UNITEXT for Physics

Series editors

Michele Cini, Roma, Italy
Attilio Ferrari, Torino, Italy
Stefano Forte, Milano, Italy
Guido Montagna, Pavia, Italy
Oreste Nicrosini, Pavia, Italy
Luca Peliti, Napoli, Italy
Alberto Rotondi, Pavia, Italy

More information about this series at http://www.springer.com/series/13351

Cosimo Bambi · Alexandre D. Dolgov

Introduction to Particle Cosmology

The Standard Model of Cosmology and its Open Problems

Cosimo Bambi
Department of Physics
Fudan University
Shanghai
China

Alexandre D. Dolgov
Dipartimento di Fisica e Scienze della Terra
Università degli Studi di Ferrara
Ferrara
Italy

and

Department of Physics
Novosibirsk State University
Novosibirsk
Russia

ISBN 978-3-662-51587-7   ISBN 978-3-662-48078-6   (eBook)
DOI 10.1007/978-3-662-48078-6

Library of Congress Control Number: 2015945603

Springer Heidelberg New York Dordrecht London
© Springer-Verlag Berlin Heidelberg 2016, corrected publication 2021
This work is subject to copyright. All rights are reserved by the Publisher, whether the whole or part of the material is concerned, specifically the rights of translation, reprinting, reuse of illustrations, recitation, broadcasting, reproduction on microfilms or in any other physical way, and transmission or information storage and retrieval, electronic adaptation, computer software, or by similar or dissimilar methodology now known or hereafter developed.
The use of general descriptive names, registered names, trademarks, service marks, etc. in this publication does not imply, even in the absence of a specific statement, that such names are exempt from the relevant protective laws and regulations and therefore free for general use.
The publisher, the authors and the editors are safe to assume that the advice and information in this book are believed to be true and accurate at the date of publication. Neither the publisher nor the authors or the editors give a warranty, express or implied, with respect to the material contained herein or for any errors or omissions that may have been made.

Printed on acid-free paper

Springer-Verlag GmbH Berlin Heidelberg is part of Springer Science+Business Media (www.springer.com)

Preface

Over the past 20 years, cosmology has emerged as a mature research field, in which it is possible to perform precise measurements and test fundamental physics. Its significance in relation to the other areas of research in physics has grown substantially. To wit, two out of ten Nobel Prizes in physics awarded during the past 10 years were conferred for studies related to cosmology (in 2006 and in 2011). Among the preceding more than 100 awards, only one-half of one Nobel Prize (in 1978) was for cosmology.

The number of researchers working in cosmology is increasing, and students taking introductory courses on cosmology include not only those who plan to work in this field, but also those with different interests, seeking to get at least a basic understanding of the subject.

The aim of this book is to provide an introduction to modern cosmology for senior undergraduate and graduate physics students, without necessarily requiring a strong background in theoretical high energy physics. Students in astronomy/astrophysics, in experimental high energy physics, or in other areas of research as well may be interested to learn some fundamental concepts of the structure and evolution of the Universe. Typically, these students are not closely familiar with General Relativity and quantum field theory, and therefore they may find it difficult to digest the existing cosmology books on the market.

This book describes the so-called Standard Cosmological Model. The model's theoretical aspects are based on General Relativity and on the Standard Model of particle physics, with the addition of the inflationary paradigm. This scenario is very successful in explaining a large amount of observational data including, in particular, the description of the Universe expansion, the primordial abundances of the light elements, and the origin and the properties of the cosmic microwave background radiation. However, there is also a plethora of observed phenomena that does not fit the frameworks of the Minimal Standard Model of particle physics and cosmology, and represents clear indications for new physics. To name just a few examples, the minimal model cannot explain the cosmological matter-antimatter asymmetry, the observed accelerated expansion of the contemporary Universe, and does not have any candidate for dark matter. The cosmological inflation is still at the level of a hypothesis. Its realization demands some new field or fields, which have not yet been discovered.

These subjects are presented here within a rather heuristic approach, which includes a needed description of observational data, and a reduction of mathematical technicalities as much as possible. Chapters 6 and 7, dealing, respectively, with inflation and baryogenesis, are more advanced and require some knowledge of quantum field theory, but students who are not familiar with those concepts can skip these chapters without affecting their comprehension of the rest of the book. The content of this book is partially based on the cosmology class given at Fudan University by one of the authors and on lectures given at a number of universities by the other.

The work of C.B. was supported by the NSFC grant No. 11305038, the Shanghai Municipal Education Commission grant No. 14ZZ001, the Thousand Young Talents Program, and Fudan University. The work of A.D.D. was supported by the grant of the Russian Federation Government 11.G34.31.0047.

Shanghai	Cosimo Bambi
Novosibirsk	Alexandre D. Dolgov
May 2015	

The original version of the book was revised: Belated corrections have been incorporated. The correction to the book is available at https://doi.org/10.1007/978-3-662-48078-6_13

Contents

1	**Introduction**		1
	1.1 Problems in Newtonian Cosmology		2
	1.2 The Standard Model of Cosmology		3
		1.2.1 Hubble's Law	6
		1.2.2 Big Bang Nucleosynthesis	7
		1.2.3 Cosmic Microwave Background	8
	1.3 Evidence for New Physics		9
		1.3.1 Inflation	10
		1.3.2 Baryogenesis	11
		1.3.3 Dark Matter	12
		1.3.4 Cosmological Constant Problems	13
	1.4 Age and Size of the Universe		14
	1.5 Cosmological Models Beyond General Relativity		16
	References		16
2	**General Relativity**		17
	2.1 Scalars, Vectors and Tensors		17
	2.2 Geodesic Equations		20
		2.2.1 Newtonian Mechanics	20
		2.2.2 Relativistic Mechanics	21
	2.3 Energy and Momentum in Flat Spacetime		23
	2.4 Energy-Momentum Tensor in Flat Spacetime		24
	2.5 Curved Spacetime		26
	2.6 Field Theory in Flat and Curved Spacetimes		27
	2.7 Einstein Equations		29
	Problems		32
	References		33

3	**The Standard Model of Particle Physics**.	35
	3.1 Fermions	37
	3.1.1 Leptons	38
	3.1.2 Quarks	38
	3.2 Bosons	39
	3.2.1 Gauge Bosons	39
	3.2.2 Higgs Particle.	41
	3.3 Feynman Diagrams	42
	3.4 Beyond the Minimal Standard Model of Particle Physics	44
	3.4.1 Supersymmetric Models.	45
	3.4.2 Grand Unification Theories	46
	3.4.3 Heavy Neutrinos	48
	3.4.4 Peccei-Quinn Model	49
	3.5 Probabilities of Reactions Among Particles	49
	Problems.	52
	References.	52
4	**Cosmological Models**	53
	4.1 Friedmann-Robertson-Walker Metric	54
	4.2 Friedmann Equations.	56
	4.3 Cosmological Models	57
	4.3.1 Einstein Universe	58
	4.3.2 Matter Dominated Universe	58
	4.3.3 Radiation Dominated Universe	60
	4.3.4 Vacuum Dominated Universe.	61
	4.4 Basic Properties of the FRW Metric	62
	4.5 Age of the Universe	64
	4.6 ΛCDM Model	65
	4.7 Destiny of the Universe	68
	Problems.	70
	References.	70
5	**Kinetics and Thermodynamics in Cosmology**	71
	5.1 Introduction	71
	5.2 Thermal Equilibrium in the Early Universe	71
	5.2.1 General Features.	71
	5.2.2 Kinetic Equation	76
	5.2.3 Plasma Heating and Entropy Conservation.	80
	5.3 Freezing of Species.	82
	5.3.1 Decoupling and Gershtein-Zeldovich Bound.	82
	5.3.2 Freezing of Non-relativistic Particles	86
	5.4 Neutrino Spectrum and Effective Number of Neutrino Species.	90
	Problems.	92
	References.	92

6	**Inflation**		93
	6.1	Introduction and History	93
	6.2	Problems of Pre-inflationary Cosmology	94
		6.2.1 Kinematics and Main Features of Inflation	95
		6.2.2 Flatness Problem	96
		6.2.3 Horizon Problem	96
		6.2.4 Origin of the Cosmological Expansion	98
		6.2.5 Smoothing Down the Universe and Creation of Primordial Density Perturbations	98
		6.2.6 Magnetic Monopole Problem	99
	6.3	Mechanisms of Inflation	101
		6.3.1 Canonical Scalar Inflaton with Power Law Potential	101
		6.3.2 Other Mechanisms of Inflation	103
	6.4	Universe Heating	105
		6.4.1 Perturbative Production	106
		6.4.2 Non-perturbative Phenomena	109
		6.4.3 Parametric Resonance	112
		6.4.4 Particle Production in a Gravitational Field	115
	6.5	Generation of Gravitational Waves	117
	6.6	Generation of Density Perturbations	121
	References		125
7	**Baryogenesis**		127
	7.1	Observational Data	127
	7.2	General Features of Baryogenesis Models	129
		7.2.1 Sakharov Principles	129
		7.2.2 CP Breaking in Cosmology	131
	7.3	Models of Baryogenesis	132
		7.3.1 Baryogenesis by Heavy Particle Decays	133
		7.3.2 Electroweak Baryogenesis	135
		7.3.3 Baryo-Through-Leptogenesis	138
		7.3.4 Evaporation of Primordial Black Holes	139
		7.3.5 Spontaneous Baryogenesis	143
		7.3.6 Baryogenesis by Condensed Scalar Baryons	149
	7.4	Cosmological Antimatter	151
	Problems		156
	References		156
8	**Big Bang Nuclesynthesis**		161
	8.1	Light Elements in the Universe	162
	8.2	Freeze-Out of Weak Interactions	163
	8.3	Electron-Positron Annihilation	165
	8.4	Deuterium Bottleneck	166

	8.5	Primordial Nucleosynthesis	168
	8.6	Baryon Abundance	170
	8.7	Constraints on New Physics	171
	Problems		173
	References		173

9 Dark Matter 175
    9.1   Observational Evidence 176
    9.2   Dark Matter Candidates 178
          9.2.1   Lightest Supersymmetric Particle 180
          9.2.2   Axion 181
          9.2.3   Super-Heavy Particles 181
          9.2.4   Primordial Black Holes 182
    9.3   Direct Search for Dark Matter Particles 183
    9.4   Indirect Search for Dark Matter Particles 187
    Problems 188
    References 188

10 Cosmic Microwave Background 191
    10.1   Recombination and Decoupling 192
    10.2   Formalism for the Description of Fluctuations 193
    10.3   Anisotropies of the CMB 198
          10.3.1 Primary Anisotropies 198
          10.3.2 Secondary Anisotropies 200
          10.3.3 Polarization Anisotropies 200
    10.4   Primordial Perturbations 203
    10.5   Determination of the Cosmological Parameters 204
    Problems 205
    References 205

11 Dark Energy 207
    11.1   Cosmological Acceleration 207
          11.1.1 Astronomical Data 209
          11.1.2 Acceleration by a Scalar Field 211
          11.1.3 Modified Gravity 212
    11.2   Problem of Vacuum Energy 213
    References 215

12 Density Perturbations 217
    12.1   Density Perturbations in Newtonian Gravity 217
    12.2   Density Perturbations in General Relativity 223
          12.2.1 Metric and Curvature 223
          12.2.2 Energy-Momentum Tensor 224
          12.2.3 Choice of Gauge 225

		12.2.4	Evolution of Perturbations in Asymptotically Flat Spacetime	227
		12.2.5	Evolution of Perturbations in Cosmology	228
		12.2.6	Concluding Remarks	235
	12.3	Density Perturbations in Modified Gravity		236
		12.3.1	General Equations	236
		12.3.2	Modified Jeans Instability	241
		12.3.3	Effects of Time Dependent Background	243
Problems				243
References				244

Correction to: Introduction to Particle Cosmology C1

Appendix A: Natural Units 245

Appendix B: Gauge Theories 247

Appendix C: Field Quantization 249

Chapter 1
Introduction

Cosmology (from Greek *kosmos*, world, and *logos*, study) is the science devoted to the study of the Universe on large scales and of its evolution. In comparison to other branches of physics, cosmology possesses some peculiar features. First, we have only one system, namely our Universe. This is not the case, for instance, in particle or nuclear physics, where one can repeat an experiment many times, exploring different samples, or even in astronomy, where one can observe different objects belonging to the same class. Second, we are observing the Universe at a specific moment of its evolution, that is, today. Despite that, we can study the early Universe by looking at regions far from us, thanks to the finite value of the speed of light. Lastly, in cosmology the observer performing the measurements is inside the system.

Modern cosmology was born after the advent of General Relativity in 1915. However, till the end of the 20th century it was not possible to perform precise measurements, and thus the approach was to use the known laws of physics, tested today and usually only on small scales, to study the Universe on large scales and at different times. In the last years of the 20th century, cosmology became a mature research field and entered a golden age with a large number of high quality observational data. The importance of cosmology in physics has grown a lot and this is proved by the recent Nobel Prizes in physics: in the last 10 years, two out of ten prizes were awarded to studies in cosmology (see Table 1.1).

Using modern advanced instruments, it is possible to measure the cosmological parameters with very high precision and therefore to use the Universe as a laboratory to test elementary particle physics. Today there is an unambiguous astronomical evidence strongly requiring new physics beyond the Minimal Standard Model (MSM) of particle physics and maybe even beyond General Relativity. Table 1.2 shows the milestones of modern cosmology, which will be briefly reviewed in this chapter and then discussed with more details in the rest of this book.

Table 1.1 Nobel Prizes in physics for studies in cosmology

Year	Winners	Motivation
1978	Penzias, Wilson[a]	Detection of the CMB
2006	Mather, Smoot	Detection of the CMB anisotropies
2011	Perlmutter, Schmidt, Riess	Discovery of the accelerating expansion of the Universe

[a]Penzias and Wilson shared the 1978 Nobel Prize in physics with Kapitsa, who was awarded for his contribution to low-temperature physics

Table 1.2 Milestones of modern cosmology

Year	Event
1915	Einstein formulated the theory of General Relativity
1922	Friedmann derived and solved the Einstein equations for cosmology (Friedmann equations)
1927	Lemaitre derived the Friedmann equations and the would-be Hubble law
1929	Hubble measured the Hubble constant
1933	Zwicky got evidence for the existence of dark matter
1946	Gamow predicted the cosmic microwave background (CMB)
1948	Alpher and Gamow published the theory of the big bang nucleosynthesis (BBN)
1960–1970s	The Standard Model of particle physics was formulated
1964	Penzias and Wilson detected the CMB
1967	Sakharov proposed the Sakharov conditions for baryogenesis
1967	Zeldovich pointed out the cosmological constant problem
1974	Two groups (Einasto, Kaasik and Saar; Ostriker, Peebles and Yahil) announced the discovery of the flat rotational curves in galaxies
1980	Kazanas, Starobinsky, and Guth proposed the inflationary paradigm
1992	The COBE satellite detected the CMB anisotropies
1998	The Supernova Cosmology Project and the High-Z Supernova Search Team discovered the accelerating expansion of the Universe
2003	The WMAP satellite measured with high precision the CMB anisotropies
2003	Gravitational lensing studies of the Bullet Cluster provided very strong evidence for the existence of dark matter
2013	The Planck satellite measured with high precision the CMB anisotropies

1.1 Problems in Newtonian Cosmology

Before the advent of General Relativity in 1915, the knowledge of the Universe was quite poor. In Newtonian mechanics, it is problematic to describe an infinite distribution of matter. Before the discovery of Hubble's law, the Universe was supposed to be static, infinitely old, and having an infinite volume. However, this was in disagreement with the darkness of the night sky. Such a paradox is commonly attributed to the German physician and astronomer Heinrich Wilhelm Olbers, who described it in 1823, and for this reason it is today known as *Olbers' paradox*. However, it seems that the problem was already known for a long time. The paradox starts from three assumptions:

1. The Universe is infinite in space.
2. The Universe is static and infinitely old.
3. Stars are uniformly distributed in the Universe and they have the same luminosity.

The assumption 3 is clearly an approximation, but it sounds reasonable if we imagine averaging over large volumes. The number of stars in a spherical shell with radius r and width dr is

$$N = 4\pi r^2 n\, dr, \tag{1.1}$$

where n is the mean stellar number density in the Universe, which is a constant according to the assumption 3. If L is the luminosity of a single star, i.e. the energy emitted per unit time, the intensity of the star radiation at the distance r is

$$I = \frac{L}{4\pi r^2}. \tag{1.2}$$

Eventually, the observed intensity from all the stars in the whole Universe should be

$$I_{\text{tot}} = \int_0^{+\infty} \frac{L}{4\pi r^2} 4\pi r^2 n\, dr, \tag{1.3}$$

which diverges. In other words, the night sky should be extremely bright rather than dark. This tells us that at least one of the initial assumptions is wrong. While we do not know today if the Universe is spatially infinite or not, the assumption 2 is surely wrong and the Universe is far from being static.

1.2 The Standard Model of Cosmology

The Standard Model of cosmology (also called Standard Cosmological Model) is currently the best theory for the description of the Universe. It is based on two fundamental ingredients: the Standard Model of particle physics, which is used for the matter content, and General Relativity, which describes gravitational interactions. It also requires the inflationary paradigm, which is an elegant mechanism to fix a few problems of the scenario. The Standard Model of cosmology very successfully explains a huge amount of observational data, including as the most remarkable ones Hubble's law, the primordial abundances of light elements, and the cosmic microwave background.

However, there are several puzzles that strongly suggest to look for new physics. It is striking that only 5% of the Universe is made of the known matter (mainly protons and neutrons). About 25% of the Universe is likely made of some weakly interactive particles not belonging to the MSM of particle physics. For the time being, this component is called *dark matter*. The other 70% of the Universe is really a mystery: it looks like a uniformly distributed substance with an unusual equation of state $P \approx -\rho$, where P is the pressure and ρ is the energy density[1] and it is responsible for the contemporary accelerated expansion of the Universe. This substance is usually called *dark energy*, but its origin is not clear at all and a breakdown of General Relativity at large scales is also a possibility. The mechanism of inflation still remains at the level of paradigm. It does not fit the framework of the Standard Model of particle physics. Lastly, we do not understand the origin of the matter-antimatter asymmetry around us. The local Universe is clearly matter dominated, but such an asymmetry cannot be created within the MSM of particle physics.

Albert Einstein published his paper on the theory of General Relativity in 1916. In 1922, a Russian physicist and mathematician, Alexander Friedmann, derived from the field equations of General Relativity the fundamental equations for the description of the evolution of the Universe, today known as the Friedmann equations. The same equations and the prediction of the expansion of the Universe were obtained independently by a Belgian priest, Georges Lemaitre, in 1927. While these fundamental equations come from the theory of General Relativity, it is useful to see a heuristic derivation of them. Let us assume that matter is uniformly distributed in the whole Universe, which is therefore homogeneous and isotropic. The motion of a particle with respect to a point P is determined by Newton's law of universal gravitation

$$m\ddot{a} = -\frac{G_N M m}{a^2}, \qquad (1.4)$$

where a is the distance between the point P and the particle, a dot indicates a derivative with respect to the time coordinate, m is the particle mass, and M is the mass enclosed in the sphere of radius a centered at the point P. If we multiply both sides of Eq. (1.4) by $\dot{a}$ and integrate it over time, we get

$$\dot{a}^2 = \frac{2G_N M}{a} - k, \qquad (1.5)$$

where k is an integration constant. If we divide both sides of this equation by a^2, we replace M by $(4/3)\pi a^3 \rho$, where ρ is the mass density, and introduce the Hubble parameter $H = \dot{a}/a$, we find

$$H^2 = \frac{8\pi G_N}{3}\rho - \frac{k}{a^2}, \qquad (1.6)$$

---

[1] In this textbook, we use the so-called natural units in which $c = \hbar = k_B = 1$, unless stated otherwise. For more details, see Appendix A.

1.2 The Standard Model of Cosmology

which is the first Friedmann equation. a is called cosmological scale factor and sets the distances in the Universe as a function of time. For instance, if at the time t_1 two objects are at a distance d_1, at the time t_2, as a consequence of the expansion/contraction of the Universe, the distance between these two objects is

$$d_2 = d_1 \frac{a(t_2)}{a(t_1)}. \quad (1.7)$$

The Hubble parameter measures the expansion rate at the time t. The value of the Hubble parameter today is called the Hubble constant and it is usually denoted as H_0. Since the value of the Hubble constant was quite uncertain for a long time, it is common to use the expression[2]

$$H_0 = 100 h_0 \frac{\text{km}}{\text{s} \cdot \text{Mpc}} \quad (1.8)$$

and to keep the parameter h_0 in all the equations. Today we know that $h_0 \approx 0.70$ with a few percent accuracy.

From the Friedmann equations, it follows the so-called *big bang model* of the Universe. In this picture, the Universe started expanding from an infinitely dense primordial plasma. The time $t = 0$ is the moment of the big bang, which, however, has to be interpreted with caution, because the Friedmann equations predict a spacetime singularity and an infinite energy density of matter, which are most likely pathologies of classical General Relativity and are believed to be fixed by unknown quantum gravity effects. In the course of the Universe expansion, the temperature and the particle density of the primordial plasma drops down. As a consequence, the particle reaction rate decreases with time and at a certain point some particle species stop interacting with the rest of the plasma. This is expected to be a very common phenomenon during the history of the Universe and naturally permits the production of relics, which, if stable, may survive till today. Table 1.3 summarizes the main events in the history of the Universe, from the big bang singularity to the present days. The physics above 200 GeV is not known, so the predictions are based on speculations and depend on the specific models. Moreover, for the time being we have no observational signatures from the Universe before the big bang nucleosynthesis (BBN) ,[3] and, though the electroweak and the QCD phase transitions can be expected from the known physics, there is no proof that the Universe was at some time at those temperatures.

---

[2]The parsec is a common astronomical unit of length and its symbol is pc. 1 pc = $3.086 \cdot 10^{16}$ m. 1 kpc = 10^3 pc, 1 Mpc = 10^6 pc, etc.

[3]An important exception is the spectrum of the primordial density perturbations, which were generated before the BBN, presumably during the period of inflation.

Table 1.3 History of the Universe

Age	Temperature	Event
0	$+\infty$	Big bang (prediction of classical General Relativity)
10^{-43} s	10^{19} GeV	Planck era (?)
10^{-35} s	10^{16} GeV	Era of Grand Unification (?)
?	?	Inflation (?)
?	?	Baryogenesis
10^{-11} s	200 GeV	Electroweak symmetry breaking[a]
10^{-5} s	200 MeV	QCD phase transition[a]
1 s –15 min	0.05–1 MeV	Big bang nucleosynthesis
60 kyr	1 eV	Matter-radiation equality
370 kyr	0.3 eV	Recombination and photon decoupling
0.2–1 Gyr	15–50 K	Reionization
1–10 Gyr	3–15 K	Structure formation
6 Gyr	4 K	Transition from a decelerating to an accelerating Universe
9 Gyr	3 K	Formation of the Solar System
13.8 Gyr	2.7 K	Today

[a]No observational evidence, but prediction based on known physics

1.2.1 Hubble's Law

Hubble's law was predicted on the basis of the Friedmann equations by Lemaitre in a paper published in 1927. In 1929, Edwin Hubble confirmed the law with astronomical observations and measured the value of the today expansion rate of the Universe, now called the *Hubble constant* and indicated by H_0. Hubble's law reads

$$v = H_0 d, \tag{1.9}$$

where v is the recession velocity of a source at the distance d.

Hubble was studying Cepheid stars, which are very luminous variable stars and their luminosity and pulsation period are strongly correlated. Thanks to this correlation, these stars can be used as distance indicators. From the brightness of the star, it is possible to infer the distance to the host galaxy d. From the spectrum of the host galaxy, one can measure the redshift $z = \Delta\lambda/\lambda$, where λ is the wavelength at the point of the emission and $\Delta\lambda$ is the difference between the wavelengths at the point of the observation and of the emission. Hubble found a proportionality between z and d. If the redshift z is interpreted as Doppler boosting, then for $v \ll 1$ we have $z \approx v$ and thus we find Eq. (1.9). However, the exact origin of the phenomenon is the expansion rate of the Universe in General Relativity, not a recession velocity in Special Relativity. The law holds for sources at a distance of a few Mpc up to a few hundreds Mpc ($z < 1$). For closer sources, the actual Doppler boosting due

1.2 The Standard Model of Cosmology

to the motion of the source is dominant. For far away sources, there are significant deviations from the simple form of Hubble's law.

The original measurement of Hubble was

$$H_0 \sim 500 \frac{\text{km}}{\text{s} \cdot \text{Mpc}}. \tag{1.10}$$

With the notation of Eq. (1.8), it corresponds to $h_0 \approx 5$, which is much higher than the accepted today value, $h_0 \approx 0.7$, probably due to systematic errors in the measurements of the distances by Hubble.

1.2.2 Big Bang Nucleosynthesis

The prediction of the primordial abundances of the light elements is another important success of the Standard Model of cosmology. Indeed, most of the ^{4}He in the Universe was produced during the first few minutes after the big bang. It cannot be explained by the production through nuclear reactions inside stars. This can be easily seen with the following rough estimate. The present day luminosity of the Galaxy is

$$L = 4 \cdot 10^{36} \text{ J/s}. \tag{1.11}$$

If we take into consideration that the production of 1 kg of ^{4}He provides an energy in electromagnetic radiation at the level of $6 \cdot 10^{14}$ J and that our Galaxy is about 10 Gyr old, we find that the amount of the produced ^{4}He is

$$M_{^4\text{He}} \approx \frac{4 \cdot 10^{36} \text{ J/s} \cdot 3 \cdot 10^{17} \text{ s}}{6 \cdot 10^{14} \text{ J/kg}} \approx 2 \cdot 10^{39} \text{ kg}. \tag{1.12}$$

This is about 1 % of the mass of our Galaxy, $M_{\text{Galaxy}} \approx 3 \cdot 10^{41}$ kg. On the other hand, we observe an abundance of ^{4}He at the level of 25 %.

The theory of the production of light elements in the early Universe was pioneered by Alpher and Gamow in 1948. In the course of the cosmological expansion, the temperature of the primordial plasma dropped down and it became possible to form bound states of nucleons. The BBN started when the Universe was about 1 s old and the plasma temperature was close to 1 MeV. However, the onset of the synthesis of the light elements took place later. The first element to be produced was the deuterium and its binding energy is about 2 MeV. An efficient production of deuterium started only when the temperature of the Universe was around 80 keV, because the number density of photons was much higher than that of protons and neutrons and therefore even though the number of high energy photons was suppressed at lower temperatures by the Boltzmann factor, their amount was still large enough to destroy deuterium nuclei. After the production of deuterium, the nucleosynthesis continued to form ^{3}He, ^{4}He, and ^{7}Li. At first approximation one can assume that all the neutrons that survived till the beginning of the deuterium synthesis eventually formed ^{4}He, because this nucleus is very strongly bound. Heavier elements were not significantly produced during the BBN, because of lack of time and the absence of stable nuclei with $A = 5$ and 6 just above ^{4}He. Therefore they were produced later inside stars.

The prediction of the primordial abundances of light elements requires numerical calculations. The first codes were written at the end of the 1960s. These calculations require the knowledge of the Standard Model of particle physics (in particular the number of the light particle species and their interaction properties) as well as a number of nuclear reaction rates that can be measured in laboratory. In the Standard Model, the matter content is known and the model has only one free parameter, namely the ratio of the cosmological number density of baryons (i.e. the sum of proton and neutron number densities) to that of photons, which is usually denoted as η. The predictions of the primordial abundances of D, ^{3}He, ^{4}He, and ^{7}Li can thus be calculated as functions of the single parameter η. From the comparison between theoretical predictions and observations, we can determine its value. The determination of the primordial abundances of these elements from astronomical observations is not an easy job, because of later reactions that changed the initial values. However, already in the 1970s it was clear that

$$\eta \sim 10^{-10} - 10^{-9}. \tag{1.13}$$

Now it is known with a few percent accuracy. This number can be eventually converted into the contribution of ordinary matter on the total mass/energy of the Universe. It turns out that protons and neutrons only represent about 5 % of the total budget.

1.2.3 Cosmic Microwave Background

When the temperature of the Universe dropped below ~ 0.3 eV, electrons and protons combined to form neutral hydrogen atoms (as well as electrons and helium nuclei formed helium atoms). Similar to the onset of the BBN, this phenomenon occurred at a temperature much lower than the ionization energy of hydrogen $E_{\text{ion}} = 13.6$ eV, because of the large number of photons with respect to electrons and protons: even though exponentially suppressed by the Boltzmann factor, the number of photons with sufficient energy in the plasma was large enough to prevent the recombination when the temperature of the Universe was higher. This event is called recombination and occurred at the redshift $z_{\text{rec}} \approx 1100$ or, equivalently, when the Universe was about 370,000 year old. Before recombination, photons were in thermal equilibrium with matter through elastic Thomson scattering off free electrons. After recombination, the absence of free electrons caused the decoupling of photons from matter. At this point, photons started freely propagating in the Universe. The events of the recombination and of the photon decoupling are clearly correlated and occurred more or less at the same time, namely $z_{\text{rec}} \approx z_{\text{dec}}$. These photons form the cosmic microwave background (CMB), reaching us today from the so-called last scattering surface, a spherical surface centered at our position where such photons interacted with matter for the last time. The spectrum of the CMB is very close to that of a black body with a temperature of 2.7 K. There are very small anisotropies at the level of 10^{-5}. The today number density of CMB photons is around 400 cm^{-3} and they only contribute ~0.005 % in the total energy budget of the Universe.

The CMB was predicted by Gamow in 1946. Later, other authors tried to estimate the present day temperature of these photons, obtaining different results due to the poor knowledge of the values of the cosmological parameters. The CMB was accidentally detected by Penzias and Wilson, who were working on a satellite communication experiment. For this discovery, Penzias and Wilson received the Nobel Prize in physics in 1978.[4] A real breakthrough in the study of the CMB and in cosmology was the launch of the COBE satellite in 1989. This experiment was able to detect temperature anisotropies at the level of one part in 10^5. These small fluctuations do depend on the values of the cosmological parameters and their detection by COBE is thought to be the birth of cosmology as a precise science. George Smoot and John Mather received the Nobel Prize in physics in 2006 as principal investigators of two instruments on COBE. At the very end of the 20th century, several balloon-borne experiments measured the CMB fluctuations on small angular scales, providing a better estimate of several cosmological parameters. The WMAP satellite was launched in 2001 and its data were released from 2003 till 2012. Its measurements of the CMB anisotropies provided precise estimates of many cosmological parameters at the level of a few percent. In particular, they confirmed the energy budget of the Universe, in which only about 5 % is ordinary matter, about 25 % is dark matter, and about 70 % is dark energy. After WMAP, the CMB anisotropies were studied by a more advanced satellite, Planck, which was launched in 2009 and the mission's all-sky map was released in 2013.

1.3 Evidence for New Physics

As it has been already stressed, the Standard Model of cosmology relies on the current understanding of the fundamental forces of Nature and the established types of the basic constituents of matter. In other words, the model is based on the Standard Model of particle physics and General Relativity.

With these two ingredients, the Standard Cosmological Model encounters several problems. Some of them can be fixed by the inflationary paradigm, which postulates a period of exponential expansion in the very early Universe. The physics of inflation cannot be cast in the Standard Model of particle physics and General Relativity. Moreover, while the inflationary scenario is usually thought to be the best candidate

---

[4] A similar, though less accurate, observation was done by Ter-Shamonov, who was calibrating the antenna prototype of the Russian radio telescope RATAN-600.

to do this job, it is not the only one and it is still at the level of speculation, without observational evidence in support of it and against alternative scenarios.

Moreover, this is not the end of the story. The Standard Model of particle physics and General Relativity cannot explain a number of observations, which therefore strongly suggests to look for new physics. The local Universe is clearly matter dominated: there is not much antimatter around us. When and how this matter-antimatter asymmetry was created? After years of investigation, we have to conclude that the MSM of particle physics cannot explain it, but a reasonable extension of it can still successfully resolve the issue.

Another fundamental open problem concerns the origin and the nature of dark matter. From the BBN, we find that only about 5 % of the total energy budget of the Universe is contributed by ordinary matter, mainly by protons and neutrons (the latter are bound in atomic nuclei). From a number of observations, like the study of the rotational curves of nearby galaxies, we infer that the amount of gravitating matter should be much higher, around 30 % the total energy density of the Universe. What is the 25 % of the gravitating matter made of? In the MSM of particle physics there are no good candidates for that. So new physics is necessary, and there are indeed potentially good candidates in theories beyond the MSM.

Lastly, from the study of the expansion rate of the Universe we find that we are in a phase of accelerated expansion. If we believe in the Friedmann equations, the phenomenon could be explained by an exotic nature of the 70 % of the energy in the Universe. The latter is called dark energy, but actually we do not know if new physics comes from the matter sector, i.e. physics beyond the Standard Model of particle physics, or from the gravity one, namely from a breakdown of the Einstein theory of General Relativity at large scales. This body of evidence coming from astronomical observations is today one of the main motivations to look for new physics outside the standard theoretical frameworks.

1.3.1 Inflation

The Standard Cosmological Model, if it is solely based on the MSM of particle physics and on General Relativity, runs into several serious problems.

First, we see that the angular anisotropy of the CMB is quite small. This suggests that different parts of the sky were causally connected at the time of the last scattering. However, this is not what one would expect from the Friedmann equations for a radiation or a matter dominated universe. This is called the horizon problem.

Second, observations show that the Universe is quite close to the geometrically flat case; that is, the term proportional to the constant k appearing in Eq. (1.6) can be neglected. On the other hand, as one can see from the Friedmann equations dominated by usual matter, an increasingly strong departure from flatness is developed with time. So the Universe had to be extremely fine-tuned to the flat one at the very beginning. This fine-tuning looks very unnatural and creates the so-called flatness problem.

1.3 Evidence for New Physics

Third, we need a mechanism to generate primordial inhomogeneities at cosmologically large scales, the seeds of future galaxies.

Fourth, beyond the framework of the MSM of particle physics, we would expect the formation of dangerous relics at early times. These relics could be some heavy particles or topological defects that would force the Universe to recollapse too soon.

A way to solve the problems of flatness, horizon, and dangerous relics is to postulate a period of exponential cosmological expansion, the so-called inflation. It was suggested in 1979 by Kazanas and Starobinsky, and, in a more complete form, by Guth in 1980. However, the concrete mechanism of inflation suggested by Guth was not realistic. This fault was cured in 1982 by Linde and, independently, by Albrecht and Steinhardt, who suggested the so-called new inflationary model and introduced a dynamical inflaton field to produce the exponential expansion of the Universe. The mechanism of generation of density perturbations at the inflationary stage was proposed by Mukhanov and Chibisov (in Russian alphabetic order). Now many different scenarios of inflation are worked out and there is common agreement that a sufficiently long period of inflation could solve all the four problems mentioned above.

At present, the inflationary paradigm is still at the level of speculation, but the inflationary prediction of the spectrum of primordial density perturbations in good agreement with the data is a strong argument in favor of the existence of an inflationary period. There is no established inflaton candidate within the MSM, though attempts to identify the inflaton and Higgs fields are pursued. In principle, other mechanisms may solve the problems of the Standard Cosmological Model. All these mechanisms require new physics.

1.3.2 Baryogenesis

The local Universe is clearly matter dominated. The amount of antimatter is very small. The latter can be explained as a result of antiparticle production by high energy collisions in space. The existence of large regions of antimatter in our neighborhood would produce high energy electromagnetic radiation created by matter-antimatter annihilation, which is not observed. On the other hand, matter and antimatter seem to have quite similar properties and thus it would be natural to expect a matter-antimatter symmetric Universe. If we believe in the inflationary paradigm, an initially tiny asymmetry cannot help, because the exponential expansion of the Universe during the inflationary period would have washed out any initial asymmetry, just like it can wash out dangerous heavy relics produced in the very early Universe. A satisfactory model of the Universe should be able to explain the origin of the matter-antimatter asymmetry.

The term baryogenesis is used to indicate the generation of the asymmetry between baryons (basically protons and neutrons) and antibaryons (antiprotons and antineutrons). In 1967, Andrei Sakharov pointed out the three ingredients, today known as Sakharov principles, to produce a matter-antimatter asymmetry from an initially symmetric universe. Actually, none of these conditions is strictly necessary, but counterexamples require some tricky or exotic mechanisms. The MSM of particle physics does have all the ingredients to meet the Sakharov principles. However, it turns out that it is impossible to generate the observed matter-antimatter asymmetry in this framework: its parameters simply do not have the right values to do it. This is a clear indication for physics beyond the MSM. Unfortunately, there is no unambiguous indication for the energy scale of the new physics. Today there are several scenarios that can potentially explain the matter-antimatter asymmetry around us. However, they typically involve physics at too high energies to be tested at particle colliders, at least for the time being.

1.3.3 Dark Matter

In 1933 Zwicky found that most of the mass in the Coma cluster seemed to be made of some form of non-luminous stuff. From the motion of the galaxies near the edge of the cluster, he got an estimate of the total mass of the Coma cluster. Such a value was significantly higher than that inferred from the estimate based on the brightness of the galaxies. Later, other observations confirmed that a significant fraction of the mass in galaxies had to be made of non-visible matter. In particular, strong evidence came from the measurement of galactic rotation curves. The issue became more intriguing in the 1970s, with the advent of accurate numerical calculations of the abundances of primordial light elements. From the comparison of theoretical predictions and observational data, the BBN theory required an amount of ordinary matter made of protons and neutrons at the level of 5% with respect to the total energy of a flat universe, namely a universe with $k = 0$ in Eq. (1.6). At the same time, the estimates of the mass in galaxies and clusters suggested something like 30%.

The study of the discrepancy between the amount of matter inferred from the BBN (capable of counting only ordinary matter, i.e. protons and neutrons) and the one inferred through the effect of the gravitational force is a very active research field. The possibility of a breakdown of Newtonian gravity at scales larger than a few kpc, and therefore of a wrong estimate of the galaxy masses with the virial theorem, has been seriously considered for a long time. In 2003, gravitational lensing studies of the Bullet Cluster provided quite a strong evidence for the existence of non-baryonic dark matter. The Bullet Cluster is a system made of a cluster and a smaller subcluster that collided about 150 Myr ago. The components of the system responded to this collision in a different way. The stars, observed at optical wavelengths, were not significantly affected by the collision, while the hot gas, observed in the X-ray band and representing most of the baryonic matter of the clusters, was strongly affected by the collision. Gravitational lensing studies show that there is a significant displacement between the center of the total mass and the center of the baryonic matter inferred from optical and X-ray observations. The interpretation is that the mass of these clusters is dominated by some dark matter objects, which presumably are weakly interactive particles that were not affected by the collision. The displacement between the two centers of mass strongly disfavors a solution based on a modification of gravity and supports the dark matter hypothesis.

Dark matter must be made of weakly interactive objects/particles and be stable or have a very long lifetime, because otherwise it would have been already decayed into something else. For sure, it is electrically neutral. In the context of the MSM of particle physics there are no good dark matter candidates. The neutrinos of the Standard Model would have the correct interaction properties, but we know today that their mass is too low to contribute a significant fraction of the cosmological dark matter. Good dark matter candidates exist in theories beyond the MSM. Until recently, the strongest candidate was the lightest supersymmetric particle of the supersymmetric extensions of the MSM. However, so far no signature of supersymmetry has been observed at the Large Hadron Collider (LHC) at CERN in Geneva. The non-observation of supersymmetric particles created serious doubts on the validity of low energy supersymmetry.

Let us mention in conclusion that dark matter is not necessarily just elementary particles. There are models of dark matter consisting of some bound states of new elementary particles or even of macroscopically large entities being either some kinds of solitons or compact stellar-like objects.

1.3.4 Cosmological Constant Problems

In addition to the first Friedmann equation given in (1.6), the evolution of the Universe is governed by the second Friedmann equation, which reads

$$\frac{\ddot{a}}{a} = -\frac{4\pi G_N}{3}(\rho + 3P), \tag{1.14}$$

where P is the matter pressure. The Friedmann equations inevitably predict that the Universe is either expanding or contracting. There is no natural way for a static system. For ordinary matter, $\rho + 3P \geq 0$ and therefore $\ddot{a} < 0$; that is, the expansion can only decelerate.

Before the Hubble discovery of the cosmological expansion, the common belief was that the Universe was static and eternal (the assumptions leading to Olbers' paradox). To resolve the contradiction between the General Relativity prediction of a non-stationary universe and this wrong belief, Einstein introduced the so-called cosmological constant Λ. With this new parameter, the Friedmann equations become

$$\frac{\dot{a}^2}{a^2} = \frac{8\pi G_N}{3}\rho + \frac{\Lambda}{3} - \frac{k}{a^2}, \tag{1.15}$$

$$\frac{\ddot{a}}{a} = -\frac{4\pi G_N}{3}(\rho + 3P) + \frac{\Lambda}{3}. \tag{1.16}$$

In this case, it is possible to have a static universe (even if unstable). After the discovery by Hubble, the constant was removed, being useless at that time. Einstein, according to some quotations, said that its introduction was the "biggest blunder" of his life. On the contrary, in 1967, Zel'dovich pointed out that the effective value of the cosmological constant should receive a huge contribution from particle physics. There is currently no satisfactory explanation for this puzzle, which is usually called "old problem of the cosmological constant".

At the end of the 20th century, the Supernova Cosmology Project and the High-Z Supernova Search Team explored type Ia supernovae at high redshift to study the expansion of the Universe. Surprisingly, they found that the Universe was accelerating rather than decelerating, as expected from the Friedmann equations with ordinary matter satisfying the condition $\rho + 3P \geq 0$. For such a discovery, the team leaders of these projects, Saul Perlmutter, Brian Schmidt, and Adam Riess, received the 2011 Nobel Prize in physics.

The origin of the phenomenon is completely unknown. Observations may be explained with a tiny but positive cosmological constant, and for this reason the puzzle is sometimes called "new cosmological constant problem". The accelerated expansion rate of the Universe might be caused by some exotic stuff uniformly distributed over the whole Universe, generically called dark energy, which behaves as a perfect fluid with the unusual equation of state $\rho \approx -P$. Alternatively, classical General Relativity may break down at large scales and not to be appropriate for the description of the Universe at large distances.

There is also the so-called "coincidence problem", concerning the reason why the possible energy density of this exotic stuff is today of the same order of magnitude as the energy density of the normally gravitating matter, despite very different evolutions during the cosmological expansion: they contribute respectively about 70 % and 30 % to the energy of the Universe and dark energy stays constant in the course of the expansion, while "normal" matter (including the dark one) drops down as $1/a^n$ with $n = 3$ or 4.

1.4 Age and Size of the Universe

By age of the Universe we mean the time that passed from the beginning of the expansion to the present days. We do not have exhaustive knowledge about the physics of the very early Universe and we do not possess observational evidence of the Universe at temperatures above a few MeV, except for the spectrum of the primordial density perturbations. However, from the Friedmann equations we see that the time since the big bang singularity predicted by General Relativity and the BBN (the first event for which we have a rigorous observational evidence) is only about 1 s and therefore completely negligible with respect to the time from the BBN to the present days. A simple estimate of the age of the Universe can be obtained assuming a constant expansion rate. In this case, the age of the Universe is simply the inverse of the Hubble constant; that is, $1/H_0 \approx 14$ Gyr. A more accurate estimate requires the

1.4 Age and Size of the Universe

knowledge of the energy content of the Universe, since, according to the Friedmann equations, the Hubble parameter depends on the energy density ρ and the constant k, see Eq. (1.6). The age of the Universe accepted today is 13.8 Gyr. Such an estimate is consistent with the age of the oldest stars in globular clusters, which are supposed to be formed 1–2 Gyr after the big bang.

Let us now get a rough idea of the typical size of the structures in the Universe. We can start from our Galaxy, the Milky Way, which is quite a standard galaxy. It has a stellar disk with a radius of about 15 kpc and an average thickness of 0.3 kpc. There are about 10^{11} stars in our Galaxy, and the Solar System is in this disk at about 8 kpc from the center. The disk is surrounded by a spheroidal halo of old stars with roughly the same radius as the disk. A larger spheroidal halo is made of dark matter. The total mass of the Milky Way is estimated to be around 10^{12} $M_\odot$.[5] Our Galaxy belongs to the Local Group, which is a group of about 50 galaxies in which the Milky Way and the Andromeda galaxy are the largest ones and most of the others are dwarf galaxies. The Local Group has a total mass of about 10^{13} $M_\odot$ and a radius of 1.5 Mpc. The Local Group belongs to the Virgo Supercluster, which contains about 100 galaxy groups and clusters. The total mass of the Virgo Supercluster is about 10^{15} $M_\odot$ and its radius is about 15 Mpc. Above the Virgo Supercluster, we find the visible Universe: it contains about 10^8 superclusters, which all together count something like 10^{11} galaxies, its total mass is 10^{23} $M_\odot$, and its radius is about 15 Gpc.

At present, we do not know the actual size of the whole Universe. In the simplest case of a homogeneous and isotropic Universe, there can be three kinds of geometries, depending on the sign of k in Eq. (1.6). $k = 0$ corresponds to a flat universe, $k > 0$ to a closed universe, and $k < 0$ to an open universe. It is usually assumed that the Universe has a trivial topology, even if this assumption could be questioned. In the case of a trivial topology, flat and open universes are spatially infinite, so they might contain an infinite number of galaxies. A closed universe is like a 3-dimensional sphere and has a finite volume. Current observations suggest that our Universe is close to be flat, which means that all the three scenarios could be possible and the Universe may either be spatially finite or spatially infinite. People have also studied the possibility of more complicated scenarios and universes with non-trivial topologies (Linde 2004; Luminet et al. 2015). In this case, even flat and open universes with $k \leq 0$ may be compact and have a finite volume. If the Universe were spatially finite and everything were within the visible Universe, we could observe electromagnetic radiation emitted by the same very distant source and coming from two different points in the sky. The study of possible similar correlations in the CMB has only provided lower bounds on the size of the Universe; that is, if the Universe has a finite size it must be anyway larger than the visible Universe (Cornish et al. 2004).

---

[5] $M_\odot \approx 2 \cdot 10^{33}$ g is the Solar mass, which is quite commonly used as a unit of mass in astronomy.

1.5 Cosmological Models Beyond General Relativity

Classical General Relativity predicts an initial spacetime singularity, where the scale factor a vanishes and the energy density of matter becomes infinitely high. However, a breakdown of the classical theory is more likely and the removal of the singularity could be achieved by unknown quantum gravity effects, which are supposed to show up at the Planck scale $M_{Pl} = 10^{19}$ GeV. In classical General Relativity, the concepts of space and time are quite different from those in Newtonian physics, as suggested by our experience in everyday life. It is likely that in the quantum gravity regime even the relativistic concept of spacetime becomes inadequate and it may be misleading to use it when talking about the beginning of the Universe, or at least this concept has to be taken with some caution. At present, we do not have any robust and predictable theory of quantum gravity capable of providing clear answers about the physics at the Planck scale and the origin of our Universe. Nevertheless, people have tried to study possible scenarios on the basis of some quantum gravity inspired models.

In many extensions of classical General Relativity it turns out that gravity may become repulsive at very high densities. In this case, going backwards in time, we still find that the Universe was smaller, denser, and hotter, but at a certain point we should reach a critical density and have a bounce. Bouncing cosmological scenarios can arise from different theories (Novello and Bergliaffa 2008). The key point is that the singularity of classical General Relativity, in which the scale factor a vanishes and the energy density diverges, is replaced by a bounce, occurring at a critical value of the energy density. From dimensional arguments, if the bounce arises due to quantum gravity effects, it is natural to expect that such a critical density is the Planck energy density $\rho_{Pl} = M_{Pl}^4$. The bouncing scenarios that remove the big bang singularity are only partially a solution, because they do not tell us anything about new concepts of the spacetime and they just move the origin of the Universe to an earlier time. An extension of this picture is the idea that our Universe was born from the gravitational collapse of a region in another universe and, more general, that there may be many universes generated in this way. In other words, if there is somewhere an overdense region that collapses, an exterior observer may see the formation of a black hole, while the collapsing matter inside may eventually bounce and expand, generating a new universe. At present, all these scenarios are at the level of speculations, and it is not clear if they could ever be tested.

References

N.J. Cornish, D.N. Spergel, G.D. Starkman, E. Komatsu, Phys. Rev. Lett. **92**, 201302 (2004). arXiv:astro-ph/0310233
A.D. Linde, JCAP **0410**, 004 (2004). arXiv:hep-th/0408164
J.P. Luminet, B.F. Roukema, arXiv:astro-ph/9901364
M. Novello, S.E.P. Bergliaffa, Phys. Rept. **463**, 127 (2008). arXiv:0802.1634 [astro-ph]

Chapter 2
General Relativity

General Relativity is our current theory for the description of the gravitational force. It is the first ingredient in the Standard Model of cosmology. General Relativity has successfully passed a large number of tests, mainly in Earth's gravitational field, Solar System, and by studying the orbital motion of binary pulsars (Will 2006). Today, the observed accelerating expansion rate of the Universe is questioning the validity of the theory at very large scales, but at present the phenomenon may be explained with a small positive cosmological constant.

The basic idea of General Relativity is that the gravitational force can be interpreted as a deformation of the geometry of the spacetime, which is not flat any more. The kinematics, namely how particles move in the spacetime, is determined by the geodesic equations and it is a relatively easy problem. The dynamics, i.e. how the energy makes the spacetime curved, is regulated by the Einstein equations. The latter are second order non-linear partial differential equations for the metric coefficients and it is highly non-trivial to find a solution. Analytical solutions are thus possible only in special cases, in which the spacetime possesses some nice symmetries.

This section provides a short review on General Relativity, focusing on the concepts necessary for an introductory course on cosmology. More details can be found in standard textbooks like Hartle 2003, Landau and Lifshitz 1975, Stephani 2004.

2.1 Scalars, Vectors and Tensors

Let us start considering the usual Euclidean space in 3 dimensions. The coordinate system can be indicated by **x** or $\{x^i\}$, with $i = 1, 2$, and 3. In the case of Cartesian coordinates, we have $\{x^i\} = \{x, y, z\}$. If we have a curve γ from the point A to the point B, its length is given by

$$I = \int_\gamma ds, \qquad (2.1)$$

where ds is the line element. The curve γ can be parametrized in terms of a chosen coordinate system as $\gamma(\lambda) = \{x(\lambda), y(\lambda), z(\lambda)\}$, where λ is an affine parameter running along the curve. Equation (2.1) becomes

$$I = \int_{\lambda_1}^{\lambda_2} \left[\left(\frac{dx}{d\lambda}\right)^2 + \left(\frac{dy}{d\lambda}\right)^2 + \left(\frac{dz}{d\lambda}\right)^2 \right]^{1/2} d\lambda, \qquad (2.2)$$

where $\gamma(\lambda_1)$ and $\gamma(\lambda_2)$ correspond, respectively, to the point A and B. Equation (2.2) can be written in a more compact way by introducing the *metric tensor* g_{ij}

$$I = \int_{\lambda_1}^{\lambda_2} \left[g_{ij} \frac{dx^i}{d\lambda} \frac{dx^j}{d\lambda} \right]^{1/2} d\lambda, \qquad (2.3)$$

where we have used the Einstein convention of summation over repeated indices; that is,

$$g_{ij} \frac{dx^i}{d\lambda} \frac{dx^j}{d\lambda} \equiv \sum_{i=1}^{3} \sum_{j=1}^{3} g_{ij} \frac{dx^i}{d\lambda} \frac{dx^j}{d\lambda}. \qquad (2.4)$$

In this case, g_{ij} is 1 for $i = j$ and 0 for $i \neq j$. In the case of spherical coordinates $\{r, \theta, \phi\}$, the line element is

$$ds^2 = dr^2 + r^2 d\theta^2 + r^2 \sin^2\theta d\phi^2, \qquad (2.5)$$

and therefore $g_{11} = 1$, $g_{22} = r^2$, $g_{33} = r^2 \sin^2\theta$, and all the off-diagonal terms vanish.

If we go from the coordinate system $\{x^i\}$ to the coordinate system $\{x'^i\}$, the infinitesimal displacements change as

$$dx^i \rightarrow dx'^i = \frac{\partial x'^i}{\partial x^a} dx^a. \qquad (2.6)$$

Since the length of the curve and the line element must be independent of the choice of the coordinate system, the metric tensor changes as

$$g_{ij} \rightarrow g'_{ij} = \frac{\partial x^a}{\partial x'^i} \frac{\partial x^b}{\partial x'^j} g_{ab}. \qquad (2.7)$$

It is easy to verify that this is indeed the case for the metric tensor in Cartesian and spherical coordinates.

In general, we call *vector* an object V with components V^is changing according to the rule

2.1 Scalars, Vectors and Tensors

$$V^i \to V'^i = \frac{\partial x'^i}{\partial x^a} V^a, \tag{2.8}$$

when we go from the coordinate system $\{x^i\}$ to the coordinate system $\{x'^i\}$.

The *dual vector* of V is the object with components given by

$$V_i = g_{ij} V^j, \tag{2.9}$$

and, under a coordinate transformation, its components change with the opposite rule; that is,

$$V_i \to V'_i = \frac{\partial x^a}{\partial x'^i} V_a. \tag{2.10}$$

Upper indices are used for components that obey the rule in Eq. (2.8), lower indices when the transformation rule is given by Eq. (2.10).

A *scalar* is an object invariant under a coordinate transformation. For instance, a scalar is

$$S = V_i V^i. \tag{2.11}$$

From the transformation rules of V_i and V^i, we see that $S \to S' = S$.

The derivative operator, $\partial_i \equiv \partial/\partial x^i$, is a dual vector because

$$\partial_i \to \partial'_i = \frac{\partial x^a}{\partial x'^i} \partial_a. \tag{2.12}$$

In general, upper indices can be lowered with the use of g_{ij}, as shown in Eq. (2.9), and lower indices can be raised with the use of g^{ij}, which is the inverse matrix of g_{ij}, so

$$V^i = g^{ij} V_j. \tag{2.13}$$

Indeed, $V^i = g^{ij} V_j = g^{ij} g_{jk} V^k = \delta^i_k V^k = V^i$, where δ^i_k is the Kronecker delta and $g^{ij} g_{jk} = \delta^i_k$ by definition of inverse matrix. We also note that the metric tensor with an upper and a lower index is the Kronecker delta, $g^i_k = \delta^i_k$.

Tensors are a multi-index generalization of vectors and dual vectors. The metric tensor g_{ij} is an example of tensor with special properties. In general, the components of a tensor can have some upper and some lower indices; examples are T_{ij}, T^{ij}, $T^i{}_{jk}$, $T_{ij}{}^k{}_l$, etc. In the case of a change of coordinates, upper indices transform according to the rule in Eq. (2.8), while lower indices follow the rule of Eq. (2.10). For example, in the case of a tensor with components $T_{ij}{}^k{}_l$, we have

$$T_{ij}{}^k{}_l \to T'_{ij}{}^k{}_l = \frac{\partial x^a}{\partial x'^i} \frac{\partial x^b}{\partial x'^j} \frac{\partial x'^k}{\partial x^c} \frac{\partial x^d}{\partial x'^l} T_{ab}{}^c{}_d. \tag{2.14}$$

Upper indices can be lowered with g_{ij}, lower indices can be raised with g^{ij}. For instance,

$$T_{ijkl} = g_{ka} T_{ij}{}^a{}_l, \quad T_{ij}{}^{kl} = g^{la} T_{ij}{}^k{}_a, \quad \text{etc.} \tag{2.15}$$

2.2 Geodesic Equations

The geodesic equations determine the kinematics, namely how test-particles move in space. With the introduction of the metric tensor, we can use the same formalism for Newtonian and relativistic mechanics.

2.2.1 Newtonian Mechanics

In Newtonian mechanics, the Principle of Least Action plays a very important role. It can be used to obtain in an elegant way the equations of motion for a system when its action is known. In the case of a free point-like particle, the Lagrangian is simply given by the kinetic energy of the particle

$$L = \frac{1}{2} m \mathbf{v}^2 = \frac{1}{2} m g_{ij} \frac{dx^i}{dt} \frac{dx^j}{dt}, \tag{2.16}$$

where m is the mass of the particle, $\mathbf{v} = (v^1, v^2, v^3)$ is the particle velocity, g_{ij} is the metric tensor, $\{x^i\}$ are the particle coordinates, and t is the time. The action is

$$S = \int L \, dt. \tag{2.17}$$

From the Principle of Least Action, we find the Euler-Lagrange equations

$$\frac{d}{dt} \frac{\partial L}{\partial \dot{x}^i} - \frac{\partial L}{\partial x^i} = 0, \tag{2.18}$$

where the dot indicates the derivative with respect to t.

If we plug the Lagrangian in Eq. (2.16) into the Euler-Lagrange equations (2.18), we obtain the *geodesic equations*

$$\ddot{x}^i + \Gamma^i_{jk} \dot{x}^j \dot{x}^k = 0, \tag{2.19}$$

where Γ^i_{jk}s are the *Christoffel symbols*

$$\Gamma^i_{jk} = \frac{1}{2} g^{il} \left(\frac{\partial g_{lk}}{\partial x^j} + \frac{\partial g_{jl}}{\partial x^k} - \frac{\partial g_{jk}}{\partial x^l} \right). \tag{2.20}$$

2.2 Geodesic Equations

We note that the Christoffel symbols are not the components of a tensor. Indeed, if we consider the coordinate transformation $\{x^i\} \to \{x'^i\}$, the Christoffel symbols change according to the rule

$$\Gamma^i_{jk} \to \Gamma'^i_{jk} = \frac{\partial x'^i}{\partial x^a} \frac{\partial x^b}{\partial x'^j} \frac{\partial x^c}{\partial x'^k} \Gamma^a_{bc} + \frac{\partial x'^i}{\partial x^a} \frac{\partial^2 x^a}{\partial x'^j \partial x'^k}. \quad (2.21)$$

Γ^i_{jk} transforms as a tensor only in the special case of linear transformations. In Cartesian coordinates, all the Christoffel symbols vanish, and therefore the geodesic equations simply reduce to $\ddot{x} = \ddot{y} = \ddot{z} = 0$ (First Newton's Law). In spherical coordinates, we have

$$\ddot{r} - r\dot{\theta}^2 - r\sin^2\theta \dot{\phi}^2 = 0, \quad (2.22)$$

$$\ddot{\theta} + \frac{2}{r}\dot{r}\dot{\theta} - \cos\theta \sin\theta \dot{\phi}^2 = 0, \quad (2.23)$$

$$\ddot{\phi} + \frac{2}{r}\dot{r}\dot{\phi} + 2\cot\theta \dot{\theta}\dot{\phi} = 0. \quad (2.24)$$

2.2.2 Relativistic Mechanics

In Special and General Relativity, time and space are not two independent entities any more and the Newtonian 3-dimensional space becomes a 4-dimensional spacetime. The coordinates are usually indicated by $\{x^\mu\}$, with $\mu = 0, 1, 2$, and 3, where the 0 component refers to the temporal one and the 1, 2, and 3 components refer to the space ones. Greek letters $\mu, \nu, \rho,\ldots$ are commonly used for the spacetime indices ranging from 0 to 3, while Latin letters $i, j, k,\ldots$ are for the space components only, ranging from 1 to 3.

In Special Relativity and in Cartesian coordinates $\{t, x, y, z\}$, or $\{t, \mathbf{x}\}$ with $\mathbf{x} = \{x, y, z\}$, the metric tensor is indicated by $\eta_{\mu\nu}$ and the line element of the spacetime ds is[1]

$$ds^2 = \eta_{\mu\nu} dx^\mu dx^\nu = dt^2 - dx^2 - dy^2 - dz^2 = dt^2 - d\mathbf{x}^2. \quad (2.25)$$

The metric coefficients are thus $\eta_{00} = 1$, $\eta_{11} = \eta_{22} = \eta_{33} = -1$, and all the off-diagonal components vanish. The Principle of Least Action can be naturally extended to relativistic mechanics. The action for a free point-like particle can now be written as

$$S = -m \int_\gamma ds, \quad (2.26)$$

---

[1] In this book, we use the metric signature convention $(+ - - -)$, which is common in particle physics. In the General Relativity community, it is more common the convention $(- + + +)$.

where m is the particle mass, γ is the particle trajectory, and ds is the line element of the spacetime. From the action (2.26) and the line element (2.25), we find that the Lagrangian is

$$S = \int L dt \quad \Rightarrow \quad L = -m\sqrt{1 - \mathbf{v}^2}, \qquad (2.27)$$

where $\mathbf{v} = d\mathbf{x}/dt$ is the particle velocity. In the non-relativistic limit $\mathbf{v}^2 \ll 1$, we recover the Newtonian Lagrangian (modulo a constant)

$$L \approx -m + \frac{1}{2}m\mathbf{v}^2, \qquad (2.28)$$

and therefore we obtain the correct Newtonian equations of motion.

In general, the metric tensor $g_{\mu\nu}$ has not the simple form of $\eta_{\mu\nu}$. The line element is an invariant; that is, it is independent of the choice of the coordinates. With the terminology of the previous section, ds^2 is a scalar. We can thus define the following coordinate independent types of trajectories:

$$\begin{aligned} ds^2 &> 0 \quad \text{time-like trajectories,} \\ ds^2 &= 0 \quad \text{light-like trajectories,} \\ ds^2 &< 0 \quad \text{space-like trajectories.} \end{aligned} \qquad (2.29)$$

In particular, massless particles like photons will follow light-like trajectories with $ds^2 = 0$; that is, massless particles move with the speed of light. The equations of motion for a massless particle can still be obtained from the action in (2.26), but now m cannot be the mass but just a constant with the dimensions of mass.

In the case of massive particles, it is convenient to use as affine parameter λ their "proper time" τ, i.e. the time measured in the rest-frame of the particle. Since ds^2 is an invariant, $d\tau^2 = ds^2$, because the coordinate system is anchored on the particle and therefore there is no motion along the spatial directions. With this choice of the affine parameter, $g_{\mu\nu}\dot{x}^\mu \dot{x}^\nu = 1$, where the dot indicates the derivative with respect to τ.

In Newtonian mechanics, the motion of a test-particle in a gravitational field can be described by adding the correct gravitational potential to the Lagrangian of the free particle. One of the key-points in General Relativity is that the gravitational field can be absorbed into the metric tensor $g_{\mu\nu}$: in other words, we have still a free particle, but now it lives in a curved spacetime and follows the geodesics of that spacetime. If the metric of the spacetime $g_{\mu\nu}$ is known, we can obtain the geodesic equations from the action in Eq. (2.26) with $\eta_{\mu\nu}$ replaced by $g_{\mu\nu}$. Equivalently, one can write the Euler-Lagrange equations for the Lagrangian

$$L = g_{\mu\nu}\dot{x}^\mu \dot{x}^\nu. \qquad (2.30)$$

2.2 Geodesic Equations

The geodesic equations have the same form as the ones in the Newtonain theory, with Latin letters replaced by Greek letters

$$\ddot{x}^\mu + \Gamma^\mu_{\nu\rho}\dot{x}^\nu\dot{x}^\rho = 0 \quad \text{with} \quad \Gamma^\mu_{\nu\rho} = \frac{1}{2}g^{\mu\sigma}\left(\frac{\partial g_{\sigma\rho}}{\partial x^\nu} + \frac{\partial g_{\nu\sigma}}{\partial x^\rho} - \frac{\partial g_{\nu\rho}}{\partial x^\sigma}\right). \quad (2.31)$$

The fact that the motion is only determined by the background geometry and not by specific features of the body meets the well-known *Weak Equivalence Principle*, which asserts that the trajectory of a test-particle is independent of its internal structure and composition (Will 2006).

It is instructive to see how we can recover the Newtonian limit. We use Cartesian coordinates and we require that: (i) the gravitational field is weak, (ii) the gravitational field is stationary, and (iii) the motion of the particle is non-relativistic. These three conditions are given, respectively, by

$$g_{\mu\nu} = \eta_{\mu\nu} + h_{\mu\nu} \quad \text{with} \quad |h_{\mu\nu}| \ll 1, \quad (2.32)$$

$$\frac{\partial g_{\mu\nu}}{\partial t} = 0, \quad (2.33)$$

$$\frac{dt}{d\lambda} \gg \frac{dx^i}{d\lambda}. \quad (2.34)$$

Within these approximations, the geodesic equations reduce to

$$\frac{d^2x^\mu}{d\lambda^2} + \Gamma^\mu_{00}\left(\frac{dt}{d\lambda}\right)^2 \approx 0 \quad \text{with} \quad \Gamma^\mu_{00} \approx \frac{1}{2}\eta^{\mu\nu}\frac{\partial h_{00}}{\partial x^\nu}. \quad (2.35)$$

After a simple integration, we find

$$\frac{d^2x^i}{dt^2} \approx -\frac{1}{2}\frac{\partial h_{00}}{\partial x^i}. \quad (2.36)$$

If we compare Eq. (2.36) with the Newtonian formula $m\ddot{\mathbf{x}} = -m\nabla\Phi$, where Φ is the Newtonian gravitational potential, and we require that the spacetime is flat at infinity, we find

$$g_{00} = 1 + 2\Phi. \quad (2.37)$$

2.3 Energy and Momentum in Flat Spacetime

From the Lagrangian in Eq. (2.27), we obtain the particle 3-momentum

$$\mathbf{p} = \frac{\partial L}{\partial \mathbf{v}} = \frac{m\mathbf{v}}{\sqrt{1-v^2}}, \quad (2.38)$$

and the particle Hamiltonian

$$\mathcal{H} = \mathbf{p}\mathbf{v} - L = \frac{m}{\sqrt{1 - \mathbf{v}^2}}. \tag{2.39}$$

$\mathcal{H}$ corresponds to the energy of the particle, i.e. $E = \mathcal{H}$. For $\mathbf{v}^2 = 0$, we get the particle *rest-energy* $E = m$. We note that, if this particle is made of a number of elementary particles, its rest-energy is not just the sum of the masses of all the elementary particles, but it includes also the kinetic energy and the interaction energy of all the constituents. In the non-relativistic limit $\mathbf{v}^2 \ll 1$, the particle energy is

$$E \approx m + \frac{1}{2}m\mathbf{v}^2. \tag{2.40}$$

The correct Newtonian kinetic energy is thus recovered subtracting the rest-energy m from the total energy E. Massive particles cannot reach the speed of light $\mathbf{v}^2 = 1$ because it would require an infinite energy.

The 4-momentum of a massive particle can be introduced as

$$p^\mu = m\dot{x}^\mu = (E, \mathbf{p}). \tag{2.41}$$

The scalar $p_\mu p^\mu = m^2$ corresponds to the well-known relativistic formula relating the energy, the mass, and the 3-momentum of a particle

$$E^2 = m^2 + \mathbf{p}^2. \tag{2.42}$$

2.4 Energy-Momentum Tensor in Flat Spacetime

Let us consider a system with action

$$S = \int L\,dt \quad \text{with} \quad L = \int \mathcal{L}\,d^3V, \tag{2.43}$$

where $\mathcal{L} = \mathcal{L}(\phi, \partial_\mu \phi)$ is the *Lagrangian density* and depends on the field $\phi(t, \mathbf{x})$ and on its first derivatives, while $d^3V = dx\,dy\,dz$ is the volume element in Cartesian coordinates. Since $\mathcal{L}$ does not explicitly depend on the coordinates x^μ, the system is closed, and its energy and momentum are conserved. If we apply the Principle of Least Action, namely we consider small variations of ϕ and $\partial_\mu \phi$ and demand that $\delta S = 0$, we get the equations of motion

$$\frac{\partial}{\partial x^\mu} \frac{\partial \mathcal{L}}{\partial (\partial_\mu \phi)} - \frac{\partial \mathcal{L}}{\partial \phi} = 0. \tag{2.44}$$

2.4 Energy-Momentum Tensor in Flat Spacetime

Using the equations of motion and the fact that $\partial_\mu \partial_\nu \phi = \partial_\nu \partial_\mu \phi$, we find

$$\begin{aligned}
\frac{\partial \mathscr{L}}{\partial x^\mu} &= \frac{\partial \mathscr{L}}{\partial \phi} \left(\partial_\mu \phi\right) + \frac{\partial \mathscr{L}}{\partial \left(\partial_\nu \phi\right)} \partial_\mu \left(\partial_\nu \phi\right) \\
&= \left[\frac{\partial}{\partial x^\nu} \frac{\partial \mathscr{L}}{\partial \left(\partial_\nu \phi\right)}\right] \left(\partial_\mu \phi\right) + \frac{\partial \mathscr{L}}{\partial \left(\partial_\nu \phi\right)} \partial_\nu \left(\partial_\mu \phi\right) \\
&= \frac{\partial}{\partial x^\nu} \left[\frac{\partial \mathscr{L}}{\partial \left(\partial_\nu \phi\right)} \left(\partial_\mu \phi\right)\right].
\end{aligned} \qquad (2.45)$$

We define the *energy-momentum tensor* of the system as

$$T_\mu^\nu = \frac{\partial \mathscr{L}}{\partial \left(\partial_\nu \phi\right)} \left(\partial_\mu \phi\right) - \eta_\mu^\nu \mathscr{L}, \qquad (2.46)$$

and Eq. (2.45) reduces to

$$\partial_\nu T_\mu^\nu = 0. \qquad (2.47)$$

A few comments are in order here. First, Eq. (2.46) looks like the Legendre transformation of the Lagrangian density, so T^{00} should be the energy density of the system and T^{0i}s should be the momentum densities of the system. Second, Eq. (2.47) is an equation of conservation. Indeed, if we integrate Eq. (2.47) over the volume V and we apply Gauss' theorem, we find

$$\frac{d}{dt} \int_V T^{00} d^3 V = -\int_\Sigma T^{0i} d^2 \sigma_i, \qquad (2.48)$$

$$\frac{d}{dt} \int_V T^{0i} d^3 V = -\int_\Sigma T^{ij} d^2 \sigma_j, \qquad (2.49)$$

where $d^2 \sigma_j$ represents the surface element of the surface Σ and it is outwardly perpendicular to Σ. Third, such a definition of the energy-momentum tensor has some ambiguity: if $T^{\mu\nu}$ is our energy-momentum tensor, the tensor

$$T^{\mu\nu} + \partial_\rho A^{\mu\nu\rho} \quad \text{with} \quad A^{\mu\nu\rho} = -A^{\mu\rho\nu} \qquad (2.50)$$

satisfies Eq. (2.47) as well. It turns out that such an ambiguity can be removed by imposing that $T^{\mu\nu}$ is a symmetric tensor, namely $T^{\mu\nu} = T^{\nu\mu}$. If the initial energy-momentum tensor is not symmetric, it is always possible to make it symmetric with a suitable choice of $A^{\mu\nu\rho}$. This requirement can be inferred by imposing the conservation of the angular momentum of the system in Special Relativity, which can be constructed from $T^{\mu\nu}$ as

$$M^{\mu\nu} = \int_V \left(x^\mu T^{\nu 0} - x^\nu T^{\mu 0}\right) d^3 V. \qquad (2.51)$$

It is easy to see that

$$\partial_\mu M^{\mu\nu} = 0 \quad \Rightarrow \quad T^{\mu\nu} = T^{\nu\mu}. \tag{2.52}$$

2.5 Curved Spacetime

In curved spacetime, but even in flat spacetime in curved coordinates, the derivative of a scalar is a vector, but the derivative of a vector is not a tensor. The generalization of the ordinary derivative ∂_μ in curved spacetime, or in curved coordinates in flat spacetime, is the *covariant derivative* ∇_μ. For a generic vector V^μ, the action of the covariant derivative is

$$\nabla_\mu V^\nu = \partial_\mu V^\nu + \Gamma^\nu_{\mu\rho} V^\rho. \tag{2.53}$$

It can be checked that for a dual vector V_μ the action of the covariant derivative is

$$\nabla_\mu V_\nu = \partial_\mu V_\nu - \Gamma^\rho_{\mu\nu} V_\rho. \tag{2.54}$$

One can see that $\nabla_\mu V^\nu$ and $\nabla_\mu V_\nu$ are tensors and that ∇_μ is the natural generalization of ∂_μ. In the case of a generic tensor with upper and lower indices, the rule is

$$\begin{aligned}\nabla_\rho T^{\mu_1 \ldots \mu_m}{}_{\nu_1 \ldots \nu_n} = &\ \partial_\rho T^{\mu_1 \ldots \mu_m}{}_{\nu_1 \ldots \nu_n} \\ &+ \Gamma^{\mu_1}_{\rho\sigma} T^{\sigma \ldots \mu_m}{}_{\nu_1 \ldots \nu_n} + \cdots + \Gamma^{\mu_m}_{\rho\sigma} T^{\mu_1 \ldots \sigma}{}_{\nu_1 \ldots \nu_n} \\ &- \Gamma^\sigma_{\rho\nu_1} T^{\mu_1 \ldots \mu_m}{}_{\sigma \ldots \nu_n} - \cdots - \Gamma^\sigma_{\rho\nu_n} T^{\mu_1 \ldots \mu_m}{}_{\nu_1 \ldots \sigma}.\end{aligned} \tag{2.55}$$

With the covariant derivative, we can introduce the *Riemann tensor* as the commutator of the derivatives

$$\nabla_\mu \nabla_\nu V_\rho - \nabla_\nu \nabla_\mu V_\rho = R^\sigma{}_{\rho\mu\nu} V_\sigma, \tag{2.56}$$

for any vector V^μ. It turns out that the Riemann tensor can be written as

$$R^\mu{}_{\nu\rho\sigma} = \frac{\partial \Gamma^\mu_{\nu\sigma}}{\partial x^\rho} - \frac{\partial \Gamma^\mu_{\nu\rho}}{\partial x^\sigma} + \Gamma^\mu_{\lambda\rho} \Gamma^\lambda_{\nu\sigma} - \Gamma^\mu_{\lambda\sigma} \Gamma^\lambda_{\nu\rho}. \tag{2.57}$$

Since it is a tensor, under a coordinate transformation $\{x^\mu\} \to \{x'^\mu\}$ it changes as

$$R^\mu{}_{\nu\rho\sigma} \to R'^\mu{}_{\nu\rho\sigma} = \frac{\partial x'^\mu}{\partial x^\alpha} \frac{\partial x^\beta}{\partial x'^\nu} \frac{\partial x^\gamma}{\partial x'^\rho} \frac{\partial x^\delta}{\partial x'^\sigma} R^\alpha{}_{\beta\gamma\delta}. \tag{2.58}$$

With the Riemann tensor, we can introduce the *Ricci tensor* $R_{\mu\nu}$ and the *scalar curvature* R

2.5 Curved Spacetime

$$R_{\mu\nu} = R^{\rho}{}_{\mu\rho\nu}, \quad R = g^{\mu\nu} R_{\mu\nu}. \tag{2.59}$$

R is a scalar, namely it is invariant under coordinate transformations. $R^{\sigma}{}_{\rho\mu\nu}$ and $R_{\mu\nu}$ are tensors. If all their components vanish in a coordinate system, they vanish in any coordinate system, as can be seen from Eq. (2.58). In particular in flat spacetime $R^{\sigma}{}_{\rho\mu\nu} = 0$.

An important issue concerns how the laws of physics formulated in Special Relativity change when we pass to General Relativity. In flat spacetime, under a coordinate transformation from Cartesian to other coordinates, it is easy to see that one has to replace $\eta_{\mu\nu}$ with $g_{\mu\nu}$ and ordinary derivatives with covariant derivatives:

$$\eta_{\mu\nu} \to g_{\mu\nu}, \quad \partial_\mu \to \nabla_\mu. \tag{2.60}$$

The integral over $d^4x = dt d^3V$ is replaced by $\sqrt{-g} d^4x$, where g is the determinant of the metric tensor

$$d^4x \to \sqrt{-g} d^4x. \tag{2.61}$$

These rules directly follow from the coordinate transformation and they are easy to check, for instance for the transformation from Cartesian to spherical or cylindrical coordinates. In the case of curved spacetime, the issue is more tricky. In principle, there may appear some interaction terms with the Riemann tensor, the Ricci tensor, or the scalar curvature. These terms are called non-minimal couplings (an example is given in Sect. 2.6). It turns out that, for the time being, experiments and observations are consistent with the simple rules of Eqs. (2.60) and (2.61). Such a prescription is not demanded by any fundamental principle and sometimes it may not be unique, but it just seems to work. Lastly, we note that the conservation of the energy-momentum tensor in Eq. (2.47) becomes

$$\nabla_\mu T^{\mu\nu} = 0 \tag{2.62}$$

in curved spacetime. However, since we have now the covariant rather than the ordinary derivative, there is no conservation of $T^{\mu\nu}$. The reason is that matter can exchange energy and momentum with the gravitational field.

2.6 Field Theory in Flat and Curved Spacetimes

In flat spacetime and Cartesian coordinates, the action of a field can be conveniently written in the form

$$S = \int \mathscr{L} d^4x, \tag{2.63}$$

where $\mathscr{L}$ is the Lagrangian density, as introduced in Sect. 2.4. In the case of the electromagnetic field, the Lagrangian density is

$$\mathscr{L} = -\frac{1}{4} F^{\mu\nu} F_{\mu\nu}, \tag{2.64}$$

where $F_{\mu\nu} = \partial_\mu A_\nu - \partial_\nu A_\mu$ is the field strength, $A^\mu = (\phi, \mathbf{A})$ is the 4-potential, ϕ is the scalar potential, and $\mathbf{A}$ is the vector potential. The electric field $\mathbf{E}$ and the magnetic field $\mathbf{B}$ are related to the 4-potential A^μ by

$$\mathbf{E} = -\partial_t \mathbf{A} - \nabla \phi, \quad \mathbf{B} = \nabla \wedge \mathbf{A}. \tag{2.65}$$

From the definition of $F_{\mu\nu}$, it follows that

$$\partial_\mu F_{\nu\rho} + \partial_\nu F_{\rho\mu} + \partial_\rho F_{\mu\nu} = 0. \tag{2.66}$$

If we write Eq. (2.66) in terms of the electric and magnetic fields $\mathbf{E}$ and $\mathbf{B}$, we find the second and the third Maxwell equations in the usual form

$$\nabla \cdot \mathbf{B} = 0, \quad \nabla \wedge \mathbf{E} = \partial_t \mathbf{B}. \tag{2.67}$$

Applying the Principle of Least Action, we consider small variations of A^μ and of its first derivatives in the action and we get the field equations of the electromagnetic field in covariant form

$$\partial_\mu F^{\mu\nu} = 0. \tag{2.68}$$

These equations in terms of the electric and magnetic fields $\mathbf{E}$ and $\mathbf{B}$ reduce to the first and the fourth Maxwell equations in vacuum, namely

$$\nabla \cdot \mathbf{E} = 0, \quad \nabla \wedge \mathbf{B} = \partial_t \mathbf{E}. \tag{2.69}$$

It is straightforward to write the action and the field equations for the electromagnetic field in curved spacetime following the recipe of Sect. 2.5. We replace ordinary derivatives with covariant derivatives. However, the field strength of the electromagnetic field is unaltered

$$F_{\mu\nu} = \nabla_\mu A_\nu - \nabla_\nu A_\mu = \partial_\mu A_\nu - \partial_\nu A_\mu, \tag{2.70}$$

because the Maxwell tensor $F_{\mu\nu}$ is antisymmetric with respect to interchange of μ and ν and symmetric Christoffel symbols disappear from the difference. $F^{\mu\nu}$ is now obtained by raising the indices with $g^{\mu\nu}$, not with $\eta^{\mu\nu}$. The Lagrangian density is still given by Eq. (2.64), while the action is

$$S = \int \mathscr{L} \sqrt{-g} d^4 x. \tag{2.71}$$

2.6 Field Theory in Flat and Curved Spacetimes

Equations (2.66) and (2.68) become, respectively,

$$\nabla_\mu F_{\nu\rho} + \nabla_\nu F_{\rho\mu} + \nabla_\rho F_{\mu\nu} = 0,$$
$$\nabla_\mu F^{\mu\nu} = 0. \tag{2.72}$$

Let us now consider a scalar field, which is the simplest field and it is widely used in cosmology. In flat spacetime and Cartesian coordinates, the action is given by Eq. (2.63) and the Lagrangian density is

$$\mathscr{L} = \frac{1}{2}\eta^{\mu\nu}\left(\partial_\mu \phi\right)\left(\partial_\nu \phi\right) - \frac{1}{2}m^2\phi^2, \tag{2.73}$$

where ϕ is the scalar field and m is the mass of the particles associated to this field. From the variation of the action, we get the field equation (Klein-Gordon equation)

$$\left(\partial_\mu \partial^\mu - m^2\right)\phi = 0, \tag{2.74}$$

where $\partial_\mu \partial^\mu = \partial_0^2 - \partial_1^2 - \partial_2^2 - \partial_3^2$ is the D'Alembert operator. In curved spacetime, the action is given by Eq. (2.71) and the Lagrangian density becomes

$$\mathscr{L} = \frac{1}{2}g^{\mu\nu}\left(\partial_\mu \phi\right)\left(\partial_\nu \phi\right) - \frac{1}{2}m^2\phi^2. \tag{2.75}$$

The field equation in curved spacetime is

$$\left(\nabla_\mu \partial^\mu - m^2\right)\phi = 0. \tag{2.76}$$

As discussed in Sect. 2.5, it is not guaranteed that the Lagrangian density in the presence of a gravitational field is given by Eq. (2.75) and there are no interaction terms. In cosmology, it is common to introduce some non-minimal couplings. In the simplest case, the Lagrangian density can be taken as

$$\mathscr{L} = \frac{1}{2}g^{\mu\nu}\left(\partial_\mu \phi\right)\left(\partial_\nu \phi\right) - \frac{1}{2}m^2\phi^2 + \xi R\phi^2, \tag{2.77}$$

where ξ is a dimensionless coupling constant. We note that in Eq. (2.77) we have $\partial_\mu \phi$ and not $\nabla_\mu \phi$. This is because ϕ is a scalar and therefore $\partial_\mu = \nabla_\mu$.

2.7 Einstein Equations

In the previous sections, we have seen that the motion of test-particles is determined by the geodesic equations and the non-gravitational laws of physics in flat spacetime can be easily translated for a curved spacetime with the prescription given in

Eqs. (2.60) and (2.61). In all these cases, we just need to know the background metric $g_{\mu\nu}$. The latter depends on the coordinate system, but it takes into account the gravitational field as well, and therefore it is determined by the matter distribution. The *Einstein equations* are the master equations of General Relativity and they relate the spacetime geometry to the matter content. They can be obtained by imposing a number of "reasonable" requirements, and *a posteriori* one can check that its predictions are consistent with observations. One can thus require that

1. They are tensor equations, to be independent of the coordinate system.
2. They are partial differential equations at most of second order in the variable of the gravitational field, namely $g_{\mu\nu}$, in analogy with the other field equations in physics.
3. They must have the correct Newtonian limit.
4. $T^{\mu\nu}$ is the source of the gravitational field.
5. If $T^{\mu\nu} = 0$, the spacetime is flat.

From the requirements 1 and 4, the Einstein equations must have the form

$$G^{\mu\nu} = \kappa T^{\mu\nu}, \tag{2.78}$$

where $G_{\mu\nu}$ is the *Einstein tensor* and κ is the Einstein constant. Since $\nabla_\mu T^{\mu\nu} = 0$, we need that

$$\nabla_\mu G^{\mu\nu} = 0. \tag{2.79}$$

From the conditions 2 and 5, it follows that the simplest choice is

$$G_{\mu\nu} = R_{\mu\nu} - \frac{1}{2} g_{\mu\nu} R, \tag{2.80}$$

where $R_{\mu\nu}$ is the Ricci tensor and R is the scalar curvature. The tensor in Eq. (2.80) satisfies the condition in (2.79), called Bianchi identity. To find the Newtonian limit, we assume the approximations (2.32) and (2.33), as well as that in our coordinate system all the components of the matter energy-momentum tensor are negligible, except the 00 one, which describes the energy density and reduces to the matter density in the Newtonian limit, so

$$T_{00} = \rho \quad T_{\mu\nu} = 0 \text{ for } \mu, \nu \neq 0. \tag{2.81}$$

After some passages, we find

$$R_{00} = \frac{1}{2} \Delta h_{00} = \kappa \rho, \tag{2.82}$$

where Δ is the Laplace operator. The Poisson equation of Newtonian gravity is recovered by replacing h_{00} with 2Φ, where Φ is the Newtonian gravitational potential, as found in Sect. 2.2.2. The Einstein constant is thus

2.7 Einstein Equations

$$\kappa = 8\pi G_{\rm N} = \frac{8\pi}{M_{\rm Pl}^2}, \tag{2.83}$$

where $G_{\rm N}$ is the Newton constant and $M_{\rm Pl} = G_{\rm N}^{-1/2}$ is the Planck mass.[2] In the next chapters, we will use the Planck mass instead of $G_{\rm N}$, as it is more common in particle cosmology.

If we relax the assumption 5, the Einstein equations can have the form

$$R^{\mu\nu} - \frac{1}{2}g^{\mu\nu}R + \Lambda g^{\mu\nu} = 8\pi G_{\rm N} T^{\mu\nu}, \tag{2.84}$$

where Λ is called *cosmological constant*. For a non-vanishing Λ, we do not recover the flat spacetime in the absence of matter. However, a sufficiently small Λ cannot be ruled out by experiments.

Lastly, like any other known field equation of physics, even the Einstein equations can be derived from the Principle of Least Action. The total action of the system has the form $S_{\rm tot} = S_{\rm EH} + S_{\rm matter}$, where $S_{\rm EH}$ is the Einstein-Hilbert action describing the gravitational field

$$S_{\rm EH} = \frac{1}{16\pi G_{\rm N}} \int R\sqrt{-g}\, d^4 x, \tag{2.85}$$

while $S_{\rm matter}$ is the action of the matter sector. If we consider small variations of the metric coefficients and of their first derivatives, we get the Einstein equations. Such a procedure allows to define the matter energy-momentum tensor as

$$T^{\mu\nu} = \frac{2}{\sqrt{-g}} \frac{\delta S_{\rm matter}}{\delta g_{\mu\nu}}, \tag{2.86}$$

which is automatically a symmetric tensor (see the discussion at the end of Sect. 2.4) and reduces to the one of Special Relativity in the absence of gravitational fields. If we consider small variations of the fundamental variables of the matter sector and of their derivatives, we get the field equations of matter (e.g. the Maxwell equations in the case of an electromagnetic field).

The covariant conservation (2.62) of the energy-momentum tensor (2.86) follows from the invariance of the action with respect to arbitrary coordinate transformations, according to the Noether theorem. This property is compatible with the Einstein equations (2.84), with $\Lambda = 0$ due to the Bianchi identity (2.79). This identity is automatically fulfilled in General Relativity, as follows from the definition of the curvature tensors and the Christoffel symbols. If $\Lambda \neq 0$, Λ must be constant, and this is why it has the name "cosmological constant".

---

[2] We remind the reader that we are using units in which $c = \hbar = 1$. If we reintroduce c and $\hbar$, we have $\kappa = \frac{8\pi G_{\rm N}}{c^4} = \frac{8\pi \hbar}{M_{\rm Pl}^2 c^3}$.

There is an interesting analogy between the automatic conservation of the left hand side of the Maxwell and the Einstein equations. The Maxwell equations in the presence of electric current have the form

$$\partial_\mu F^{\mu\nu} = J^\nu \tag{2.87}$$

Because of the antisymmetry of $F^{\mu\nu}$ with respect to the interchange of μ and ν, the derivative of the left hand side vanishes, $\partial_\nu \partial_\mu F^{\mu\nu} = 0$. So it implies the current conservation.

Problems

2.1 The exterior gravitational field of a spherically symmetric object is described by the Schwarzschild solution. The line element is

$$ds^2 = \left(1 - \frac{2G_N M}{r}\right) dt^2 - \left(1 - \frac{2G_N M}{r}\right)^{-1} dr^2 - r^2 d\theta^2 - r^2 \sin^2\theta d\phi^2. \tag{2.88}$$

Here M is the gravitational mass of the object.

(a) Write the geodesic equations. [Hint: write the Euler-Lagrange equations for the Lagrangian in (2.30) with $g_{\mu\nu}$ of the Schwarzschild solution and then arrange these equations in the form (2.31).]
(b) Find the non-vanishing Christoffel symbols. [Hint: they can be gotten from the geodesic equations rather than from their definition in terms of the metric tensor.]
(c) What is the value of the Riemann tensor, Ricci tensor, and scalar curvature for $r \to +\infty$?

2.2 The Friedmann-Robertson-Walker metric describes the spacetime geometry of a homogeneous and isotropic universe. The line element is given by

$$ds^2 = dt^2 - a^2(t) \left(\frac{dr^2}{1 - kr^2} + r^2 d\theta^2 + r^2 \sin^2\theta d\phi^2\right), \tag{2.89}$$

where $a(t)$ is called scale factor and it is a function of t only, while k is a constant.

(a) Answer the questions (a) and (b) of the previous problem for the metric in (2.89).
(b) Is the energy of a test-particle conserved? And its angular momentum?

References

J.B. Hartle, *Gravity: an Introduction to Einstein's General Relativity*, 1st edn. (Addison-Wesley, San Francisco, 2003)

L.D. Landau, E.M. Lifshitz, *The Classical Theory of Fields*, 4th edn. (Pergamon, Oxford, 1975)

H. Stephani, *Relativity: an Introduction to Special and General Relativity*, 3rd edn. (Cambridge University Press, Cambridge, 2004)

C.M. Will, Living Rev. Rel. **9**, 3 (2006) [gr-qc/0510072]

Chapter 3
The Standard Model of Particle Physics

The Standard Model of particle physics currently represents the best framework for the description of all the known elementary particles and all the fundamental forces of Nature except gravity, namely the electromagnetic force, the strong nuclear force, and the weak nuclear force. Matter is described by fermions, spin-1/2 particles, which are grouped into two classes: leptons and quarks. Forces are described by gauge theories and are mediated by gauge bosons, spin-1 particles. The Standard Model of particle physics includes also a spin-0 particle, the Higgs boson, which provides a mass to the other fundamental particles (charged leptons, quarks, weak gauge bosons). Particles are classified according to their quantum numbers, which are related to the invariance of the theory under certain symmetries. Figure 3.1 shows the fundamental particles of the Standard Model of particle physics and their basic properties (Olive et al. 2014).[1] Starting from a small number of assumptions, we can write the most general Lagrangian, finding that the model depends on 19 free parameters (9 fermion masses, 3 quark mixing angles, 1 quark CP violating phase, 3 gauge couplings, 1 Higgs vacuum expectation value, 1 weak mixing angle, and 1 CP violating parameter of the strong interaction), and their numerical value has to be determined in experiments.

Interactions between matter particles are created (mediated) by the exchange of the so-called gauge bosons. The electromagnetic interaction is induced by the exchange of photons. The corresponding theory is called quantum electrodynamics, or QED. The strong interaction is mediated by eight gluons, which interact with the strong charge of quarks, called color, so the theory has the name quantum chromodynamics, or QCD. The weak interaction is induced by the exchange of heavy intermediate bosons, the electrically charged $W^\pm$ or the neutral Z^0.

The model is very successful in explaining a large number of observations and its predictions well agree with particle collider experiments. In some cases, it is possible to perform very precise measurements, and the agreement with theoretical calculations is excellent. For instance, the electron anomalous magnetic dipole moment has

---

[1] We have to distinguish the MSM and a minimally extended model, which includes Supersymmetry, possibly Grand Unification, and sometimes even something more, see below.

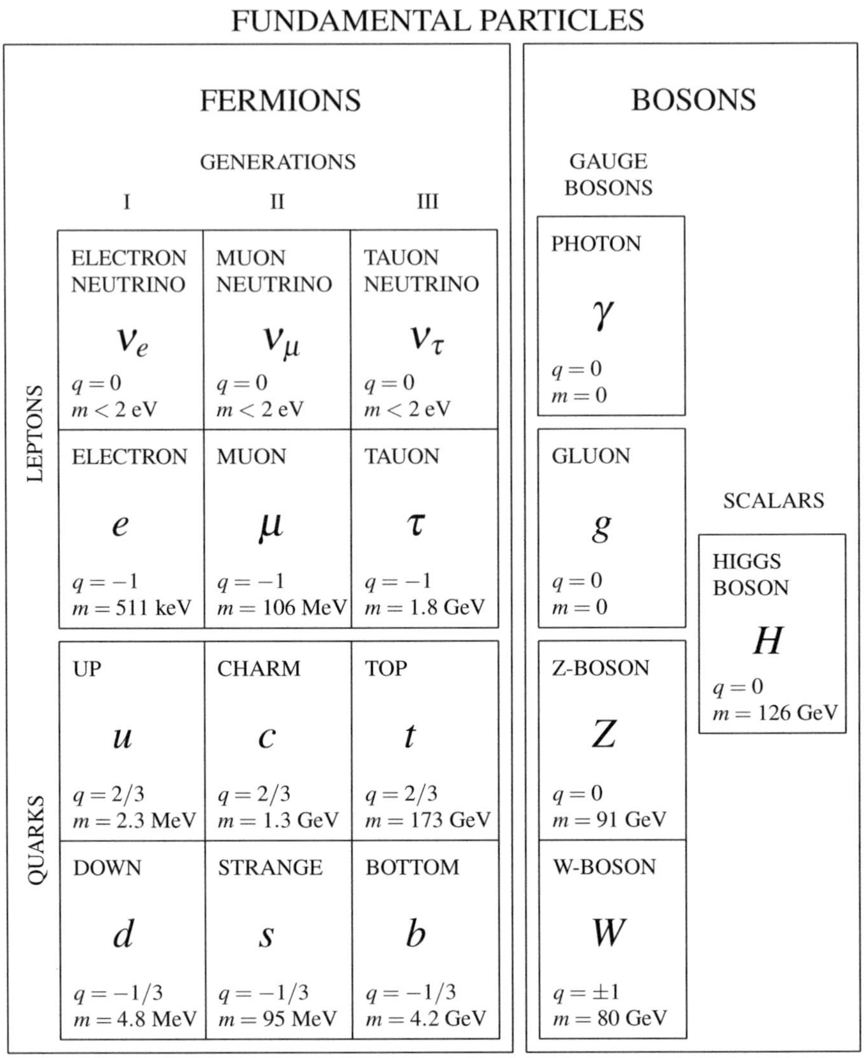

FUNDAMENTAL PARTICLES

	FERMIONS			BOSONS	
	GENERATIONS			GAUGE BOSONS	
	I	II	III		

		I	II	III	GAUGE BOSONS	
LEPTONS	ELECTRON NEUTRINO ν_e $q=0$ $m<2$ eV	MUON NEUTRINO ν_μ $q=0$ $m<2$ eV	TAUON NEUTRINO ν_τ $q=0$ $m<2$ eV	PHOTON γ $q=0$ $m=0$	
	ELECTRON e $q=-1$ $m=511$ keV	MUON μ $q=-1$ $m=106$ MeV	TAUON τ $q=-1$ $m=1.8$ GeV	GLUON g $q=0$ $m=0$	SCALARS
QUARKS	UP u $q=2/3$ $m=2.3$ MeV	CHARM c $q=2/3$ $m=1.3$ GeV	TOP t $q=2/3$ $m=173$ GeV	Z-BOSON Z $q=0$ $m=91$ GeV	HIGGS BOSON H $q=0$ $m=126$ GeV
	DOWN d $q=-1/3$ $m=4.8$ MeV	STRANGE s $q=-1/3$ $m=95$ MeV	BOTTOM b $q=-1/3$ $m=4.2$ GeV	W-BOSON W $q=\pm 1$ $m=80$ GeV	

Fig. 3.1 Building blocks of the Standard Model of particle physics. Matter is described by fermions, forces are mediated by bosons. For every particle, we show the name, symbol, electric charge q, and mass m

been tested at the level of 10^{-8}. Nevertheless, there are both theoretical problems and observational data that require new physics. From the theoretical side, the problem to make the Higgs mass stable against huge quantum corrections suggests that new physics is not far from the electroweak energy scale. A possible solution is the supersymmetric extension of the MSM, in which every Standard Model particle has a supersymmetric partner. The Standard Model of particle physics assumes that neutrinos are massless. Today we know this is not the case, but there are several ways to provide neutrinos with a mass. From cosmology, new physics is required because the Standard Model of particle physics has no good dark matter candidate, it cannot generate the matter-antimatter asymmetry observed around us, and there is no way to produce an early period of inflation.

This section provides a short review on the Standard Model of particle physics. More details can be found in many textbooks on this subject, like Griffiths (2008).

3.1 Fermions

Fermions are spin-1/2 particles and therefore obey the Fermi-Dirac statistics. They are grouped into two classes: leptons and quarks. Leptons do not interact through the strong nuclear force. Quarks interact through all the fundamental forces of Nature. Leptons and quarks are the basic building blocks of matter. Around us, there are no free quarks, as a consequence of the confining property of the strong nuclear force. Free quarks were likely to be present in the primordial plasma of the early Universe, when the temperature was above the QCD phase transition, $T_{QCD} \sim 200$ MeV. Today we observe bound states: baryons, consisting of three quarks, and mesons, which are bound states of a quark and an antiquark. Baryons and mesons are called hadrons, which was a term coined to indicate particles subject to the strong nuclear force and therefore different from leptons. Protons and neutrons are baryons made of the lightest quarks. Electrons are the lightest leptons with a non-vanishing electric charge.

There are three generations of leptons and three generations of quarks, called first, second, and third generation. In every generation, there are two particles with quite similar properties but different electric charge. The members of the first generation are light particles, those in the second generation are heavier, those in the third generation are much heavier, though we do not know if this is true for neutrinos. We do not know why there are three generations and not four or more, but there are arguments suggesting that there may not be heavier generations with relatively light new neutrinos. An indication comes from the study of the primordial abundances of light elements (see Sect. 8.7). Another argument is based on the study of the decay of the Z-boson in collider experiments. Z-bosons decay into a fermion and the corresponding antiparticle. If the decay product is a charged lepton-antilepton or a quark-antiquark pair, it is seen in the detector. If it is a neutrino-antineutrino, it is not detected. However, it is possible to measure the decay rate into "invisible" particles, namely particles that cannot be seen by the detector. It turns out that the measured decay rate into invisible particles is consistent with the theoretical predictions for three light neutrinos, not four or more. If there were a fourth generation, the corresponding neutrino should be quite heavy ($2\,m_\nu > M_Z$, to make the decay kinematically forbidden).

3.1.1 Leptons

Leptons interact through the weak nuclear force and, if they possess a non-vanishing electric charge, they have also electromagnetic interactions. In any generation, there is an electrically charged lepton and the corresponding electrically neutral neutrino. The three electrically charged leptons are the electron (e^-), the muon (μ^-), and the tauon (τ^-), and each of them has its antiparticle, namely the positron (e^+), the antimuon (μ^+), and the antitauon (τ^+). Since all these particles have spin 1/2, the total number of degrees of freedom is 12. In the MSM of particle physics, neutrinos are massless and only interact through the weak nuclear force. Since the latter only acts on left-handed particles and right-handed antiparticles, we have three massless left-handed neutrinos (electron neutrino ν_e, muon neutrino ν_μ, tauon neutrino ν_τ), and three massless right-handed antineutrinos (electron antineutrino $\bar{\nu}_e$, muon antineutrino $\bar{\nu}_\mu$, tauon antineutrino $\bar{\nu}_\tau$).[2] Overall, there are 6 degrees of freedom. The Lagrangian of the Standard Model is invariant under a global transformation of the kind $L \to e^{i\alpha} L$, where L is called weak isospin doublet and groups the Dirac spinors of the two leptons of the same generation, while α is a constant. Such a symmetry is associated to the conservation of the lepton number of every generation, usually indicated by L_e, L_μ, L_τ.

Today we know that neutrinos "oscillate"; that is, they can transform to neutrinos of another generation. Such a phenomenon clearly violates the lepton number of the generation, but not the total lepton number $L = L_e + L_\mu + L_\tau$. Neutrino oscillations occur because the neutrino mass eigenstates and the neutrino flavor eigenstates (they are also called interaction or gauge eigenstates) do not coincide. When a neutrino is generated or interacts with another particle, it falls into a flavor eigenstate, which is a linear combination of mass eigenstates. In the process of free neutrino propagation, the relative phases of the mass eigenstates change, if masses are different, and therefore neutrino stops being a single flavor eigenstate but becomes a mixture of different flavors. When the neutrino interacts with matter again, it may create a charged lepton of different flavor. This is possible only in the case of a non-vanishing mass (at least for two neutrinos), and therefore neutrinos must have a mass, though very small.

3.1.2 Quarks

Quarks interact through the electromagnetic, strong, and weak nuclear forces. As leptons, they are grouped into three generations, with the first generation made of the lightest quarks and the third one with the heaviest quarks. Every generation has a U-type quark (up quark u, charm quark c, top quark t) with electric charge $+2/3$, and a D-type quark (down quark d, strange quark s, bottom quark b) with

---

[2]Particles with spin in the same (opposite) direction as their momentum are called right-handed (left-handed). In the case of massless particles, this is independent of the reference system. For very light particles, this classification is approximately valid too.

electric charge $-1/3$. Every quark has its antiparticle, called antiquark (antiquark up $\bar{u}$, antiquark down $\bar{d}$, antiquark charm $\bar{c}$, etc.). The strong interaction among quarks is generated by the so-called color charges: every quark can have three different color charges, which are conventionally called red, blue, and green. Overall, the total number of degrees of freedom is $2 \times 2 \times 2 \times 3 \times 3 = 72$. Adding the 18 leptonic degrees of freedom, the total number of fermionic degrees of freedom is 90. Like leptons, heavier quarks can decay into lighter ones, but, in contrast to charged leptons, the flavor is not conserved, i.e. quarks of different generations can transform to each other, though the probability of such processes is suppressed with respect to transformations between the members of the same generation.

Free quarks have never been observed. Only bound states made of three quarks or three antiquarks (baryons or antibaryons) or of a quark and an antiquark (mesons) have been registered. This is due to the confining property of the strong interaction. Only colorless states are allowed to propagate as free particles at low temperatures, below the QCD phase transition, which presumably takes place at $T \sim 200$ MeV. Protons are the lightest baryons and they are a bound state uud. Neutrons are the next to the lightest baryons and they are a bound state udd. The state uuu is not the lightest one because quarks are fermions and they have spin 1/2, so we can at most arrange two fermions with the same quantum numbers but opposite spin (+1/2 and −1/2) in the same energy level. Antibaryons are bound states of three antiquarks (e.g. antiproton, antineutron, etc.), while mesons (e.g. pion π, kaon K, etc.) are colorless bound states made of a quark and an antiquark. The theory predicts even more complex colorless bound states, like penta-quarks made of four quarks and an antiquark, but, despite some claims of their observation in the past, there is no clear detection of them.

3.2 Bosons

Bosons have integer spin, so they obey the Bose-Einstein statistics. In the MSM of particle physics, elementary bosons are the gauge bosons, which are the force carries, and the Higgs scalar, which plays a special role providing a mass to quarks, charged leptons, and weak gauge bosons. The interactions among the fundamental particles of the Standard Model are summarized in Fig. 3.2.

3.2.1 Gauge Bosons

In the Standard Model of particle physics, interactions are introduced by imposing the principle of gauge invariance. We start from a Lagrangian invariant under the global transformation $\psi \to G\psi$, where ψ is a fermionic field and G is a space-time independent transformation (global transformation) belonging to some group. Global invariance can be generalized by demanding that the symmetry transformation

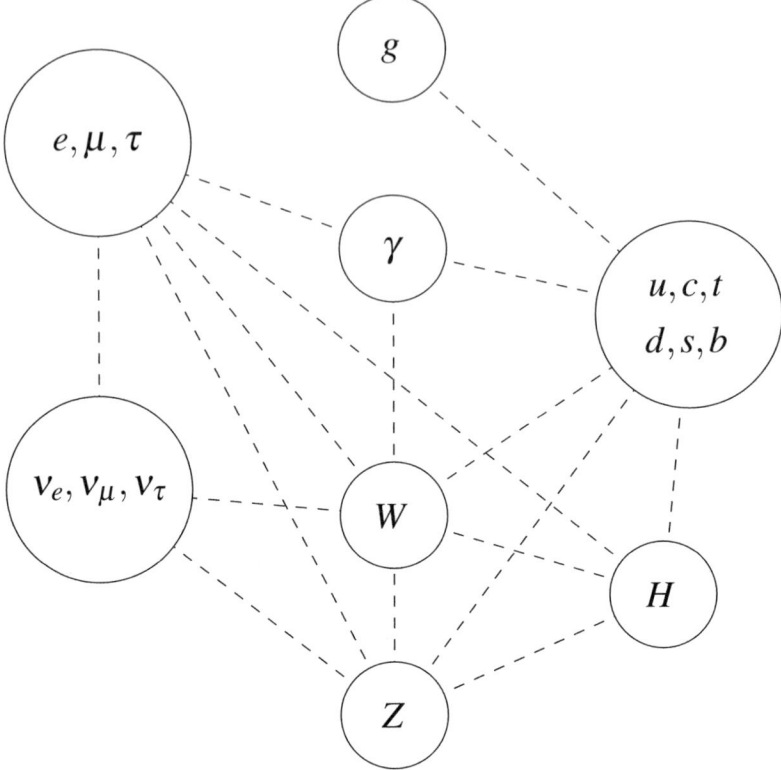

Fig. 3.2 Interactions among the particles of the Standard Model

depends upon the spacetime coordinates, $G = G(x)$. In this case, the symmetry is called local or gauge symmetry. Kinetic terms of matter fields in the Lagrangian are nor invariant with respect to coordinate dependent transformations. The compensation of this non-invariance requires the introduction of a new vector field, say A^μ. The particles associated to the new field A^μ are called gauge bosons. They must be massless, otherwise the gauge symmetry would be broken. The principle of gauge invariance is a very elegant way to introduce interactions, because with a simple assumption we obtain a well defined Lagrangian, generally leading to a renormalizable theory. A few more details on gauge theories can be found in Appendix B.

The Standard Model of particle physics is described by the gauge theory $U_Y(1) \times SU_L(2) \times SU(3)$.[3] Below the electroweak symmetry scale $\sim$200 GeV, the elec-

---

[3] We note that the Y in $U_Y(1)$ stands for hyper-charge. It is used to distinguish the $U(1)$ symmetry above the electroweak symmetry breaking from the $U(1)$ symmetry below the electroweak symmetry breaking; the latter is indicated by $U_{em}(1)$ and describes the usual Maxwell electrodynamics. The L in $SU_L(2)$ is used to indicate that the $SU(2)$ symmetry only acts on left-handed particles and right-handed antiparticles. In some extensions of the Standard Model there is also the symmetry $SU_R(2)$, which acts on right-handed particles and left-handed antiparticles.

3.2 Bosons

troweak symmetry $U_Y(1) \times SU_L(2)$ is broken, and there is the residual symmetry $U_{em}(1)$ which describes the electromagnetic force. The $U_{em}(1)$ sector is called quantum electrodynamics (QED). The photon (γ) is the massless particle associated to this residual symmetry. The W- and Z-bosons mediating the weak interaction are instead the other gauge bosons that acquire a mass because the symmetry is broken. $SU(3)$ describes the strong interaction force, which is called quantum chromodynamics (QCD). Since $SU_L(2)$ and $SU(3)$ are non-abelian gauge theories, their gauge bosons carry a non-vanishing charge and couple each other. Gauge bosons are spin-1 particles. The total number of degrees of freedom is thus 2(γ, because photons are massless and have two spin states) + 3 (Z, because it has a mass and therefore there are three possible spin states, namely +1, 0, –1) + 2 × 3 (W^+ and W^-, which are massive) + 2 × 8 (gluons g) = 27.

As we have already mentioned, QCD is described by $SU(3)$. It has two peculiar features, namely confinement and asymptotic freedom. Confinement means that the force between two quarks increases as their distance increases. In the end, it is not possible to have an isolated quark, because the energy of separation becomes so high that the process creates quark-antiquark pairs. Asymptotic freedom means that the interaction becomes weaker as the energy increases. At high energy, QCD becomes a perturbative theory, because the coupling constant is small and perturbative calculations, similar to those in QED, are applicable. At low energies, the QCD coupling constant is not a small parameter and the calculations require non-perturbative techniques.

3.2.2 Higgs Particle

The Higgs boson is the only spin-0 particle in the Standard Model of particle physics. Its existence was confirmed by collider experiments only in 2013. It plays a special role, because it provides a mass to the other particles through the so-called Higgs mechanism and it is responsible for the electroweak symmetry breaking.

The phenomenon of spontaneous symmetry breaking is known also in other fields in physics. For instance, in a ferromagnetic material, when the temperature is above the Curie temperature, T_{Curie}, the magnetic moments of the constituent particles point in all directions, so magnetic domains are absent and the magnetization of the material is zero. Below T_{Curie}, magnetic domains are spontaneously formed, stochastically choosing a certain direction, because it is energetically more favorable for the particles to align their spins along the same line.

Something similar happens with the Higgs field. Here the critical temperature is the electroweak scale $T_{ew} \sim 200$ GeV. Below the electroweak scale, the Higgs field acquires a non-vanishing vacuum expectation value, which breaks the $U_Y(1) \times SU_L(2)$ gauge symmetry. The gauge bosons of $U_Y(1) \times SU_L(2)$ mix together and the result is the weak nuclear force, mediated by the massive gauge bosons W and Z, and the electromagnetic force, mediated by the massless photon and representing the residual symmetry $U_{em}(1)$.

The Yukawa terms in the Lagrangian of the form $H\bar{\psi}\psi$, where H is the Higgs field and ψ is a fermionic field, become the fermionic mass terms when the Higgs field has a non-vanishing vacuum expectation value. In the unbroken symmetry phase, the $U_Y(1) \times SU_L(2)$ symmetry forbids any mass term, which is thus possible only when the electroweak symmetry is broken.

3.3 Feynman Diagrams

Particle physics experiments usually involve scattering processes, in which some particles collide with a target or with other particles and we want to study the products created in the collision. The initial and the final states can be approximated by free particles and we can estimate the probability amplitude within the Standard Model. In most cases, we can do it by using a perturbative expansion in the coupling constants. In the case of the electroweak sector $U_Y(1) \times SU_L(2)$, the gauge coupling constants are small and the perturbative approach works very well. In QCD at low energies, this is not true and we have to proceed with other techniques, while the perturbative approach can be used for high energy processes.

The graphical representation of the perturbative approach is the Feynman diagram. Figure 3.3 shows the elastic scattering of two electrons. The initial and the final states are two free electrons. The grey blob represents the interaction area. Since we are not able to get the exact solution of the field equations describing the process, we adopt

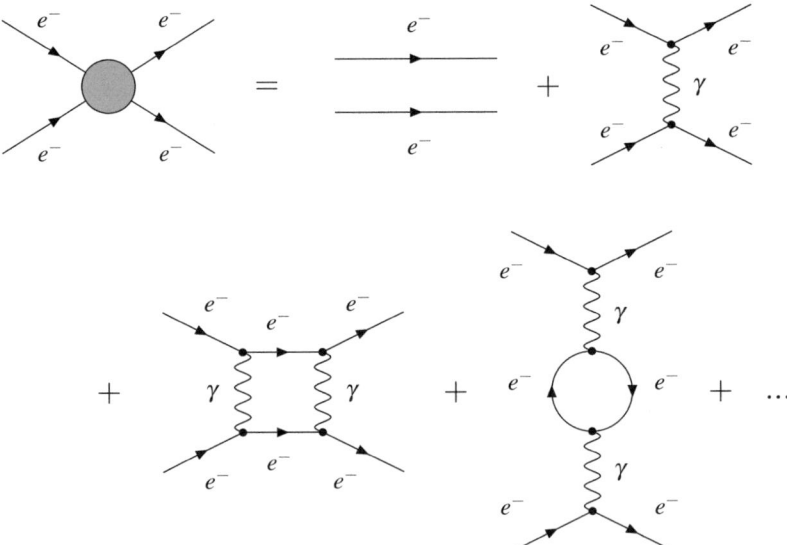

Fig. 3.3 Feynman diagrams for electron-electron scattering. The perturbative approach works because the expansion parameter is $\sqrt{\alpha_{em}} \ll 1$. Every vertex is suppressed by the factor $\sqrt{\alpha_{em}}$

3.3 Feynman Diagrams

a perturbative approach. At the zeroth order, there is no interaction. At the first order, we have the exchange of a photon. Every vertex in the diagram is suppressed by the factor $\sqrt{\alpha_{em}}$, where $\alpha_{em} = e^2/4\pi \approx 1/137 \ll 1$ is the fine structure constant and the expansion parameter. At the second order, we have several diagrams. Figure 3.3 shows the exchange of two photons and the exchange of a photon with the production of an electron-positron pair.

The Feynman diagrams are a convenient graphical representation of the perturbation theory approach. They can be easily obtained from the basic interaction vertices. The fundamental vertices of the Standard Model of particle physics (except those involving the Higgs boson) are reported in Fig. 3.4. These are the building blocks for the perturbative calculations. Every vertex must conserve the sum of the 4-momenta of incoming and outgoing particles, the electric charge, and any other quantum numbers respected by the theory. For instance, the baryonic and/or the leptonic numbers may not be conserved if they are broken in the fundamental Lagrangian. The 4-momentum of intermediate (virtual) particles is also fixed by energy-momentum conservation. However, these particles are normally off-mass-shell, i.e. their momentum does not satisfy the usual condition for free particles $p^2 = m^2$. This allows, for instance, that two light particles like electrons can interact through an exchange of a

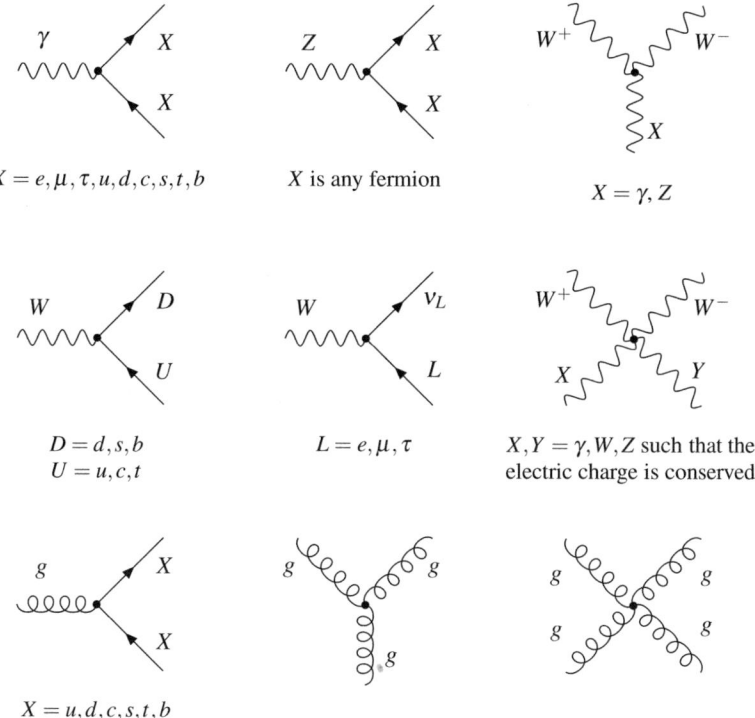

Fig. 3.4 Standard Model interactions mediated by gauge bosons

heavy boson, like Z^0. For example, a e^+e^- pair can go into a virtual Z-boson, which e.g. can then transform into a $\bar{\nu}\nu$ pair.

3.4 Beyond the Minimal Standard Model of Particle Physics

Despite the excellent agreement between theoretical predictions and experimental measurements, the MSM of particle physics has a few open problems that clearly indicate the existence of new physics. From the pure theoretical side, the main issue is the stability of the mass of the Higgs boson. Indeed, one can expect huge contributions to the square of the Higgs mass m_H^2 from the coupling with massive fermions at the level of

$$\Delta m_H^2 = -\frac{|\lambda_f|^2}{8\pi^2}\left(\Lambda_{UV}^2 + \cdots\right), \tag{3.1}$$

where λ_f is the Yukawa coupling between the Higgs field and the fermion f and Λ_{UV} is an ultraviolet energy cut-off. Unlike the other massive particles,[4] the Higgs mass is not protected by any symmetry at high energies, and therefore one could expect that such a contribution is huge, with Λ_{UV} of order the Planck scale $M_{Pl} \sim 10^{19}$ GeV, where quantum field theory presumably breaks down. Since the quark top has the largest Yukawa coupling, it should provide the main contribution to a huge Higgs mass. Its contribution to the Higgs boson mass is represented by the Feynman diagram in the left panel in Fig. 3.5. While it would be possible to renormalize the physical Higgs mass to an acceptable value of order of the electroweak scale, this would sound as an *ad hoc* fine-tuning. In other words, it would not be a natural solution. The issue is usually called hierarchy problem.

The phenomenon of neutrino oscillation is another clear evidence for physics beyond the MSM. In the MSM of particle physics, neutrinos are massless. The observed phenomenon of neutrino oscillation, in which neutrinos of one flavor can turn into neutrinos of another flavor, is only possible if the mass eigenstates and the flavor eigenstates are not the same; that is, at least two neutrinos must be massive. While it would be easy to provide neutrinos with a mass, neutrinos are special, because, in contrast to quarks and charged leptons, only the left hand component of neutrinos is interactive, while the the right hand component is sterile. Neutrinos are also electrically neutral, and this allows to introduce a mass only to left-handed neutrinos (the so-called Majorana mass).

The general Lagrangian of the MSM allows for a CP violating term in the QCD sector. Charge-conjugation transformation, C, changes a particle into the corresponding antiparticle. Parity transformation, P, is simply mirror reflection. In the MSM, we expect, on rather general grounds, the presence of a term in the QCD sector that

---

[4]Leptons, quarks, and massive gauge bosons are protected by the electroweak symmetry, which is restored at high energies and forbids mass terms.

3.4 Beyond the Minimal Standard Model of Particle Physics

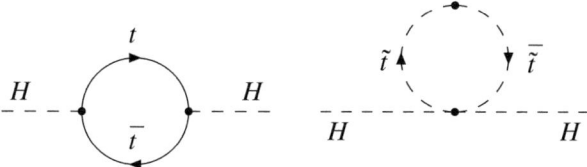

Fig. 3.5 *Left* quadratically divergent contribution from the *top* quark to the square of the Higgs mass. *Right* the similar contribution from the scalar supersymmetric partner of the *top* quark in a supersymmetric extension of the Standard Model

is not invariant under a CP transformation, the so called θ-term. However, no CP violation in QCD is observed. Thus the CP violating parameter should be very small, but there is no natural explanation for that. This is called the strong CP problem.

Physics beyond the MSM is also required by cosmology. First, there is no good dark matter candidate within the Standard Model. Dark matter particles must be massive, stable, and cannot have strong and electromagnetic interactions. The neutrinos of the Standard Model cannot do the job because their mass is too low, but in the past, when their mass constraints were much weaker, they were considered as possible candidates.

Second, within the MSM of particle physics it is impossible to generate the observed matter-antimatter asymmetry. Some time ago, the electroweak baryogenesis scenario was very popular. In this framework, the cosmological matter-antimatter asymmetry would be generated at the electroweak symmetry breaking in the MSM. However, it was later understood that the mechanism does not work for a number of reasons.

Third, there is no way to arrange inflation. Again, the scenario of a Standard Model inflation with the Higgs field in the role of inflaton field was investigated, but it was eventually found that the mechanism encounters serious problems.

We may also add a fourth problem, namely the explanation of the present accelerated expansion rate of the Universe. In this case, however, we do not know if new physics is necessary in the matter or in the gravity sector, or, in other words, if we need to modify the MSM of particle physics or Einstein's theory of General Relativity.

3.4.1 Supersymmetric Models

Supersymmetry is a symmetry relating fermions, particles with half-integer spin, and bosons, particles with integer spin. Every particle in one of the two groups must have a "super-partner" in the other group. In any supersymmetric extension of the Standard Model of particle physics any known particle should have a super-partner not yet discovered. The super-partner should exactly have the same properties as the original particle, except for the spin. Of course, the world around us is not supersymmetric, because we do not see these super-partners. For instance, there is no scalar particle with the properties of electron. So Supersymmetry, if it exists, must be broken at low energies, just like the electroweak gauge symmetry. In this case, the super-partners should get masses of order the Supersymmetry breaking scale and it is possible that such an energy scale is high enough that the super-partners of the Standard Model particles have not yet been created in particle colliders.

The first appealing feature of Supersymmetry is that it can stabilize the mass of the Higgs boson. Indeed, for any Standard Model diagram like that in the left in Fig. 3.5, we should have a diagram involving the super-partner, like that in the right in Fig. 3.5. It turns out that the contribution to the square of the Higgs mass from a scalar particle has the form

$$\Delta m_H^2 = 2\frac{\lambda_s}{16\pi^2}\left(\Lambda_{\text{UV}}^2 + \cdots\right). \tag{3.2}$$

If every fermion of the Standard Model has two scalar super-partners (fermions have spin-1/2, so they have two degrees of freedom, while scalars are spin-0 particles, just one degree of freedom) the quantum corrections to the Higgs mass can cancel. Since the Higgs mass is about 126 GeV, this is a strong argument to expect that the Supersymmetry scale is not too higher than the electroweak one, say $\sim$1 TeV. Another interesting possibility offered by supersymmetric models is that they often provide good dark matter candidates.

3.4.2 Grand Unification Theories

In the Standard Model of particle physics, the electromagnetic and the weak forces are not really unified. Even above the electroweak symmetry breaking scale, there are two different gauge groups and thus two different coupling constants. On the other hand, it is appealing to have a unified description of all the forces. Grand Unifications Theories (GUTs) have been proposed with this goal. The simplest possibility is a theory based on the symmetry group $SU(5)$. In $SU(5)$, we have 24 gauge bosons. In this scenario, $SU(5)$ should be spontaneously broken to $U_Y(1) \times SU_L(2) \times SU(3)$ at "low" energies. Below the $SU(5)$ breaking scale, the masses of 12 gauge bosons would be of the order the $SU(5)$ breaking scale, while $U_Y(1) \times SU_L(2) \times SU(3)$ would remain the residual symmetry above the electroweak scale with 12 massless gauge bosons.

In quantum field theory, the numerical value of the parameters of the theory depends on the energy scale involved in the measurement process. This is the direct consequence of quantum corrections. The coupling constants are not an exception and therefore they "run"; that is, their numerical value depends on the energy scale of the physical process. The dependence of the coupling constants on the energy scale is determined by a few factors, including the particle content. From very precise measurements at the electroweak scale performed at the Large Electron-Positron collider (LEP) at CERN, in Geneva, in the 1990s, one can see that the coupling constants of the MSM do not converge to a single value at high energies, see the top

3.4 Beyond the Minimal Standard Model of Particle Physics

panel in Fig. 3.6. However, it turns out that the unification is possible in the case of a supersymmetric extension of the Standard Model with the super-partners having masses around 1–10 TeV, as shown in the bottom panel in Fig. 3.6. This is at present the only indication in favor of a low energy Supersymmetry and of a GUT with a

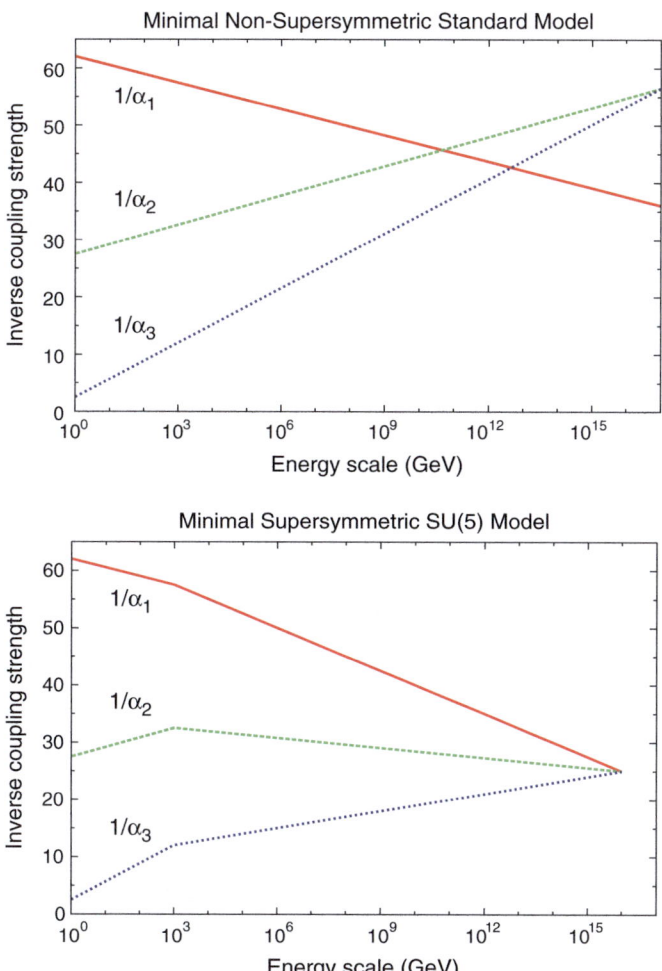

Fig. 3.6 Extrapolation of the inverse of the gauge couplings of the Standard Model to high energies on the basis of high precision measurements at LEP at CERN. α_1, α_2, and α_3 are, respectively, the gauge couplings of $U_Y(1)$, $SU_L(2)$, and $SU(3)$. In the case that the matter content is determined by the MSM particles only, we do not see any unification of the gauge couplings at high energies (*top panel*). If we assume that for every MSM particle there is a new particle with the same interaction properties and a mass in the TeV range, the gauge couplings converge at $M_{\text{GUT}} \sim 10^{16}$ GeV (*bottom panel*). The latter is interpreted as the GUT scale at which the $SU(5)$ symmetry is broken. Its precise value depends on the assumed mass of the new super-partner particles

large energy scale, around $10^{14}-10^{16}$ GeV. GUTs may have important implications in cosmology.

3.4.3 Heavy Neutrinos

In the Standard Model of particle physics, we only have massless left-handed neutrinos and massless right-handed antineutrinos. Neutrinos are the only elementary fermions without electric charge. They interact through the weak nuclear force, which only acts on left-handed particles and right-handed antiparticles.

Today we know that neutrinos are not massless, but their masses are unnaturally small in comparison with those of the other Standard Model particles. The other fermions get a mass through the interaction with the Higgs field. The Lagrangian density has a term of the form $\lambda_f H \bar{\psi} \psi$, where λ_f is the Yukawa coupling between the Higgs field H and the fermionic field ψ. When the Higgs field acquires a non-vanishing vacuum expectation value v, the former Yukawa term in the Lagrangian density becomes the mass term of the fermionic field ψ, leading to the mass $m = \lambda_f v$. This mass term mixes left-handed and right-handed fermions and it is called Dirac mass. This is the only possibility of introducing masses to electrically charged fermions. Such a mass term violates the $U_Y(1) \times SU_L(2)$ symmetry and therefore fermions can have a mass only after the electroweak symmetry breaking. Since $v \sim 250$ GeV, for a "natural" value of the Yukawa coupling λ_f the masses of the Standard Model fermions should be of order the electroweak scale. From this point of view, only the quark top has a natural mass. In the case of neutrinos, the mass constraint is $m_\nu < 2$ eV and such a small value sounds very unnatural.

There are several ways to provide neutrinos with a mass. An appealing scenario is the so-called see-saw model, which has links to very high energy physics, possibly to the GUT scale. The starting point is a mass matrix of the form

$$\begin{pmatrix} 0 & m \\ m & M \end{pmatrix}. \quad (3.3)$$

m is the neutrino Dirac mass coming from the electroweak symmetry breaking, which is expected to be of order of 100 GeV and requires right-handed neutrinos and left-handed antineutrinos. M is the neutrino Majorana mass of right-handed neutrinos and left-handed antineutrinos. Since the latter have no charge with respect to the Standard Model gauge symmetries, the mass term involving M can be generated by high energy physics and M may be huge, even of order the GUT scale $\sim 10^{14}-10^{16}$ GeV. The mass eigenvalues of the mass matrix in (3.3) are

$$m_\pm = \frac{M \pm \sqrt{M^2 + 4m^2}}{2}. \quad (3.4)$$

3.4 Beyond the Minimal Standard Model of Particle Physics

If $M \gg m$, we have

$$m_+ \approx M, \quad m_- \approx -\frac{m^2}{M}. \tag{3.5}$$

The negative mass eigenvalue does not present any problem, because for fermions the sign of the mass can be changed by the so-called γ_5 transformation of the spinor. Moreover, in all physical effects there appears only m^2, never m.

If M increases, m_+ increases too, while m_- decreases. This is the reason for the name see-saw mechanism. For $m \sim 100$ GeV and $M \sim 10^{14}$–10^{16} GeV, we find $m_- \sim 0.001 - 0.1$ eV, which is consistent with the present mass constraints.

3.4.4 Peccei-Quinn Model

Peccei and Quinn suggested a possible solution to the strong CP problem, i.e. to the explanation why the CP violating parameter of the strong interaction is unnaturally close to zero. The model introduces a new global $U(1)$ symmetry under which some complex scalar field has a non-vanishing charge. At low energies, the symmetry is spontaneously broken and the vacuum expectation value of the scalar field automatically happens to be at the point where the CP violation related to the QCD θ-term disappears.

As a result of the spontaneous symmetry breaking, there appears a very light scalar boson called axion. This is a consequence of a general theorem by Goldstone that the spontaneous breaking of a global symmetry leads to a massless scalar field. A non-zero mass of the axion appears due to additional explicit symmetry breaking related to the so called *instantons*. Depending on the value of the axion mass, these particles may also be good dark matter candidates.

3.5 Probabilities of Reactions Among Particles

The reaction rates of scattering processes can be calculated in the framework of quantum field theory. They cannot be rigorously derived within an introductory course on cosmology. In this section, we just provide a recipe to get a rough estimate of cross sections of scattering processes on the basis of the leading order Feynman diagram. More details can be found in any introductory textbook on quantum field theory, like Mandl and Shaw (2010).

As discussed in Sect. 3.3, all the quantum numbers must be conserved at every vertex, according to the conservation laws determined by the Lagrangian of the theory. The fundamental vertices including only fermions and gauge bosons are shown in Fig. 3.4. Every vertex is suppressed by the corresponding gauge charge, say g, or, equivalently, by the square root of the gauge coupling $\alpha_g = g^2/4\pi$. The perturbative approach is valid if $g \ll 1$. Such a condition holds in the electroweak sector, while it is satisfied only in the case of high energy collisions in QCD. Internal lines connecting two vertices introduce an additional factor (propagator) in the amplitude described by the diagram. In the case of a boson of mass M, the factor is

$$\sim \frac{1}{|M^2 - q^2|}, \qquad (3.6)$$

where q is the 4-momentum transferred to the virtual boson. If the internal line describes a fermion, the factor is

$$\sim \frac{M + \gamma \cdot q}{|M^2 - q^2|}. \qquad (3.7)$$

where the product $\gamma \cdot q$ is the scalar product of the 4-momentum q by some spin matrices (Dirac matrices). For an order of magnitude estimate, $q^2 \sim E^2$, where E is the characteristic energy scale of the process.

The probability of a reaction is given by the square of the amplitude described by the diagram. The cross section can be evaluated from dimensional arguments, remembering that in natural units it has the dimensions of inverse square of energy.

Example 3.1 Let us consider the scattering $e^- \nu_e \to e^- \nu_e$ at energies $E \ll M_W$. The left diagram in Fig. 3.7 is one of the leading order Feynman diagrams for this process (there is also a diagram with the exchange of a Z-boson). Every vertex introduces the factor $g \sim 0.1$, while the propagator introduces the factor $1/M_W^2$, since $E \ll M_W$. The amplitude corresponding to this diagram is thus of the order of g^2/M_W^2. The cross section is proportional to the square of the amplitude. Since, as we have mentioned above, the cross section has the dimension of inverse square of energy, it can be estimated as

$$\sigma(e^- \nu_e \to e^- \nu_e) \sim \frac{g^4}{M_W^4} E^2. \qquad (3.8)$$

The factor E^4 comes from the square of the product of four Dirac spinors; the factor $1/E^2$ comes from the particle flux by which the amplitude squared should be divided to obtain the cross section.

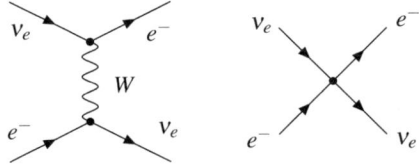

Fig. 3.7 *Left* first order Feynman diagram for the scattering $e^- \nu_e \to e^- \nu_e$ mediated by a W-boson. *Right* the same process in the Fermi theory

3.5 Probabilities of Reactions Among Particles

In the Fermi theory of the weak interactions, the scattering $e^- \nu_e \to e^- \nu_e$ is described by a 4-fermion interaction term of the form $G_F(\bar{\psi}\psi)(\bar{\psi}\psi)$ in the Lagrangian density (right diagram in Fig. 3.7). Here $G_F \approx 10^{-5}$ GeV^{-2} is the Fermi constant. From dimensional arguments, it is easy to conclude that the cross section must be

$$\sigma(e^- \nu_e \to e^- \nu_e) \sim G_F^2 E^2 \,. \tag{3.9}$$

The Fermi theory holds at low energies, $E \ll M_W$. In this regime it agrees with the Standard Model predictions, since $G_F \sim g^2/M_W^2$.

Example 3.2 Let us now consider the scattering $e^- e^+ \to \mu^- \mu^+$ at energies $m_\mu \ll E \ll M_Z$. The leading order Feynman diagram is the left one in Fig. 3.8 (the diagram with the exchange of a Z-boson is suppressed for $E \ll M_Z$). The two vertices contribute with the factor α and the propagator introduces the factor $1/E^2$. So the amplitude is of order α/E^2 and the square of the amplitude is approximately α^2/E^4. The cross section is thus

$$\sigma(e^- e^+ \to \mu^- \mu^+) \sim \frac{\alpha^2}{E^2} \,. \tag{3.10}$$

We note that the assumption $m_\mu \ll E \ll M_Z$ makes the only energy scale of the problem equal to the scattering energy in the center of mass E.

Example 3.3 Lastly, we want to estimate the cross section of the elastic scattering $e^- \gamma \to e^- \gamma$ at low energies $E \ll m_e$, which is called Thomson scattering. The leading order Feynman diagram is the right one in Fig. 3.8. The two vertices contribute with the factor α and the propagator gives the factor $1/m_e$, since $E \ll m_e$. So the cross section is of the order

$$\sigma(e^- \gamma \to e^- \gamma) \sim \frac{\alpha^2}{m_e^2} \,. \tag{3.11}$$

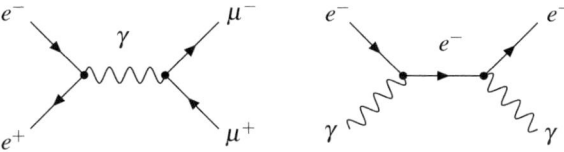

Fig. 3.8 *Left* first order Feynman diagram for the scattering $e^- e^+ \to \mu^- \mu^+$ mediated by a photon. *Right* first order Feynman diagram for the scattering $e^- \gamma \to e^- \gamma$

Problems

3.1 Estimate the cross section of the scattering $\nu_e \nu_\mu \to \nu_e \nu_\mu$, both in the case $E \ll M_Z$ and $E \gg M_Z$.

3.2 Draw the Feynman diagram for the muon decay. [Hint: since the lightest baryons and mesons are heavier than the muon, the latter can only decay into leptons.]

3.3 Draw the Feynman diagram for the neutron decay $n \to pe^-\bar{\nu}_e$. [Hint: the neutron is a bound state udd and the proton is a bound state uud. A d quark in the neutron decays into a u quark with the emission of a virtual W-boson.]

References

D. Griffiths, *Introduction to elementary particles*, 2nd edn. (Wiley-VCH, Weinheim, 2008)
F. Mandl, G. Shaw, *Quantum field theory*, 2nd edn. (Wiley, Chichester, 2010)
K.A. Olive et al., Particle data group collaboration. Chin. Phys. C **38**, 090001 (2014)

Chapter 4
Cosmological Models

The Einstein equations relate the geometry of the spacetime, encoded in $G^{\mu\nu}$, to the matter content, described by the matter energy-momentum tensor $T^{\mu\nu}$. If the matter content and its distribution are known, we can determine the spacetime geometry. In general, it is highly non-trivial to solve the Einstein equations, because they are second order non-linear partial differential equations for ten components of the metric tensor. Analytic solutions can be found only if some "helpful" symmetries of the spacetime are present. In the Standard Model of cosmology, we assume the so-called *Cosmological Principle*:

$$\text{The Universe is homogeneous and isotropic.} \tag{4.1}$$

This is clearly an approximation, because we observe a lot of structures and the Universe is far from being homogeneous and isotropic. However, if we average over large volumes, say over scales larger than 10 Mpc, the assumption sounds reasonable. Moreover, it essentially relies on the fact there are no preferred points or preferred directions in the Universe, a sort of Copernican Principle. Nevertheless, it is also possible that high precision data cannot be treated with this simplification and there is today a debate on the realm of validity of the Cosmological Principle and even on the fact that current measurements of the cosmological parameters might deviate from their correct values because they are inferred under the assumption of homogeneity and isotropy (see e.g. Fleury et al. 2013; Marra et al. 2007).

If the Cosmological Principle is valid, the geometry of the Universe is described by the Friedmann-Robertson-Walker (FRW) metric. The latter only depends on a constant parameter k and a function of time $a(t)$. The constant k may be positive, negative, or zero, respectively for the case of closed, open, or flat universe. If $k \neq 0$, it is usually normalized to unity, $k = \pm 1$. $a(t)$ is the scale factor and determines the evolution of distances between distant (not bound) objects in the Universe. If we plug the FRW metric into the Einstein equations, we obtain the Friedmann equations, which determine how $a(t)$ depends on time for different forms of cosmological matter. In the old Friedmann cosmology, it was assumed that all physically relevant models started from an initial singularity, where the scale factor was zero. The Uni-

---

The original version of this chapter was revised: The errors in this chapter have been corrected. The correction to this chapter can be found at https://doi.org/10.1007/978-3-662-48078-6_13.

verse expansion was supposed to start from this initial singularity, called big bang, or near it. The expansion rate and the final destiny of the Universe depend on its 3-dimensional geometry and the matter content. In the simplest cases of matter dominated or radiation dominated models, the Universe first expands and then recollapses if it is closed ($k > 0$), while it expands forever if it is open ($k < 0$) or flat ($k = 0$). In the presence of vacuum energy, the picture is more complicated and the fate of the Universe depends on the specific contribution of its components. Current observations support the so-called Λ cold dark matter model (ΛCDM model), in which the Universe is almost flat and today it is dominated by vacuum (or vacuum-like) energy. The latter represents about 70% of the total energy density, while the other 30% is made of non-relativistic matter. The contribution from other components is much smaller and irrelevant for the current expansion regime. The age of the Universe, namely the time interval from the beginning of its expansion up to today, is around 14 Gyr.

4.1 Friedmann-Robertson-Walker Metric

If we assume the Cosmological Principle, the background geometry is strongly constrained, independently of the Einstein equations of General Relativity. The Cosmological Principle requires indeed that there are no preferred points (homogeneity, namely invariance under spatial translations) and no preferred directions (isotropy, or invariance under spatial rotations) in the 3-dimensional space. However, it is still allowed that the spacetime geometry depends on time. It turns out that the only background compatible with these requirements is the FRW metric (Weinberg 1972), and its line element is

$$ds^2 = dt^2 - a^2(t)\left(\frac{dr^2}{1 - kr^2} + r^2 d\theta^2 + r^2 \sin^2\theta d\phi^2\right), \quad (4.2)$$

where $a(t)$ is the *scale factor*, which is independent of the spatial coordinates $\{r, \theta, \phi\}$, while k is a constant. The latter can be positive, negative, or zero, but it is always possible to properly rescale the coordinate r and have $k = \pm 1$ or 0. For $k = 1$, we have a *closed universe*, for $k = 0$ a *flat universe*, and for $k = -1$ an *open universe*. If we consider a gravity theory different from General Relativity, but we keep the assumption of the Cosmological Principle in a $3 + 1$ dimensional spacetime, the background geometry is still given by the FRW metric. General Relativity can only determine $a(t)$ and k if the properties of the matter in the universe are known. We note that a flat universe is not necessarily a flat (Minkowski) spacetime, but when a is independent on t and $k = 0$, we recover the flat spacetime of Special Relativity.

It is instructive to compute some invariants of the FRW metric. This can be easily done, for instance, with some specific Mathematica packages, but it would be a good exercise to make these calculations by hand, to gain a better insight into the formality and the spirit of the Riemann geometry. The scalar curvature turns out to be

4.1 Friedmann-Robertson-Walker Metric

$$R = -6\frac{k + \dot{a}^2 + \ddot{a}a}{a^2}, \quad (4.3)$$

and it vanishes when $k = 0$ and a is independent of t (flat spacetime of Special Relativity). We also note that R diverges when $a \to 0$. If we multiply the left and the right hand sides of the Einstein equations (2.78) with $\Lambda = 0$ by $g_{\mu\nu}$, we get

$$R = -\frac{8\pi}{M_{\text{Pl}}^2} T, \quad (4.4)$$

where $T = T_\mu^\mu$ is the trace of the matter energy-momentum tensor. The divergence of R thus implies the divergence of T. The square of the Riemann tensor is given by

$$R_{\mu\nu\rho\sigma} R^{\mu\nu\rho\sigma} = 12\frac{k^2 + 2k\dot{a}^2 + \dot{a}^4 + \ddot{a}^2 a^2}{a^4}, \quad (4.5)$$

and it also diverges for $a \to 0$.

Lastly, we note that the Einstein equations are local equations. They cannot tell us anything about global properties of the spacetime like its volume. The same metric can indeed describe topologically different universes. This point was already noticed by Friedmann (1999). If we assume that the Universe has a trivial topology, the 3-volume is finite if $k = 1$ and infinite for $k = 0$ and -1. From the FRW metric we have

$$V_{\text{universe}} = \int_V \sqrt{-^3 g}\, d^3 x = a^3(t) \int_0^{2\pi} d\phi \int_0^\pi \sin\theta\, d\theta \int_0^{R_k} \frac{r^2 dr}{\sqrt{1-kr^2}}, \quad (4.6)$$

where $^3 g$ is the determinant of the spatial 3-metric, while $R_k = 1$ for $k = 1$ and $+\infty$ for $k = 0$ and -1. The integration gives

$$V_{\text{universe}} = \begin{cases} \pi^2 a^3(t) & \text{for } k = 1 \\ +\infty & \text{for } k = 0, -1 \end{cases}. \quad (4.7)$$

Closed universes are always finite, but in the case of non-trivial topology even flat and open universes may have a finite volume. As we will see in Chap. 10, current CMB data suggest that the Universe is close to be flat, allowing both $k = 1$ and $k = -1$. This means that even assuming a trivial topology we cannot say if our Universe has a finite or an infinite volume. In the case of non-trivial topology, the size of the Universe may be evaluated by looking for "ghost images" of astronomical sources, because in a multi-connected universe the radiation emitted by a source should be detected from different directions. For the time being, there is no evidence of any ghost image and therefore we can only get a lower bound on the possible size of a topologically non-trivial universe.

4.2 Friedmann Equations

The assumption of the Cosmological Principle requires that the spacetime geometry is described by the FRW metric. The scale factor $a(t)$ and the constant k can be obtained from the Einstein equations if the matter content of the Universe is known. The latter can be described with good approximation by the energy-momentum tensor of a perfect fluid

$$T^{\mu\nu} = (\rho + P) u^\mu u^\nu - P g^{\mu\nu}, \qquad (4.8)$$

where ρ and P are, respectively, the energy density and the pressure of the fluid. Since in the coordinate system of the FRW metric the Universe is manifestly homogeneous and isotropic, we have to consider the rest frame of the fluid in which $u^\mu = (1, 0, 0, 0)$. If we plug the FRW metric and the energy-momentum tensor of this perfect fluid into the Einstein equations, from the 00 component we find

$$H^2 = \frac{8\pi}{3 M_{\rm Pl}^2} \rho - \frac{k}{a^2}, \qquad (4.9)$$

where $H = \dot{a}/a$ is the Hubble parameter. Equation (4.9) is called *first Friedmann equation*. In the general case, we can expect that the Universe is made of different components, say dust, radiation, etc. ρ and P have thus to be seen as, respectively, the total energy density and the total pressure

$$\rho = \sum_i \rho_i, \quad P = \sum_i P_i, \qquad (4.10)$$

where the sum is taken over all the relevant types of matter. The 11, 22, and 33 components of the Einstein equations provide the same equation, which reads

$$\frac{\ddot{a}}{a} = -\frac{4\pi}{3 M_{\rm Pl}^2} (\rho + 3P). \qquad (4.11)$$

This is the *second Friedmann equation*.

In the FRW metric, the covariant conservation of the energy-momentum tensor $\nabla_\mu T^{\mu 0} = 0$ becomes

$$\dot{\rho} = -3H (\rho + P). \qquad (4.12)$$

Equation (4.12) is not independent and it can indeed be derived from the first and the second Friedmann equations (see the discussion at the end of Chap. 2). At this point, we have three unknown functions of the time t, namely ρ, P, and the scale factor a, and two equations, the two Friedmann equations or one of the Friedmann equations together with the covariant conservation of the energy-momentum tensor (4.12).

4.2 Friedmann Equations

One more equation is thus necessary. One typically uses the matter equation of state, expressing the pressure in terms of the energy density, $P = P(\rho)$. In the simplest and practically important case, the relation between the energy density and the pressure is linear

$$P = w\rho, \tag{4.13}$$

with w constant. For instance, dust is described by $w = 0$, radiation by $w = 1/3$ (this is true for any kind of ultra-relativistic matter, not only for photons), and vacuum energy by $w = -1$. However, if we consider other forms of matter, w may not be a constant any more and, in general, the linear relation between the energy density and the pressure may not be valid.

For $k = 0$, the first Friedmann equation is

$$H^2 = \frac{8\pi}{3M_{\text{Pl}}^2}\rho_c \Rightarrow \rho_c = \frac{3M_{\text{Pl}}^2 H^2}{8\pi}, \tag{4.14}$$

which defines the *critical density* ρ_c. The value of the critical energy today is

$$\begin{aligned}\rho_c^0 &= \frac{3M_{\text{Pl}}^2 H_0^2}{8\pi} = 1.878 \cdot 10^{-29}\, h_0^2\, \text{g/cm}^{-3} \\ &= 1.054 \cdot 10^{-5}\, h_0^2\, \text{GeV/cm}^{-3}.\end{aligned} \tag{4.15}$$

4.3 Cosmological Models

If we know the matter content of the Universe, we can solve the Friedmann equations and find the spacetime geometry of the Universe at any time. A different matter content provides a different cosmological model. If we assume an equation of state in the form (4.13), the covariant conservation law (4.12) becomes

$$\frac{\dot{\rho}}{\rho} = -3(1+w)\frac{\dot{a}}{a}, \tag{4.16}$$

and therefore

$$\rho \sim a^{-3(1+w)}. \tag{4.17}$$

The energy density thus scales as $1/a^3$ for dust, as $1/a^4$ for radiation, and it is constant in the case of vacuum energy. If we plug this result into the first Friedmann equation and we neglect the k/a^2 term, we find (for $w \neq -1$)

$$a(t) \sim t^\alpha \quad \text{with } \alpha = \frac{2}{3(1+w)}. \tag{4.18}$$

This is a good approximation at early times, because, if we go backward in time, $a \to 0$ and the k/a^2 term is subdominant with respect to those of dust and radiation, which scale, respectively, as $1/a^3$ and $1/a^4$.

4.3.1 Einstein Universe

If we write the Einstein equations with a non-vanishing cosmological constant Λ as in Eq. (2.84), the first and the second Friedmann equations become

$$H^2 = \frac{8\pi}{3M_{\text{Pl}}^2}\rho + \frac{\Lambda}{3} - \frac{k}{a^2}, \tag{4.19}$$

$$\frac{\ddot{a}}{a} = -\frac{4\pi}{3M_{\text{Pl}}^2}(\rho + 3P) + \frac{\Lambda}{3}. \tag{4.20}$$

The Einstein universe is a cosmological model in which matter is described by dust ($P = 0$) and a positive cosmological constant is introduced to make the universe static, in accordance with the common belief before the discovery of Hubble's law. Imposing the condition $\dot{a} = \ddot{a} = 0$ in Eqs. (4.19) and (4.20), we find

$$\rho = \frac{\Lambda M_{\text{Pl}}^2}{4\pi}, \quad a = \frac{1}{\sqrt{\Lambda}}, \quad k = 1. \tag{4.21}$$

We note that the Einstein universe is unstable, in the sense that small perturbations would force it either to collapse or expand, and therefore it does not lead to a static universe as it was its original purpose.

4.3.2 Matter Dominated Universe

Non-relativistic matter with the equation of state $P = 0$ or, equivalently, $w = 0$ is usually called dust. From Eq. (4.12), we find

$$\rho a^3 = \text{constant} \equiv A. \tag{4.22}$$

The first Friedmann equation becomes

$$\dot{a}^2 = \frac{8\pi A}{3M_{\text{Pl}}^2}\frac{1}{a} - k. \tag{4.23}$$

Let us introduce the new variable η, related to t by

4.3 Cosmological Models

$$\frac{d\eta}{dt} = \frac{1}{a(t)}. \tag{4.24}$$

The new time variable η is called conformal time. In terms of η, Eq. (4.23) can be rewritten as

$$a'^2 = \frac{8\pi A}{3M_{\text{Pl}}^2} a - ka^2, \tag{4.25}$$

where the prime denotes the derivative with respect to the conformal time η. Equation (4.25) can be integrated by separation of variables. If we choose the initial condition $a = 0$ for $t = 0$, we find the following parametric solutions for t and a in the case of closed ($k = 1$), flat ($k = 0$), and open ($k = -1$) universes

$$k = 1 \quad a = \frac{4\pi A}{3M_{\text{Pl}}^2}(1 - \cos\eta), \quad t = \frac{4\pi A}{3M_{\text{Pl}}^2}(\eta - \sin\eta), \tag{4.26}$$

$$k = 0 \quad a = \frac{2\pi A}{3M_{\text{Pl}}^2}\eta^2, \quad t = \frac{2\pi A}{9M_{\text{Pl}}^2}\eta^3, \tag{4.27}$$

$$k = -1 \quad a = \frac{4\pi A}{3M_{\text{Pl}}^2}(\cosh\eta - 1), \quad t = \frac{4\pi A}{3M_{\text{Pl}}^2}(\sinh\eta - \eta). \tag{4.28}$$

Figure 4.1 shows the scale factor $a(t)$ as a function of the cosmological time t for the three scenarios. A closed universe expands up to a critical point and then recollapses. An open universe expands forever. A flat universe is the boundary case between

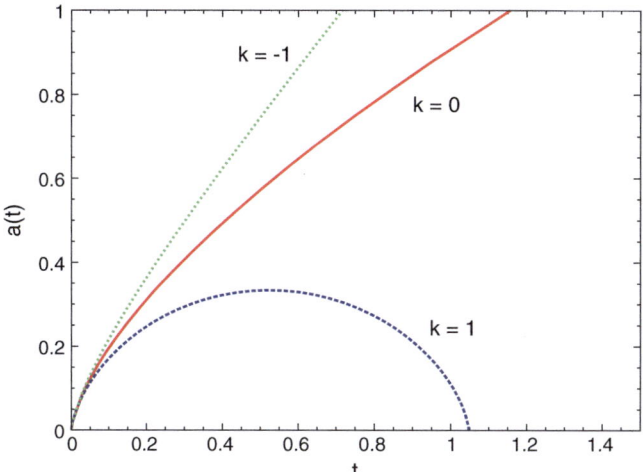

Fig. 4.1 Scale factor a as a function of the cosmological time t for the three types of matter dominated universes: closed universe ($k = 1$), flat universe ($k = 0$), and open universe ($k = -1$). Here t and a are presented in units in which $8\pi A/M_{\text{Pl}}^2 = 1$

open and closed universes: it expands forever, but the expansion rate asymptotically approaches zero, namely $H \to 0$ for $t \to +\infty$. These statements are valid if the cosmological constant is zero.

4.3.3 Radiation Dominated Universe

The equation of state of relativistic matter is $P = \rho/3$, so $w = 1/3$. It describes a gas of massless non-interacting particles, but it is also valid in the case of an ultra-relativistic gas, in which the particles' rest-energy is negligible in comparison to their total energy. From Eq. (4.12), we find that

$$\rho a^4 = \text{constant} \equiv B, \tag{4.29}$$

and the first Friedmann equation can be written as

$$\dot{a}^2 = \frac{8\pi B}{3M_{\text{Pl}}^2}\frac{1}{a^2} - k. \tag{4.30}$$

If we impose the initial condition $a(t = 0) = 0$, we find the following solutions for the scale factor $a(t)$ in the case of closed ($k = 1$), flat ($k = 0$), and open ($k = -1$) universes:

$$k = 1 \quad a = \left[2\sqrt{\frac{8\pi B}{3M_{\text{Pl}}^2}}t - t^2\right]^{1/2}, \tag{4.31}$$

$$k = 0 \quad a = \left[2\sqrt{\frac{8\pi B}{3M_{\text{Pl}}^2}}t\right]^{1/2}, \tag{4.32}$$

$$k = -1 \quad a = \left[2\sqrt{\frac{8\pi B}{3M_{\text{Pl}}^2}}t + t^2\right]^{1/2}. \tag{4.33}$$

As in the matter dominated universe, when $k = 1$ the universe first expands, then the scale factor reaches a maximum value, and eventually the universe recollapses to $a = 0$. An open universe ($k = -1$) expands forever, while a flat universe represents the critical case separating the closed and the open models. Figure 4.2 shows the scale factor $a(t)$ as a function of the cosmological time t for these three scenarios.

4.3 Cosmological Models

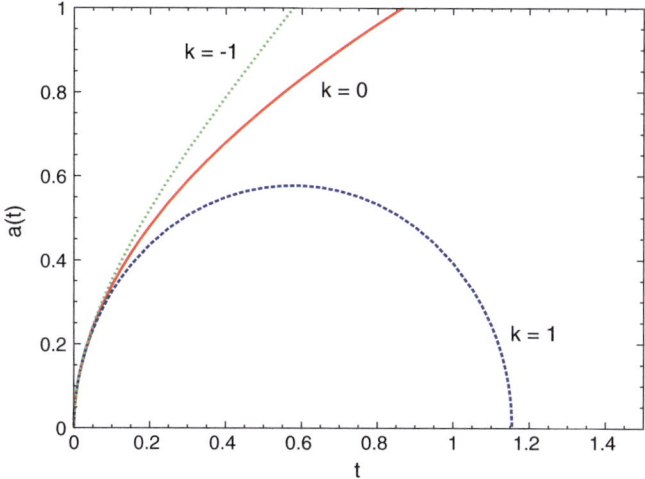

Fig. 4.2 Scale factor a as a function of the cosmological time t for the three types of radiation dominated universes: closed universe ($k = 1$), flat universe ($k = 0$), and open universe ($k = -1$). The scale factor is expressed in units in which $8\pi B/M_{\text{Pl}}^2 = 1$

4.3.4 Vacuum Dominated Universe

In a vacuum dominated universe, there is no matter, so $\rho = P = 0$, but we allow for a non-vanishing cosmological constant. As we will show in Chap. 11, in quantum field theory the vacuum is not empty and in General Relativity it should act as an effective cosmological constant or, more precisely, the vacuum energy is equivalent, up to a constant factor, to a cosmological constant.

In the case of $\Lambda > 0$, the Friedmann equations provide the following solutions

$$k = 1 \quad a = \sqrt{\frac{3}{\Lambda}} \cosh\left(\sqrt{\frac{\Lambda}{3}}t\right), \tag{4.34}$$

$$k = 0 \quad a = a(0)\exp\left(\sqrt{\frac{\Lambda}{3}}t\right), \tag{4.35}$$

$$k = -1 \quad a = \sqrt{\frac{3}{\Lambda}} \sinh\left(\sqrt{\frac{\Lambda}{3}}t\right). \tag{4.36}$$

If $\Lambda < 0$, the solution is

$$k = -1 \quad a = \sqrt{-\frac{3}{\Lambda}} \cos\left(\sqrt{-\frac{\Lambda}{3}}t\right), \tag{4.37}$$

and there is no solution for $k = 0$ and 1. Lastly, when $\Lambda = 0$, we recover the flat spacetime with $k = 0$ and a constant.

4.4 Basic Properties of the FRW Metric

As it is discussed in Sect. 2.2, the motion of test-particles in curved spacetime can be studied by considering the Lagrangian

$$L = g_{\mu\nu} x'^{\mu} x'^{\nu}, \qquad (4.38)$$

where now we use the prime to indicate the derivative with respect to the affine parameter λ, while the dot is reserved for the derivative with respect to the cosmological time t. Since the FRW metric depends on t through the scale factor $a(t)$, the particle's energy is not a constant of motion. Let us now consider a photon and a coordinate system in which its motion is only along the radial direction. In this case, the relation $g_{\mu\nu} x'^{\mu} x'^{\nu} = 0$ becomes

$$t'^2 = \frac{a^2}{1 - kr^2} r'^2. \qquad (4.39)$$

If we write the Euler-Lagrange equations for $x^{\mu} = t$ and we use Eq. (4.39), we find

$$t'' = -a\dot{a}\frac{r'^2}{1-kr^2} = -\frac{\dot{a}}{a} t'^2 = -\frac{a'}{a} t'. \qquad (4.40)$$

t' is proportional to the photon's energy E and therefore $t''/t' = E'_\gamma/E_\gamma$. We thus find that a photon propagating in a FRW background redshifts as the inverse of the scale factor

$$E_\gamma \sim 1/a. \qquad (4.41)$$

The phenomenon is called *cosmological redshift*, to be distinguished from the Doppler redshift due to the relative motion of a source and from the gravitational redshift due to the climbing in a gravitational potential. It is also responsible for the behavior of the energy density of radiation that scales as $1/a^4$: the photon number density scales as the inverse of the volume, $1/a^3$, while the photon's energy scales as $1/a$.

The cosmological redshift of a photon can also be derived in the following way. We consider the emission of monochromatic electromagnetic radiation emitted by a source at the origin $r = 0$. The wavefront emitted at the time $t = t_e$ is detected at the time $t = t_o$ at the radial coordinate $r = r_o$. Since $ds^2 = 0$, we can write

$$\int_{t_e}^{t_o} \frac{d\tilde{t}}{a} = \int_0^{r_o} \frac{d\tilde{r}}{\sqrt{1 - k\tilde{r}^2}}. \qquad (4.42)$$

The right hand side does not depend on the time t. If we consider the wavefront after, which is emitted by the source at the time $t = t_e + \delta t_e$ at $r = 0$ and it is detected at the time $t = t_o + \delta t_o$ at $r = r_o$, we have

4.4 Basic Properties of the FRW Metric

$$\int_{t_e+\delta t_e}^{t_o+\delta t_o} \frac{d\tilde{t}}{a} = \int_0^{r_o} \frac{d\tilde{r}}{\sqrt{1-k\tilde{r}^2}} = \int_{t_e}^{t_o} \frac{d\tilde{t}}{a}, \quad (4.43)$$

and therefore

$$\int_{t_e}^{t_e+\delta t_e} \frac{d\tilde{t}}{a} = \int_{t_o}^{t_o+\delta t_o} \frac{d\tilde{t}}{a} \Rightarrow \frac{\delta t_e}{a(t_e)} = \frac{\delta t_o}{a(t_o)}. \quad (4.44)$$

δt_e and δt_o are, respectively, the wavelengths of the radiation measured at the time t_e and at the time t_o. It follows that the photon wavelength scales as a and the photon's energy as $1/a$, in agreement with Eq. (4.41).

An important concept is that of *particle horizon*, which is the distance travelled by a photon from the moment of the big bang to a certain time t. It defines the radius of causally connected regions at the time t, in the sense that two points at a distance larger than the particle horizon have never exchanged any information. For a flat universe ($k = 0$), from the equation for light propagation $ds^2 = 0$, we find

$$r = \int_0^r d\tilde{r} = \int_0^t \frac{d\tilde{t}}{a} = \frac{1}{a(t)} \frac{t}{(1-\alpha)}, \quad (4.45)$$

where in the last passage we have used the fact that $a \sim t^\alpha$ for $w \neq -1$, see Eqs. (4.18), (4.27), and (4.32). The proper distance at the time t between the origin and a point with radial coordinate r is $d(t) = a(t)r$ and therefore the particle horizon is

$$d = \frac{t}{1-\alpha}. \quad (4.46)$$

In the case of a universe made of dust, we find $d = 3t$, while for a radiation dominated universe we have $d = 2t$. We note that the particle horizon increases linearly with time, while $a \sim t^\alpha$ with $\alpha < 1$ in the case of non-exotic matter with $w \geq 0$, which means that more and more points of the Universe become causally connected at later times.

The case of a universe filled with vacuum energy has to be treated separately. Here the scale factor is given by Eq. (4.35) and the particle horizon is

$$d = \sqrt{\frac{3}{\Lambda}} \left[\exp\left(\sqrt{\frac{\Lambda}{3}} t\right) - 1 \right]. \quad (4.47)$$

Since the scale factor also grows exponentially, if two regions are at a coordinate distance larger than $\sqrt{3/\Lambda}$, they will never been able to exchange any information.

4.5 Age of the Universe

From the Friedmann equations, one finds that a universe begins with a spacetime singularity (the scale factor a vanishes) and expands (a increases). At the singularity, the energy density diverges. Actually, we do not expect that standard physics can work above the Planck scale $M_{Pl} \sim 10^{19}$ GeV, where quantum gravity effects should become important. However, in the absence of a robust and reliable theory of quantum gravity we can only describe the Universe within classical General Relativity. In this framework, the Universe was born at the time in which the energy density diverged and the age of the Universe is the time interval measured with respect to the temporal coordinate of the FRW metric from this initial time till today. The age of the Universe can be estimated from the value of the Hubble constant and its energy content.

We note that the initial moment is absolutely inessential for the calculation of the age of the Universe. The difference between the Universe age estimated from the initial Planck time and from the moment of the BBN is approximately one second, to be compared with about 10 billion years.

If we multiply and divide the left hand side of the first Friedmann equation by the critical density ρ_c, we obtain

$$H^2 = \frac{8\pi}{3 M_{Pl}^2} \rho_c \left(\sum_i \frac{\rho_i}{\rho_c} - \frac{\rho_k}{\rho_c} \right), \qquad (4.48)$$

where we have introduced the effective energy density corresponding to a possible non-vanishing spatial curvature, $\rho_k = 3 M_{Pl}^2 / 8\pi \; k/a^2$. Today the Universe is mainly filled with non-relativistic matter, which has energy density scaling as $1/a^3$, and vacuum energy with constant density. In the estimate of the age of the Universe, at first approximation we can neglect the period of radiation dominated regime, because this time interval is much shorter than the period of matter dominated and vacuum energy dominated regimes.

To do the calculations, we introduce the dimensionless ratios of the energy densities of different forms of matter to the critical energy density, $\Omega_i = \rho_i / \rho_c$, and we rewrite Eq. (4.48) as

$$H^2 = H_0^2 \left[\Omega_m^0 (1+z)^3 + \Omega_\Lambda^0 + \Omega_k^0 (1+z)^2 \right], \qquad (4.49)$$

where the indices 0 indicate the today values. Ω_m^0 and Ω_Λ^0 are, respectively, the contributions from non-relativistic matter (dust) and from a non-vanishing cosmological constant. To take into account the evolution of these energy densities in the course of the cosmological expansion, we have introduced the *redshift factor*

$$1 + z \equiv \frac{a_0}{a}. \qquad (4.50)$$

4.5 Age of the Universe

As it can be easily checked, the factor $(1+z)^n$ properly takes into account the expansion of the Universe. For instance $\Omega_m(z) = \Omega_m^0 (1+z)^3$, because the energy density of dust scales as $1/a^3$. From the definition of Hubble parameter, we see that

$$H = \frac{d}{dt}\ln\frac{a}{a_0} = \frac{d}{dt}\ln\frac{1}{1+z} = -\frac{1}{1+z}\frac{dz}{dt}, \qquad (4.51)$$

which can be plugged into Eq. (4.49) to obtain

$$\frac{dt}{dz} = -\frac{1}{1+z}\frac{1}{H_0\sqrt{\Omega_m^0(1+z)^3 + \Omega_\Lambda^0 + \Omega_k^0(1+z)^2}}. \qquad (4.52)$$

With the substitution $\Omega_k^0 = 1 - \Omega_m^0 - \Omega_\Lambda^0$, we find the time difference between today ($z = 0$) and the time at which the redshift of the Universe was z in terms of H_0, Ω_m^0, and Ω_Λ^0

$$\Delta t = \frac{1}{H_0}\int_0^z \frac{d\tilde{z}}{1+\tilde{z}}\frac{1}{\sqrt{\left(1 + \Omega_m^0 \tilde{z}\right)(1+\tilde{z})^2 - \tilde{z}(2+\tilde{z})\Omega_\Lambda^0}}. \qquad (4.53)$$

The total age of the Universe is obtained for $z \to +\infty$. To be more accurate, we need to take into account the contribution of relativistic matter, but, as we have mentioned above, this leads to a very small correction.

It is easy to see that the integral is typically of order unity, so the time scale is set by $1/H_0 \sim 14$ Gyr. In the simple case of flat universe with no vacuum energy, namely $\Omega_m^0 = 1$ and $\Omega_\Lambda^0 = 0$, the age of the Universe would be

$$\tau_U = \frac{1}{H_0}\int_0^{+\infty}\frac{d\tilde{z}}{(1+\tilde{z})^{5/2}} = \frac{2}{3}\frac{1}{H_0} \sim 10 \text{ Gyr}. \qquad (4.54)$$

In other cases, one can calculate the integral numerically. The results for a flat Universe ($\Omega_\Lambda^0 = 1 - \Omega_m^0$) and for a Universe with no vacuum energy ($\Omega_\Lambda^0 = 0$) as a function of Ω_m^0 are shown in Fig. 4.3.

4.6 ΛCDM Model

Ω_m^0 and Ω_Λ^0 can be estimated by measuring the apparent luminosity of *standard candles*, namely sources with a known intrinsic luminosity. If there were no expansion of the Universe, the flux density of the radiation emitted by a similar source and measured by a detector on Earth would simply be $\Phi = L/4\pi d^2$, where L is the intrinsic luminosity (power) of the source and d is its distance from us. In an expanding universe, the flux Φ scales as the area of the spherical shell at the detection

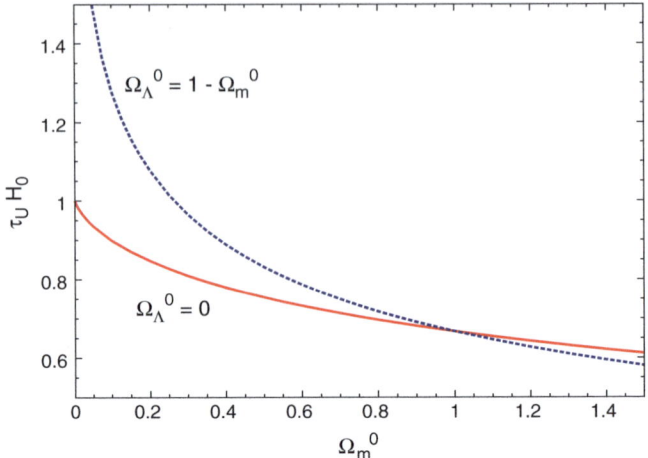

Fig. 4.3 Age of the Universe τ_U as a function of Ω_m^0 if the Universe is made of dust only ($\Omega_\Lambda^0 = 0$) and if it is flat and made of dust and vacuum energy ($\Omega_\Lambda^0 = 1 - \Omega_m^0$)

time, so d should be replaced by $a_0 r$, where r is the radial coordinate of the FRW metric and the detector is assumed to be at $r = 0$. Moreover, photons are redshifted by the factor $1 + z = a_0/a_e$, where a_e is the scale factor at the emission time, and any time interval at the emission time is measured by us to be longer by the same redshift factor $1 + z$. Eventually, the flux density of the source is

$$\Phi = \frac{L}{4\pi a_0^2 r^2 (1+z)^2} = \frac{L}{4\pi d_L^2}. \tag{4.55}$$

where we have introduced the luminosity distance d_L

$$d_L = a_0 r (1+z) = \sqrt{\frac{L}{4\pi \Phi}}. \tag{4.56}$$

For radial photon trajectories $g_{00} dt^2 + g_{11} dr^2 = 0$, and therefore

$$(1+z) dt = a_0 \frac{dt}{a} = a_0 \frac{dr}{\sqrt{1-kr^2}}. \tag{4.57}$$

Using Eq. (4.52), Eq. (4.57) can be recast in the following form

$$a_0 \int_0^r \frac{d\tilde{r}}{\sqrt{1-k\tilde{r}^2}} = \int_0^z \frac{d\tilde{z}}{H_0 \sqrt{(1+\tilde{z})^2 (1+\tilde{z}\Omega_m^0) - \tilde{z}(2+\tilde{z})\Omega_\Lambda^0}}, \tag{4.58}$$

4.6 ΛCDM Model

where

$$\int_0^r \frac{d\tilde{r}}{\sqrt{1-k\tilde{r}^2}} = \begin{cases} \arcsin r & \text{if } k = 1, \\ r & \text{if } k = 0, \\ \text{arcsinh } r & \text{if } k = -1. \end{cases} \quad (4.59)$$

If we combine Eq. (4.56) with Eq. (4.58), we can write the luminosity distance as a function of z, H_0, Ω_m^0, and Ω_Λ^0. For $k \neq 0$, we find

$$d_L\left(z, H_0, \Omega_m^0, \Omega_\Lambda^0\right) = \frac{(1+z)}{H_0\sqrt{|\Omega_k^0|}} \mathscr{S}\left(\sqrt{|\Omega_k^0|} \int_0^z F(\tilde{z})d\tilde{z}\right) \quad (4.60)$$

where

$$\mathscr{S}(x) = \begin{cases} \sin x & \text{if } k = 1, \\ x & \text{if } k = 0, \\ \sinh x & \text{if } k = -1. \end{cases} \quad (4.61)$$

$$F(z) = \frac{1}{\sqrt{(1+z)^2\left(1+z\Omega_m^0\right) - z(2+z)\Omega_\Lambda^0}}. \quad (4.62)$$

If the Universe is flat, $k = 0$, the luminosity distance is given by

$$d_L\left(z, H_0, \Omega_m^0, \Omega_\Lambda^0\right) = \frac{(1+z)}{H_0} \mathscr{S}\left(\int_0^z F(\tilde{z})d\tilde{z}\right). \quad (4.63)$$

In recent years, high redshift supernovae are efficiently used for the determination of the cosmological parameters. Type Ia supernovae (SNe Ia) are thought to occur in binary systems in which one of the stars is a carbon-oxygen white dwarf. While white dwarfs are the remnants of stars that have ceased nuclear fusion, carbon-oxygen white dwarfs can restart nuclear reactions if the temperature of their core is raised and exceeds a critical value. In a type Ia supernova, a carbon-oxygen white dwarf should accrete matter from a companion star. This would force the star to contract, leading to an increase of the core temperature. Since white dwarfs are unable to regulate the burning process as normal stars do, they undergo a runaway reaction, with the subsequent release of a large amount of energy in a short time interval. This leads to a supernova explosion. The efficiency of the mechanism is determined by the temperature of the core and therefore by the mass of the white dwarf. After some corrections for every source, the peak luminosity can be used as a standard candle (Leibundgut 2001). From the study of low-redshift type Ia supernovae, it is possible to measure the Hubble constant H_0. The study of high redshift type Ia supernovae led to the discovery of the accelerated expansion rate of the Universe (Perlmutter et al. 1999; Riess et al. 1998). Current data support

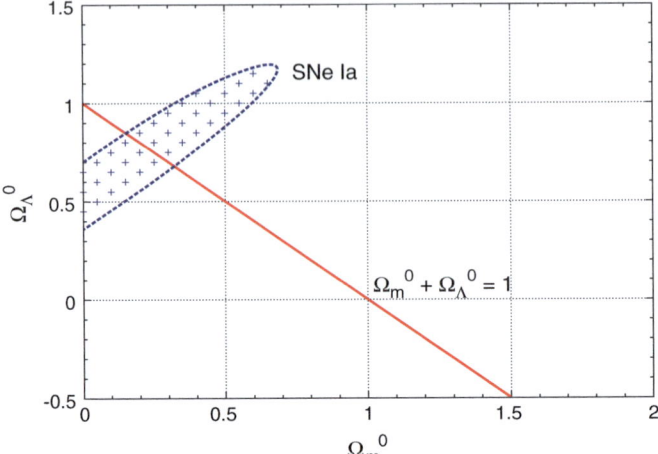

Fig. 4.4 Constraint from type Ia supernovae on the plane (Ω_m^0, Ω_Λ^0). In combination with the CMB data that require $\Omega_m^0 + \Omega_\Lambda^0 \approx 1$, observations favor the so-called ΛCDM model with $\Omega_m^0 \approx 0.3$ and $\Omega_\Lambda^0 \approx 0.7$

the so-called ΛCDM model, in which the present day Universe is dominated by vacuum energy plus some non-relativistic matter (see Fig. 4.4). When the supernova measurement is combined with CMB data, which suggest $\Omega_m^0 + \Omega_\Lambda^0 \approx 1$, the best fit values are

$$\Omega_m^0 \approx 0.3, \quad \Omega_\Lambda^0 \approx 0.7. \tag{4.64}$$

4.7 Destiny of the Universe

The geometry of a universe is determined by the sign of k, which can be properly rescaled to be 0 or ± 1. If a universe is only filled with dust ($P = 0$) and vacuum energy ($P = -\rho$), the condition to be flat is

$$\Omega_m + \Omega_\Lambda = 1, \tag{4.65}$$

while the universe would be closed (open) if $\Omega_m + \Omega_\Lambda > 1 \ (< 1)$.

If $\Lambda = 0$, a closed universe first expands and then recollapses, as shown by Eq. (4.26). Open and flat universes expand forever, see Eqs. (4.28) and (4.27). In the presence of vacuum energy, the situation is more complicated. If $\Omega_m \leq 1$, the fate of the universe is determined by the sign of Λ: for $\Lambda = 0$, the universe would expand

4.7 Destiny of the Universe

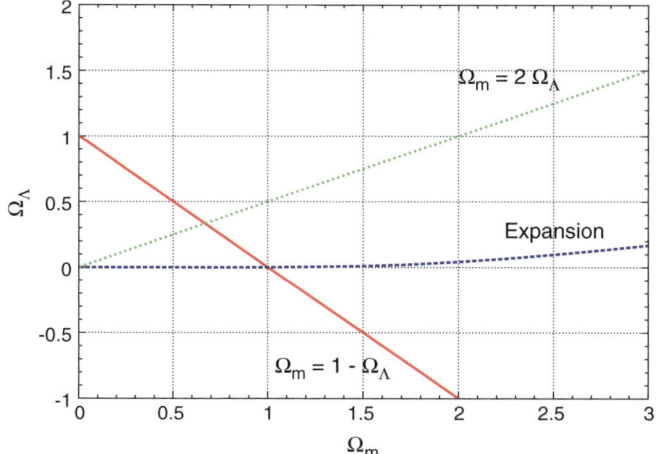

Fig. 4.5 Properties of universes filled with dust and vacuum energy. Flat universes are situated on the line $\Omega_m = 1 - \Omega_\Lambda$. This line separates closed universes ($\Omega_m + \Omega_\Lambda > 1$) from open universes ($\Omega_m + \Omega_\Lambda < 1$). Universes at the moment of their evolution in which the cosmological acceleration turns to zero, $\ddot{a} = 0$, are situated on the line $\Omega_m = 2\Omega_\Lambda$. This line separates accelerating ($\ddot{a} > 0$) and decelerating ($\ddot{a} < 0$) phases of the cosmological evolution, which are situated, respectively, above and below this line. The line "Expansion" separates universes that expand forever (*above*) from universes that first expand and then recollapse (*below*)

forever, but since $\Omega_m/\Omega_\Lambda \sim 1/a^3$, at a sufficiently late time even a tiny cosmological constant becomes dominant and, if it is negative, it terminates the expansion and the universe recollapses. For $\Omega_m > 1$, the universe expands forever if the effect of a positive vacuum energy becomes important before the universe starts recollapsing. The line on the (Ω_m, Ω_Λ) plane separating eternally expanding universes from universes having initially an expanding phase followed by a contraction is given by

$$\Omega_\Lambda = \begin{cases} 0 & \text{for } \Omega_m \leq 1, \\ 4\Omega_m \sin^3\left[\frac{1}{3}\arcsin\left(\frac{\Omega_m-1}{\Omega_m}\right)\right] & \text{for } \Omega_m > 1. \end{cases} \quad (4.66)$$

Lastly, we can distinguish accelerating ($\ddot{a} > 0$) and decelerating ($\ddot{a} < 0$) universes. From Eq. (4.11), we see that the condition for $\ddot{a} = 0$ is

$$\rho + 3P = 0 \quad \Rightarrow \quad \Omega_m = 2\Omega_\Lambda. \quad (4.67)$$

A universe is thus accelerating (decelerating) if $\Omega_m < 2\Omega_\Lambda$ ($\Omega_m > 2\Omega_\Lambda$).

The geometry, the final destiny, and the acceleration of different types of universes are illustrated on the (Ω_m, Ω_Λ) plane in Fig. 4.5.

Problems

4.1 Verify Eqs. (4.3) and (4.5).

4.2 Reconsider the discussion on the age of the Universe presented in Sect. 4.5 in the case of a non-negligible radiation component Ω_γ.

4.3 Let us consider universes only filled with radiation and vacuum energy. Determine the line $\ddot{a} = 0$ and the one separating universes in eternal expansion from those that will recollapse.

References

P. Fleury, H. Dupuy, J.P. Uzan, Phys. Rev. Lett. **111**, 091302 (2013). arXiv:1304.7791 [astro-ph.CO]
A. Friedmann, Z. Phys. **21**, 326 (1924) [Gen. Rel. Grav. **31**, 2001 (1999)]
B. Leibundgut, Ann. Rev. Astron. Astrophys. **39**, 67 (2001)
V. Marra, E.W. Kolb, S. Matarrese, A. Riotto, Phys. Rev. D **76**, 123004 (2007). arXiv:0708.3622 [astro-ph]
S. Perlmutter et al., [Supernova Cosmology Project Collaboration], Astrophys. J. **517**, 565 (1999) [astro-ph/9812133]
A.G. Riess et al., [Supernova Search Team Collaboration], Astron. J. **116**, 1009 (1998) [astro-ph/9805201]
S. Weinberg, *Gravitation and Cosmology*, 1st edn. (Wiley, San Francisco, 1972)

Chapter 5
Kinetics and Thermodynamics in Cosmology

5.1 Introduction

At the early epochs of the cosmological evolution, the state of matter in the Universe was very close to the thermal equilibrium one. In a sense, the situation was opposite to normal thermodynamics, where thermal equilibrium is established after sufficiently long time. In cosmology, the younger is the Universe, the faster are reactions among particles, and less time is necessary to reach the equilibrium. In the early Universe, the reaction rates were typically higher than the expansion rate, $H = \dot{a}/a$, despite the rise of H as $1/t$ at short cosmological times. This condition was fulfilled almost till the GUT epoch, $T_{\text{GUT}} \sim 10^{15}$ GeV, if it ever existed in the Universe.

In thermal equilibrium, the state of matter is described by very few parameters: temperature and chemical potentials of different particle species. The distribution of particles over their energies is determined by the equilibrium and is given by the canonical Fermi-Dirac or Bose-Einstein forms (including possible formation of Bose-Einstein condensates).

The equilibrium state of the primeval plasma grossly simplifies theoretical considerations. However, the most interesting phenomena appear because of deviations from the equilibrium, such as, e.g., freezing of massive species, distortion of massless neutrino spectrum, baryogenesis (considered in Chap. 7), etc.

5.2 Thermal Equilibrium in the Early Universe

5.2.1 General Features

Observing the Universe today, we can understand what were the cosmological physical conditions in the past. The first evident conclusion is that the Universe was denser and hotter. According to Eq. (4.41), the temperature of radiation drops down as the inverse of the cosmological scale factor. So, if we travel backward in time, we

---

The original version of this chapter was revised: The errors in this chapter have been corrected. The correction to this chapter can be found at
https://doi.org/10.1007/978-3-662-48078-6_13.

would observe the corresponding temperature rise. According to the results of the previous chapter, the energy density of radiation and of non-relativistic matter scale, respectively, as $1/a^4$ and $1/a^3$. Going backward in time to smaller and smaller a, we would see that the density of matter rises in accordance with these laws. This rise should change or terminate when the equation of state of the cosmological matter changes or quantum gravity effects become operative. As for the latter, we do not expect them at temperatures below M_{Pl}, so it may be safe to extrapolate back to such high temperatures. Another possibility is a change of equation of state of the cosmological matter. This is what is realized in inflationary models. Roughly speaking, the equation of state became $P = -\rho$ and thus the energy density stayed constant. In the simplest versions of the inflationary models, it was the vacuum-like energy of the inflaton field.

According to the Standard Cosmological Model, the matter in the Universe was created by the inflaton decay. Prior to that, the Universe was an exponentially expanding darkness with only the inflaton field present there. At some stage, this quasi-empty dark state exploded, producing all the elementary particles, mostly with masses smaller than the inflaton mass. This picture surprisingly closely reminds the biblical account of Creation. Initially "Earth was without form and void, and darkness was over the face of the deep". Then all of a sudden there was a powerful explosion, the big bang, exactly in the spirit of "let there be light".

The theory of particle production by an external time dependent (oscillating) field is well established. The process is reliably described by mathematical equations, so we have a rigorous description of this phenomenon. The energy spectra of the created particles depend upon the details of the production and might be rather complicated. Fortunately (for theorists), thermal equilibrium in the primeval plasma was soon established and the whole production history forgotten.

The thermal equilibrium state is realized when the characteristic reaction rate, Γ, is larger than the Universe expansion rate, H. For particle scattering with cross section σ, the reaction rate can be estimated as

$$\Gamma_r = \frac{\dot{n}}{n} \sim \sigma v n, \tag{5.1}$$

where n is the number density of particles in the plasma and v is their relative velocity. The typical magnitude of cross sections at high energies is $\sigma \sim g^4/s$, where g is the coupling constant of particle interactions and $s = (p_1 + p_2)^2$ is the total energy of the colliding particles in their center of mass system. In GUT models, $g^2 \sim \alpha \sim 0.01$. If thermal equilibrium is established, then $s \sim T^2$ and $n \sim T^3$, where T is the plasma temperature. If the essential processes are the particle decay and the inverse decay, then the characteristic reaction rate is simply the decay width

$$\Gamma_d \sim \frac{g^2 m}{\gamma}, \tag{5.2}$$

where m is the decaying particle mass and γ is its Lorenz γ-factor.

5.2 Thermal Equilibrium in the Early Universe

The reaction rates should be compared with the Hubble parameter, which, from Eq. (4.9), turns out to be $H \sim T^2/M_{\text{Pl}}$ for a radiation dominated universe, see below. Since the Planck mass is much larger than the typical mass parameters in elementary particle physics, we could expect that at temperatures below $10^{-3} - 10^{-4}\, M_{\text{Pl}}$ thermal equilibrium was well established:

$$g^4 T \gtrsim \frac{T^2}{M_{\text{Pl}}} \Rightarrow T \lesssim g^4 M_{\text{Pl}}. \tag{5.3}$$

It is interesting that, in thermodynamics, equilibrium is usually established after sufficiently long time, while in cosmology it takes place at short times in the early Universe, because of the high density of particles there. More details are presented in the concrete examples studied below.

Now we discuss some properties of the thermal equilibrium plasma in the cosmological background. Equilibrium distributions in the homogeneous case have the well known universal form, but somewhat different for bosons and fermions

$$f_j^{(eq)}(E_j, T, \mu_j) = \frac{1}{\exp\left[(E_j - \mu_j)/T\right] \pm 1}, \tag{5.4}$$

where j denotes the particle type, $E_j = \sqrt{p^2 + m_j^2}$ is the particle energy, T is the plasma temperature (common to all species in equilibrium), and μ_j is the chemical potential for the j-type particle. Plus sign in front of unity corresponds to fermions and minus sign corresponds to bosons.

The number density n and the energy density ρ of particles are expressed through the integrals of the distribution function over momentum, namely

$$n_j = \sum_s \int \frac{d^3 p}{(2\pi)^3} f_j, \tag{5.5}$$

$$\rho_j = \sum_s \int \frac{d^3 p}{(2\pi)^3} E(p) f_j. \tag{5.6}$$

Let us present also an expression for the pressure, which will be used in what follows:

$$P_j = \sum_s \int \frac{d^3 p}{(2\pi)^3} \frac{p^2}{3E} f_j. \tag{5.7}$$

The summations here and in Eqs. (5.5) and (5.6) are made over all the spin states of the particles in question. If, as it is often the case, all the spin states are equally populated, the summation is reduced to the multiplication of the integrals by the number of spin states, g_s. There is an exception for this rule: for neutrinos, only left-handed states are populated, while right-handed states are (almost) absent, see below Sect. 5.3.1.

For massless particles with zero chemical potentials, the integrals (5.5) and (5.6) can be taken leading to the following results for bosons

$$n_b^0 \approx 0.122 g_s T^3, \quad \rho_b^0 = \frac{\pi^2}{30} g_s T^4. \tag{5.8}$$

The analogous expressions for fermions can be obtained using the following rule expressing integrals from fermionic distribution functions through the bosonic ones

$$\int dp p^n f_f = (1 - 1/2^n) \int dp p^n f_b. \tag{5.9}$$

Correspondingly, $n_f = (3/4)n_b$ and $\rho_f = (7/8)\rho_b$ for massless particles with $\mu = 0$.

In the non-relativistic case, when $T \ll m$ and $\mu = 0$, the bosonic and fermionic distributions are almost the same and have the form

$$n_b^m = n_f^m \equiv n^m = g_s e^{-m/T} \left(\frac{mT}{2\pi}\right)^{3/2}, \quad \rho_b^m = \rho_f^m = mn^m \tag{5.10}$$

During practically all the history of the early Universe, the primeval plasma was dominated by relativistic matter. Indeed, the contribution of non-relativistic particles, i.e. of the particles with large masses, $m > T$, was Boltzmann suppressed as $\exp(-m/T)$, while for $m \leq T$, the particles were relativistic. Let us express the Hubble parameter through the plasma temperature at the onset of the BBN epoch, when the plasma temperature was about 1 MeV. According to the equations presented above, the energy density can be written as

$$\rho_{rel} = \frac{\pi^2}{30} g_* T^4, \tag{5.11}$$

where g_* includes the contribution of all the relativistic particle species in the plasma, i.e. the number of their spin states for bosons and 7/8 of the number of spin states for fermions.[1] So the contribution from photons is $g_\gamma = 2$; the contribution from e^+e^- pairs is 7/2, and the contribution from neutrinos and antineutrinos is $(7/4)(3 + \Delta N_\nu)$, where ΔN_ν is the effective number of any other relativistic particle species with the energy density normalized to the equilibrium energy of one neutrino species. Summing up all the contributions we find

---

[1] We can still use Eq. (5.11) in the case of one or more decoupled relativistic components with possible different temperatures, but now g_* is

$$g_*(T) = \sum_{bosons} g_i \left(\frac{T_i}{T}\right)^4 + \frac{7}{8} \sum_{fermions} g_i \left(\frac{T_i}{T}\right)^4, \tag{5.12}$$

where the first summation is over all the boson species and the second one over all the fermion species, g_i is the number of spin states of the species i, and T_i is the temperature (or effective temperature in the case of decoupled particles) of the species i.

5.2 Thermal Equilibrium in the Early Universe

$$g_* = 10.75 + \frac{7}{4}\Delta N_\nu. \tag{5.13}$$

In addition to the already included particle species, there were baryons and dark matter particles in the plasma at that time. However, the energy densities of these non-relativistic contributions were about 10^{-6} with respect to the relativistic matter.

Now we can express the Hubble parameter through the temperature of the relativistic primeval plasma. From Eqs. (4.9), (5.11), and (5.13), one can find

$$H = 5.44\sqrt{\frac{g_*(T)}{10.75}}\frac{T^2}{M_{\text{Pl}}}. \tag{5.14}$$

Since at the relativistic stage $H = 1/(2t)$, the law of the Universe cooling with time turns out to be $tT^2 = const$ and, keeping in mind that $a(t) \sim \sqrt{t}$, we can see that

$$T \sim \frac{1}{a(t)}. \tag{5.15}$$

The temperature of radiation drops according to this law for any expansion regime if the energy exchange with massive particles can be neglected. Elastic scattering of radiation off massive particles would lead to a somewhat faster cooling. On the other hand, massive particle annihilation to relativistic species would heat up the plasma and result in a slower cooling. The origin of such cooling or heating is described in Sect. 5.2.3. Note that, for a quick estimate of the temperature of the plasma, a simple approximate relation is helpful: the temperature in MeV is equal to the square root of the Universe age in seconds. In the calculations made above we neglected possible chemical potentials of the relativistic particles. They are supposed to be small and, if they are known, they are small indeed. For example, the baryonic chemical potential of quarks is about $10^{-9}\,T$ and the leptonic chemical potential of neutrinos is smaller than $0.1\,T$.

Evidently, for vanishing chemical potentials, the number and energy densities of particles and antiparticles are the same in equilibrium, since the masses of particles and antiparticles are supposed to be equal. This equality is a consequence of the CPT theorem. So to describe the case when there is an asymmetry between particles and antiparticles it is necessary to introduce for them unequal chemical potentials. Let us stress that chemical potentials do not vanish in equilibrium if the difference of number densities of particles and antiparticles is conserved. It happens if the particles in question possess a conserved quantum number, such as, e.g., baryonic or leptonic number which are supposed to be conserved at low energies.

As we see in what follows, in equilibrium the chemical potentials for particles and antiparticles have equal magnitude and opposite signs, $\bar{\mu} = -\mu$. We can find the difference between the densities of massless fermions and antifermions using Eqs. (5.4) and (5.5):

$$n - \bar{n} = g_s T^3 \frac{\xi^3 + \pi^2 \xi}{6\pi^2}, \tag{5.16}$$

where $\xi = \mu/T$ is the dimensionless chemical potential, see e.g. Landau and Lifshitz (1980). As we show below, if there is no entropy release, e.g. by massive particle annihilation, the temperature drops down as the inverse scale factor, $T \sim 1/a$, and $\xi = const$. The last condition describes the conservation of particle number difference in the comoving volume, i.e. in the volume which expands together with the Universe, $V \sim a^3$.

5.2.2 Kinetic Equation

Here we present the kinetic equations governing the evolution of the distribution functions and their approach to equilibrium. In the case of weak interactions among particles in the plasma, the equation can be written in the following form

$$\frac{df^i}{dt} = \left[\partial_t + \dot{\mathbf{p}} \partial_\mathbf{p} + \dot{\mathbf{r}} \nabla \right] f^i = I^i_{coll}, \tag{5.17}$$

where f^i is the distribution function for particles of type i and I^i_{coll} is the collision integral describing the particle interactions. It will be specified below. In general, the function f^i depends upon time, space coordinate $\mathbf{r}$, and the vector of particle momentum $\mathbf{p}$. Sometimes we omit the index i.

In the homogeneous and isotropic case, the distribution functions only depend on the time t and on the absolute value of the particle momentum p, while they do not depend on the space coordinate $\mathbf{r}$, so the last term in the left hand side of this equation vanishes. Following the approach discussed in Sect. 4.4, we can see that $\dot{p} = -Hp$ in the FRW background. We can thus rewrite the left hand side as

$$\frac{df^i}{dt} = \left[\partial_t + \dot{p} \partial_p \right] f^i = Hx \partial_x f^i(x, y^j) = I^i_{coll}, \tag{5.18}$$

where $x = m_0 a$, $y_i = p_i a$, and m_0 is an arbitrary normalization parameter with dimension of mass. The temperature of the primeval plasma usually drops as the cosmological scale factor, see Eq. (5.15), and during this regime it is convenient to take $x = m_0/T$ and $y = p/T$.

The collision integral for the process $i + Y \leftrightarrow Z$, where Y and Z are some arbitrary, generally multi-particle, states, has the form

$$I^i_{coll} = \frac{(2\pi)^4}{2E_i} \sum_{Z,Y} \int dv_Z \, dv_Y \delta^4(p_i + P_Y - P_Z)$$

$$\times \left[|A(Z \to i + Y)|^2 \prod_Z f \prod_{i+Y} (1 \pm f) - |A(i + Y \to Z)|^2 f_i \prod_Y f \prod_Z (1 \pm f) \right], \tag{5.19}$$

where $A(i + Y \to Z)$ is the transition amplitude from the state $i + Y$ to the state Z, $\prod_Y f$ is the product of the distribution functions forming the state Y, the sign in $\prod(1 \pm f)$ is $+$ for bosons and $-$ for fermions, $P_{Z,Y}$ is the total momentum of the state Z or Y, and

$$d\nu_Y = \prod_Y \overline{dp} \equiv \prod_Y \frac{d^3p}{(2\pi)^3 2E}. \tag{5.20}$$

All the terms in I_{coll} can be easily understood. The factors $1/2E$ is related to the relativistic invariant normalization of the amplitudes. The δ-function ensures the energy-momentum conservation in the reactions. The integration is taken over the phase space of all participating particles except for the particle i. The reaction probability is evidently proportional to the particle densities in the initial state and to the Fermi suppression or Bose enhancement factors in the final state, which is realized by the products of f_{in} and of $(1 \pm f_{fin})$.

By definition, the equilibrium distribution functions are the functions annihilating the collision integral, i.e. $I_{coll}[f_{eq}] = 0$. We can easily check this if the invariance with respect to time reversal holds and the approximation of the Boltzmann statistics, for which $f_{eq} = \exp(\mu - E)/T$, is valid. Because of the invariance with respect to time inversion, the absolute values of the amplitudes of direct and inverse reactions are equal after time reversal of the kinematic variables, i.e. the change of the signs of momenta and spins. Since in the collision integral the summation over all the spin states and the integration over momenta is made, this change of signs is not essential and we can take the direct and inverse amplitudes to be equal and thus the collision integral would contain the factor

$$\Pi_{in} f_{in} \, \Pi_{fin}(1 \pm f_{fin}) - \Pi_{fin} f_{fin} \, \Pi_{in}(1 \pm f_{in}). \tag{5.21}$$

Since the Boltzmann statistics is valid for $f \ll 1$, this factor is reduced to $\Pi_{in} f_{in} - \Pi_{fin} f_{fin}$. This difference evidently vanishes due to the energy conservation and if the usual equilibrium condition among chemical potentials is fulfilled, namely

$$\sum \mu_{in} = \sum \mu_{fin}. \tag{5.22}$$

We leave as an exercise to prove that the difference (5.21) is also zero for the case of quantum (Bose and/or Fermi) statistics and that the collision integral vanishes even if the T-invariance is not fulfilled. The last exercise may be too difficult and the readers can find the prove in Dolgov (1979).

The energy conservation condition is automatically fulfilled in the absence of external time dependent fields, but the conservation of chemical potentials (5.22) is only valid in the equilibrium case. The system evolves to the state with conserved μ if the efficiency of inelastic reactions is high enough. Elastic reactions do not help to enforce the condition (5.22) because for elastic reactions this condition is automatically fulfilled, so chemical potentials do not evolve in elastic reactions.

Efficient elastic scattering reactions induce the canonical form of the distribution over energy, so, in the case of equilibrium with respect to elastic reactions, the dependence of f on the particle energy takes the form $f \sim \exp[(\mu - E)/T)]$ in the case of the Boltzmann statistics or $f = 1/[\exp(E - \mu)/T) \pm 1]$, but the parameters μ may be arbitrary. Only inelastic reactions with non-conservation of the number of the particles in question lead to the equilibrium condition (5.22).

To understand the role of inelastic processes, let us consider the electron-positron annihilation into two and three photons. If these reactions are sufficiently fast, they enforce the following conditions

$$\mu_{e^-} + \mu_{e^+} = 2\mu_\gamma = 3\mu_\gamma. \tag{5.23}$$

From these equations, it follows that in equilibrium

$$\mu_\gamma = 0, \quad \mu_{e^+} = -\mu_{e^-}. \tag{5.24}$$

These equations demonstrate the general situation that in equilibrium the chemical potential of a particle species is zero if the particle's number is not conserved, and that the chemical potential of particles and antiparticles have equal absolute values but opposite signs.

Note that chemical potentials of bosons are bounded from above by their masses, i.e. $\mu \leq m$, otherwise the distribution function at some low momenta, $p^2 < \mu^2 - m^2$, would be negative, which is physically senseless. Now an interesting problem arises. As it is mentioned above, chemical potentials are introduced to describe an asymmetry between the number densities of particles and antiparticles. If we take a plasma with an excess of bosons with respect to antibosons, this difference, if it is sufficiently small, can be described by a properly chosen chemical potential. But what happens if the asymmetry increases? The chemical potential should also increase till it reaches the maximum allowed value $\mu = m$. What happens after that? The answer is that a larger asymmetry would induce the formation of a Bose condensate and the equilibrium distribution would take the form

$$f^B_{eq}(E, T, m, C) = \frac{1}{\exp\left[(E - m)/T\right] - 1} + \frac{C}{(2\pi)^3} \delta^{(3)}(\mathbf{p}). \tag{5.25}$$

The last term proportional to the δ-function of the momentum describes the condensed part of the distribution with the constant C being the amplitude of the condensate. It can be shown that this function, $f^B_{eq}(E, T, m, C)$, is an equilibrium distribution, i.e. it annihilates the collision integral if and only if $\mu = m$. We leave this problem as an exercise.

Note that the equilibrium distribution functions are always determined by two parameters: the temperature, which is common for all particles, and the chemical potential, which is generally different for different particle species. If a chemical potential is fixed by its maximum value $\mu = m$, then there appears another parameter, the amplitude of condensate, C.

5.2 Thermal Equilibrium in the Early Universe

Since the collision integral vanishes at $f = f_{eq}$, it is often approximated as

$$I_{coll} \approx -\Gamma(f - f_{eq}), \quad (5.26)$$

where Γ is the effective reaction rate. This is not a very accurate approximation and it is not always applicable, but in some cases it works reasonably well. We use this equation to estimate deviations from equilibrium of massive particles in the FRW background. We assume that the deviation is small and the distribution function can be presented as

$$f = f_{eq} + \delta f. \quad (5.27)$$

Using the variables x and y, introduced after Eq. (5.18), we write f_{eq} in the form

$$f_{eq} = \frac{1}{\exp(\sqrt{x^2 + y^2}) \pm 1}, \quad (5.28)$$

where we neglected possible chemical potentials.

Substituting this expressions into the kinetic equation (5.18) with the collision integral in the form (5.26), we find

$$\frac{\delta f}{f_{eq}} = \frac{Hx}{\Gamma} \frac{\partial_x f_{eq}}{f_{eq}} = -\frac{Hx^2}{\Gamma\sqrt{x^2+y^2}} = -\frac{m^2 H}{T E \Gamma}, \quad (5.29)$$

where we took the normalization mass m_0 equal to the mass of the particles under scrutiny. With H from Eq. (5.14) and $T \sim m$, one estimates the deviation from equilibrium as $\delta f / f_{eq} \sim m^2/(M_{Pl}\Gamma)$. Typically $\Gamma \sim \alpha^n m$, where $\alpha \sim 10^{-2}$, $n = 1$ for decays and $n = 2$ for two-body reactions. So for the particle masses below 10^{16} GeV the violation of thermal equilibrium is quite small.

As one can see from Eq. (5.29), the equilibrium of massless particles is not destroyed by the cosmological expansion. Even if the interaction is switched off, the distribution remains of the equilibrium form with the temperature and the possible chemical potential dropping down as $1/a$. Indeed, the left hand side of the kinetic equation can be written as (here we return to the standard variables, p and t)

$$(\partial_t - Hp\partial_p) f_{eq}\left(\frac{E - \mu(t)}{T(t)}\right) = \left[-\frac{\dot T}{T}\frac{E-\mu}{T} - \frac{\dot\mu}{T} - \frac{Hp}{T}\right] f'_{eq}, \quad (5.30)$$

where the prime means the derivative of f over its argument $(E - \mu)/T$. The factor in the square brackets vanishes if $\dot\mu = \dot T/T = -H$ and if $E(\dot T/T) = -Hp$, which can be and is true only for $E = p$, i.e. for $m = 0$. It explains, in particular, the observed perfect equilibrium spectrum of the CMB. Note that $\mu/T = const$ implies the particle conservation in the comoving volume, i.e. $n \sim T^3$; however, it is not of importance for the CMB, because $\mu/T \ll 1$.

5.2.3 Plasma Heating and Entropy Conservation

It is shown above that the temperature of a plasma consisting of massless or very light ($m \ll T$) particles drops down as $1/a(t)$. However, a noticeable fraction of massive particles with $m \sim T$ would destroy this simple law. There are several instructive examples that help to understand the reasons for deviations from the $1/a$-law.

Let us first consider light particles in thermal equilibrium with the rest of the plasma. Their distribution had the equilibrium form (5.4), with $\exp(E/T) = \exp(\sqrt{x^2 + y^2})$. If these particles completely stopped interacting at some temperature T_d, then the distribution would become

$$\exp(E/T) = \exp\left(\sqrt{x_d^2 + y^2}\right), \tag{5.31}$$

where $x_d = m/T_d$. A realistic example of such particles is presented by neutrinos, for which $T_d \sim 1$ MeV and $x_d \ll 1$ (see Sect. 5.3.1). So we would expect that the neutrino distribution after their decoupling should be

$$f_\nu = \frac{1}{\exp(p/T_\nu) + 1}, \tag{5.32}$$

where T_ν does not have a meaning of temperature because the temperature is determined for the equilibrium case when the distribution depends upon E/T. For non-interacting particles, the parameter T_ν always drops as $1/a$. In the case of massless particles, it is a real temperature.

The situation would be different if the massive particles are in mutual equilibrium but do not interact with any other matter. In this case, the distribution should have the equilibrium form (5.4), but the temperature would not follow the law $T \sim 1/a$. A simple and practically interesting example is a collection of non-relativistic particles that have sufficiently strong elastic scattering maintaining the canonical distribution over their energy but their annihilation is switched-off, so the number density in comoving volume remains constant. To achieve this, an effective chemical potential should be developed, $\mu \sim m$, the same for particles and antiparticles, if the latter exist and the number densities of particles and antiparticles are equal. Since in this conditions f_{eq} depends upon $p^2/(2mT)$, the constant number density in the comoving volume implies $T \sim 1/a^2$.

If there is a mixture of relativistic and non-relativistic matter with an energy exchange between them, then if the massive particles do not annihilate, the plasma cooling rate would be between $T \sim 1/a$ and $T \sim 1/a^2$. The presence of massive non-annihilating particles acts as an extra cooling agent leading to a faster than $1/a$ drop-off of the temperature.

If, however, the annihilation is essential, the cooling would proceed slower than $1/a$ due to the release of energy accumulated in the particle masses. One can determine the law of cooling in this case using the entropy conservation law per comoving

5.2 Thermal Equilibrium in the Early Universe

volume

$$\frac{dS}{dt} \equiv \frac{d}{dt}\left(a^3 \frac{\rho + P}{T}\right) = 0, \qquad (5.33)$$

which can be derived as follows. The expressions for the total pressure and energy density are given by Eqs. (5.6) and (5.7) after summation over all particle species j. For relativistic (i.e. essentially massless) particles

$$s = \frac{\rho + P}{T} = \frac{2\pi^2}{45} g_*^s T^3, \qquad (5.34)$$

where $g_*^s = g_b + (7/8)g_f$, see the comment below Eq. (5.9).[2] The entropy is conserved for the equilibrium distributions with zero chemical potentials. In fact, a weaker condition is sufficient, namely that f is a function of E/T, where T is an arbitrary function of time. To derive Eq. (5.33), we use the covariant energy conservation $\dot{\rho} = -3H(\rho + P)$, Eq. (4.12), and the law of the pressure evolution

$$\dot{P} = \frac{\dot{T}}{T}(\rho + P), \qquad (5.36)$$

which can be obtained by differentiation of Eq. (5.7) over time and integration of the result by parts.

Let us consider now an example of plasma populated by photons and electron-positron pairs with the initial temperature exceeding the electron mass, $T_{in} > m_e$, and the final temperature $T_{fin} \ll m_e$, at which practically all electrons and positrons are annihilated. In the real situation, there is a small excess of electrons over positrons, but we neglect it here. The initial entropy of the plasma fully populated by γ, e^+, and e^- is $(11\pi^2/45)(a_{in}T_{in})^3$ and the final entropy is $(4\pi^2/45)(a_{fin}T_{fin})^3$. Using the conservation law (5.33), we find that the photon temperature drops slower than $1/a$

$$\frac{T_{fin}}{T_{in}} = \left(\frac{11}{4}\right)^{1/3} \frac{a_{in}}{a_{fin}}. \qquad (5.37)$$

This result is used in the next subsection for the calculations of the neutrino to photon temperature ratio.

---

[2] If all the components of the plasma have the same temperature, $g_* = g_*^s$. If this is not the case, we have the counterpart of Eq. (5.12) and g_*^s is

$$g_*^s(T) = \sum_{bosons} g_i \left(\frac{T_i}{T}\right)^3 + \frac{7}{8} \sum_{fermions} g_i \left(\frac{T_i}{T}\right)^3, \qquad (5.35)$$

which may be different from g_*.

5.3 Freezing of Species

According to the simple estimates presented above, in a certain temperature range there is good contact among particles in the primeval plasma and thermal equilibrium is established. When the temperature drops down, interactions of some particles with the rest of the plasma become too weak to maintain equilibrium and they start to live their own free life. This process is called *freezing of species*. There are two possibilities to terminate the interaction. The first one is realized if the interaction strength drops down with energy. In this case, the decoupled particles are generally relativistic at the moment of decoupling and their frozen number density is close to the number density of the CMB photons. This happens with weakly interacting particles such as neutrinos.

The second type of freezing occurs even with strongly interacting particles. They may be in strong elastic contact with the plasma but they stop annihilating because their number becomes exponentially small and it is impossible to find a partner to commit a pairwise suicide. Correspondingly, the number density of such particle becomes constant in the comoving volume. This is the case of non-relativistic freezing. By such a mechanism, the number density of dark matter particles was fixed.

5.3.1 Decoupling and Gershtein-Zeldovich Bound

An example of particles that decoupled while being relativistic is represented by neutrinos. At low energies, below the masses of the W- and Z-bosons, neutrinos possess 4-fermion interactions determined by the Fermi coupling constant $G_F \approx 10^{-5}\,\text{GeV}^{-2}$. Correspondingly, the processes of neutrino scattering on charged leptons and neutrinos and $\bar{\nu}\nu$-annihilation all have the cross section of the same order of magnitude, namely $\sigma_W \sim G_F^2 s$, where $s = (p_1 + p_2)^2$ is the total energy of the colliding particles in the center of mass system. The corresponding reaction rate $\Gamma_W = \sigma_W n \sim G_F^2 T^5$ is to be compared with the Hubble parameter $H \sim T^2/M_{\text{Pl}}$. Thus the equilibrium would be broken below the temperature

$$T_f \sim m_N \left(\frac{10^{10} m_N}{M_{\text{Pl}}} \right)^{1/3} \sim \text{MeV}\,, \tag{5.38}$$

where $m_N \sim 1\,\text{GeV}$ is the nucleon mass. More accurate calculations are necessary to establish if T_f is larger or smaller than m_e, which is important for the calculations of the number density of relic neutrinos n_ν at the present time and for the cosmological bound on their mass m_ν.

For more accurate calculations of the decoupling temperature from the electron-positron plasma, we can use the kinetic equation in Boltzmann approximation with only the direct reactions with electrons, i.e. νe elastic scattering and $\nu\bar{\nu}$ annihilation taken into account

5.3 Freezing of Species

$$Hx \frac{\partial f_\nu}{f_\nu \, \partial x} = -\frac{80 G_F^2 \left(g_L^2 + g_R^2 \right) y}{3\pi^3 x^5}, \tag{5.39}$$

where the exact expression for the amplitude of the neutrino interaction is used with g_L and g_R being some constants of order unity and different for ν_e and $\nu_{\mu,\tau}$ (see any textbook on weak interactions). The dimensionless parameter x is the neutrino energy in units of temperature, $x = E/T$. It is clear from this equation that the freeze-out temperature, T_f, depends upon the neutrino momentum $y = p/T$, and this can distort the spectrum of the decoupled neutrinos, as we see in what follows. For the average value of the neutrino momentum, $y = 3$, the temperature of decoupling of neutrinos from $e^\pm$ is

$$T_{\nu_e} = 1.87 \text{ MeV}, \quad \text{and} \quad T_{\nu_\mu, \nu_\tau} = 3.12 \text{ MeV}. \tag{5.40}$$

The decoupling temperatures with respect to the annihilation $\nu \bar{\nu} \leftrightarrow e^+ e^-$ which can change the number density of neutrinos are equal to $T_{\nu_e} \approx 3$ MeV and $T_{\nu_\mu, \nu_\tau} \approx 5$ MeV, respectively for ν_e and $\nu_{\mu,\tau}$.

To take into account all the reactions experienced by neutrinos, including elastic νe and all $\nu \nu$ scattering, we need to make the substitution $(g_L^2 + g_R^2) \leftrightarrow (1 + g_L^2 + g_R^2)$ in Eq. (5.39). In this way, we can find that neutrinos completely decoupled and started to propagate freely in the Universe when the temperature dropped below

$$T_{\nu_e} = 1.34 \text{ MeV} \quad \text{and} \quad T_{\nu_\mu, \nu_\tau} = 1.5 \text{ MeV}. \tag{5.41}$$

An accurate calculation of the neutrino decoupling temperature is presented in Dolgov (2002).

To summarize, the decoupling of all the neutrino flavors took place at a temperatures higher than the electron mass, when e^+ and e^- were abundant in the plasma. So the $e^+ e^-$ annihilation heated up the photon gas, while the neutrino temperature remained intact and dropped down as $1/a$. At high temperatures, the ratio of neutrino and photon densities was $(n_\nu + n_{\bar{\nu}})/n_\gamma = 3/4$. It is assumed here that the neutrino chemical potentials are zero and so $n_\nu = n_{\bar{\nu}}$. For non-zero μ_ν, the total number density of neutrinos plus antineutrinos would be higher, with an increase proportional to μ_ν^2. Another important assumption is that only one spin state of neutrinos is present in the primeval plasma. If neutrinos were massless or had a Majorana mass, then only the left-handed spin state would exist. However, massive neutrinos with Dirac mass have altogether four states: left-handed neutrinos and antineutrinos plus right-handed ones. Neutrinos participating in the usual weak interactions are left-handed, i.e. their helicity (spin projection on the direction of the neutrino momentum) is equal to $-1/2$, while antineutrinos have helicity equal to $+1/2$. Right-handed neutrinos must be produced by the weak interactions but with a small probability proportional to $\Gamma_R \sim (m_\nu/E)^2 \Gamma_L \sim G_F^2 m_\nu^2 T^3$, where Γ_L is the probability of left-handed neutrino reactions due to the usual weak interactions. Rescaling the equilibrium condition (5.38) for left-handed neutrinos to the right-handed ones, we find that ν_R would have equilibrium density at $T_R \sim (G_F^2 m_\nu^2 M_{\text{Pl}})^{-1} \geq 10^{10}$ GeV, if $m_\nu < 1$ eV.

However, at energies comparable to, or larger than, the masses of the weak bosons, the weak interaction cross section stops rising and instead drops down as $\sigma_W \sim \alpha^2/T^2$, where the coupling constant is $\alpha \sim 0.01$. The interaction has a sharp maximum near the W or Z resonance, when the energies of the colliding leptons at the center of mass are equal to half of the W or Z mass. The rate of ν_R production is strongly enhanced at the resonance but still the equilibrium ν_R density can be reached only for $m_\nu > 2$ keV. Details can be found in Dolgov (2002), Sect. 6.4. The ν_R number density may be unsuppressed if they were produced by a new interactions with the strength similar to that of the electroweak one. As we will see in Sect. 8.7, an analysis of light elements abundances created at the BBN allows to put restrictive bounds on the strength of such interactions.

Even if ν_R were abundant in the Universe at some high temperatures, their number density with respect to ν_L would be suppressed at ~ 1 MeV due to the entropy release by massive particle annihilation on the way. This effect is analogous to the photon heating by e^+e^- annihilation discussed in Sect. 5.2.3. Thus we conclude that ν_R were practically absent in the cosmological plasma at MeV temperatures.

The ratio of the number densities of the usual left-handed neutrinos and photons after the neutrino decoupling and e^+e^- annihilation is given by the ratio of the cube of their temperatures, $n_\nu/n_\gamma = (3/4)(T_\nu/T_\gamma)^3$ with $T_\nu/T_\gamma = (4/11)^{1/3}$, see Eq. (5.9). This is true for any neutrino flavor, ν_e, ν_μ, and ν_τ, or, as it would be better to say, for any of the three neutrino mass eigenstates, $\nu_{1,2,3}$. Let us note, however, that the eigenstates of the neutrino Hamiltonian in a hot cosmic plasma differ from those in vacuum.

The temperature of the decoupled neutrinos dropped down as $T \sim 1/a$, while T_γ dropped slower, as given by Eq. (5.37). Correspondingly

$$\frac{n_{\nu_L} + n_{\bar{\nu}_L}}{n_\gamma} = \frac{3}{11} \tag{5.42}$$

and the ratio of the neutrino temperature to the photon temperature is

$$\frac{T_\nu}{T_\gamma} = \left(\frac{4}{11}\right)^{1/3} = 0.714. \tag{5.43}$$

In fact we can speak about the neutrino temperature after their decoupling only when $T \gg m_\nu$, see the end of Sect. 5.4.

In the Standard Cosmological Model, the number densities of neutrinos and photons remained constant in the comoving volume, so we can calculate the present day number density of neutrinos knowing the today number density of CMB photons, $n_\gamma = 0.2404\, T^3 = 411(T/2.725\text{K})^3\, \text{cm}^{-3}$, see Eq. (10.19):

$$n_\nu = 56\,\text{cm}^{-3} \Sigma_{species} = 336\,\text{cm}^{-3}, \tag{5.44}$$

where the summation is done over all the neutrino and antineutrino species, assuming that all the mass eigenstates are equally populated and that the densities of neutrinos and antineutrinos are equal, i.e. $\mu_\nu = \mu_{\bar\nu} = 0$.

The present day energy density of neutrinos must be smaller than the total energy density of matter, ρ_m. This leads to the following upper bound on the sum of the masses of the three mass eigenstates of neutrinos

$$\sum m_{\nu_j} < 94 \text{ eV } \Omega_m h^2 . \qquad (5.45)$$

Since $h^2 \approx 0.5$, $\Omega_m \approx 0.25$, and the masses of different neutrinos are nearly equal, as follows from the data on neutrino oscillations, we find $m_\nu < 5$ eV for any neutrino mass eigenstate. This bound was derived in 1966 in a seminal paper by Gershtein and Zeldovich (1966). Somewhat similar result was obtained 6 years later by Cowsik and McClelland (1972), who, however, assumed that all the spin states of neutrino (left- and right-handed) were equally populated and did not take into account the photon heating by e^+e^- annihilation, with the result that they overestimated the neutrino density by roughly a factor seven. The result in Eq. (5.42) is a cornerstone of the cosmological bounds on m_ν, which are being obtained nowadays with better and better precision. The Gerhstein-Zeldovich (GZ) limit (5.45) may be immediately further strengthened by taking into account that cosmological structure formation would be inhibited at small scales if $\Omega_{HDM} > 0.3 \Omega_{CDM}$ (see Sect. 9.2). This gives $m_\nu < 1.7$ eV. Recently, on the basis of detailed studies of the contemporary data on the large scale structure of the Universe and on the spectrum of angular fluctuations of the CMB, the bound is $m_\nu < 0.3$ eV, which is almost an order of magnitude stronger than direct laboratory measurements (Olive et al. 2014).

The next question is how robust the GZ bound is. Is it possible to modify the standard picture to avoid or weaken it? The bound is based on the following assumptions:

1. Thermal equilibrium between ν, $e^\pm$, and γ at $T \sim 1$ MeV. If the Universe never was at $T \geq 1$ MeV, neutrinos might be under-abundant and the bound would be much weaker. However, a successful description of light element production at the BBN makes it difficult or impossible to eliminate the equilibrium neutrinos at the MeV phase in the Universe evolution.
2. Negligible lepton asymmetry. A non-zero lepton asymmetry would result in a larger number/energy density of neutrinos plus antineutrinos and the bound would be stronger.
3. No extra production of CMB photons after the neutrino decoupling. Strictly speaking, this is not excluded but strongly restricted. If the extra photons were created before the BBN terminated, they might have distorted the abundances of light elements. Late time creation of extra photons, after the BBN, would lead to distortions of the energy spectrum of the CMB, so there remains only very small freedom, not sufficient to change n_ν/n_γ essentially.
4. Neutrino stability on the cosmological scale, $\tau_\nu > t_U$. If neutrinos decay into other normal neutrinos, e.g. $\nu_\mu \to \nu_e + X$, the total number of neutrinos does not change and the limit on the mass of the lightest neutrino remains undisturbed, but

heavier neutrinos are allowed. If the decay goes into a new lighter fermions, e.g. sterile neutrino, the bound may be weakened for all neutrino species.
5. No late-time annihilation of $\nu + \bar{\nu}$ into a pair of (pseudo) Goldstone bosons, e.g. majorons. For noticeable annihilation, a too strong coupling of neutrinos to majorons is necessary, which is probably excluded by astrophysics.

So we have to conclude that there is no trivial way to weaken the GZ bound.

5.3.2 Freezing of Non-relativistic Particles

Heavy and sufficiently strongly interacting particles may have decoupled from the primordial plasma at a temperature that was much smaller than their mass, $T_f < m_h$. After decoupling, the number density of these particles stopped falling down as $n_h \sim \exp(-m_h/T)$, according the Boltzmann suppression law, but remained constant in the comoving volume. This happened because the number density had already turned very small, dropping down as $\exp(-m_h/T_f)$, and it became very difficult to find a partner for self-destruction through mutual annihilation. It is assumed, of course, that these particles are stable. This phenomenon is called freezing in English literature. In Russian literature, it originally got the name "quenching" as was suggested by Zeldovich who pioneered the study of this process.

The number density of heavy particles at the decoupling is given by

$$n_h/n_\gamma \sim (m_h/T_f)^{3/2} e^{-m_h/T_f} \ll 1, \tag{5.46}$$

so such particles may have masses much larger than those allowed by the GZ bound and can make cosmologically interesting cold dark matter with $\Omega_h \sim 1$. The frozen number density of such particles is determined by the cross section of their annihilation and it is given by a simple expression, see the derivation below and, e.g., Dolgov and Zeldovich (1981)

$$\frac{n_h}{n_\gamma} \approx \frac{(m_h/T_f)}{\langle \sigma_{ann} v \rangle M_{\text{Pl}} m_h}, \tag{5.47}$$

where $m_h/T_f \approx ln(\langle \sigma_{ann} V \rangle M_{\text{Pl}} m_h) \sim 10 - 50$. To derive this expression, one has to numerically solve the kinetic equation governing the evolution of the number density of heavy particles, but it is anyway instructive to make some analytic calculations. Moreover, the analytic results are pretty accurate. The calculations of frozen abundances are usually done under the following assumptions:

1. Boltzmann statistics. It is usually a good approximation for heavy particles at $T < m$, since their distribution function is small, namely $f_h \ll 1$.
2. It is usually assumed that the number densities of the heavy particles and their antiparticles under scrutiny are equal, so in full thermal equilibrium at high temperatures, namely for $T \geq m$, their chemical potential vanishes.

5.3 Freezing of Species

3. At lower temperatures, however, the heavy particles were in kinetic, but not chemical, equilibrium, i.e. their distribution function has the form

$$f_h = e^{-E/T + \xi(t)}, \qquad (5.48)$$

where ξ is the effective chemical potential normalized to temperature, $\xi = \mu/T$. Chemical equilibrium is enforced by the annihilation for which an antipartner is needed. However, its number density at $T < m$ is exponentially suppressed. On the other hand, kinetic equilibrium demands encounter with abundant light particles. That is why the chemical equilibrium stops being maintained much earlier than the kinetic one.

4. The light particles, produced by the annihilation of h and $\bar{h}$ and producing $h\bar{h}$ pairs by the inverse process, are supposed to be in complete thermal equilibrium.

5. By assumption, the charge asymmetry of heavy particles is negligible and thus the effective chemical potentials for particles and antiparticles are equal, $\xi = +\bar{\xi}$. However, sometimes this restriction is lifted, and the asymmetry is allowed to be essential. In this case, the annihilation proceeds much more efficiently and the survived abundance of h is determined by the magnitude of their charge asymmetry.

The kinetic equation under this assumption turns into an ordinary differential equation, which was derived in 1965 by Zeldovich (1965) and used for the calculation of the frozen number density of non-confined massive quarks in Zeldovich et al. (1965). In 1978, the equation was applied to the calculations of the frozen number densities of stable heavy leptons in Lee and Weinberg (1977); Vysotsky et al. (1977), and after that it got the name Lee-Weinberg equation, though it would be more proper to call it Zeldovich equation.

The equation has the following Riccati type form

$$\dot{n}_h + 3Hn_h = \langle \sigma_{ann} v \rangle (n_h^{eq \, 2} - n_h^2), \qquad (5.49)$$

where n_h is the number density of these heavy particles, n_h^{eq} is its equilibrium value, and $\langle \sigma_{ann} v \rangle$ is the thermally averaged annihilation cross section multiplied by velocity of the annihilating particles

$$\langle \sigma_{ann} v \rangle = \frac{(2\pi)^4}{(n_h^{eq})^2} \int \overline{dp_h} \overline{dp_{\bar{h}}} \overline{dp_f} \overline{dp_{f'}} \delta^4(P_{in} - P_{fin}) |A_{ann}|^2 e^{-(E_p + E_{p'})/T}, \qquad (5.50)$$

where $\overline{dp} = d^3 p / [2E (2\pi)^3]$. The annihilation (and the inverse annihilation) is assumed to be a simple two-body process, namely $h + \bar{h} \leftrightarrow f + \bar{f}$.

The integration in Eq. (5.50) can be taken down to one variable and we find (Gondolo and Gelmini 1991)

$$\langle \sigma_{ann} v \rangle = \frac{x}{8 m_h^5 K_2^2(x)} \int_{4m_h^2}^{+\infty} ds \, (s - 4m_h^2) \sigma_{ann}(s) \sqrt{s} K_1\left(\frac{x\sqrt{s}}{m_h}\right), \qquad (5.51)$$

where $x = m_h/T$, $s = (p + \bar{p})^2$, and K_1 and K_2 are the modified Bessel functions. Usually $x \gg 1$ and $\sigma_{ann}v \to const$ near the threshold, so the thermally averaged product $\langle \sigma_{ann}v \rangle$ is reduced just to the threshold value of $\sigma_{ann}v$. The expression above can be useful if the annihilation cross section noticeably changes near the threshold, e.g. in the case of resonance annihilation.

For the derivation of Eq. (5.49), we start with the general kinetic equation

$$\partial_t f - H p \partial_p f = I_{el} + I_{ann}, \tag{5.52}$$

where we take into account only two-body processes with heavy particles. The elastic scattering is governed by the I_{el} term in the collision integral and the two-body annihilation is governed by I_{ann}. When $T < m_h$, the former is much larger than the latter because of the exponential suppression of the density of heavy particles, $f_h \sim \exp(-m_h/T)$. Since I_{el} is large, it enforces kinetic equilibrium, i.e. the canonical Boltzmann distribution over energy

$$f_h = \exp[-E/T + \xi(t)]. \tag{5.53}$$

With such a form of f_h, we can integrate both sides of Eq. (5.52) over $d\bar{p}$ and the large elastic collision integral disappears, but the impact of it remains in the distribution (5.53).

As the last step, we express $\xi(t)$ through n_h

$$\exp(\xi) = \frac{n_h}{n_h^{eq}} \tag{5.54}$$

and we arrive at Eq. (5.49). This equation can be solved analytically, approximately, but quite accurately. At high temperatures, $T \geq m_h$, the annihilation rate is usually high

$$\sigma_{ann} v n_h / H \gg 1, \tag{5.55}$$

and thus the equilibrium with respect to annihilation is approximately maintained. We can thus write $n_h = n_h^{eq} + \delta n$, where δn is small. It is convenient to introduce the dimensionless ratio of the number density to the entropy $r = n_h/s$, so the effects of the expansion of the Universe disappear from the equation

$$\dot{n} + 3Hn = s\dot{r}, \tag{5.56}$$

since, according to Eq. (5.33), the entropy is conserved in comoving volume, $\dot{s} = -3Hs$.

Introducing the new variable $x = m_h/T$ and assuming that $\dot{T} = -HT$, which is approximately true if the entropy release by massive particle annihilation is small, we arrive at the equation

5.3 Freezing of Species

$$r' = Qx^{-2}\left(r_{eq}^2 - r^2\right), \quad (5.57)$$

where the prime denotes the derivative over x, $Q = g_{s*}\sqrt{90/8\pi^3 g_*}\,\sigma_{ann}vm_hM_{Pl} \gg 1$, g_{s*} is defined in Eq. (5.34), the Hubble parameter is given by Eq. (5.14), and according to Eqs. (5.8) and (5.10) we have

$$r_{eq}(x) = \frac{g_h}{g_{s*}}e^{-x}\left(\frac{x}{2\pi}\right)^{3/2}. \quad (5.58)$$

Since the coefficient Q in front of the brackets in Eq. (5.57) is huge, r should weakly deviate from equilibrium, so we can write $r = r_{eq}(1 + \delta r)$, where $\delta r \ll 1$. It is the so-called stationary point approximation, which implies that the factor that is multiplied by Q in the right hand side of the equation vanishes with an accuracy of $1/Q$ terms. It follows that

$$\delta r \approx -\frac{r'_{eq}x^2}{2Q\,r_{eq}^2}. \quad (5.59)$$

Since r_{eq} exponentially drops down, δr rises, and the approximation breaks down roughly speaking when δr reaches unity. After that, we use another approximation: neglecting r_{eq}^2, we analytically integrate the equation for r and obtain the final result (5.47). For its derivation, we assume that the annihilation proceeded in s-wave, so the product $\sigma_{ann}v$ tends to a non-vanishing constant. The annihilation cross section in higher partial wave vanishes as a power of the center of mass 3-momentum of the colliding particles. The result can be easily generalized for this case and we leave its derivation as an exercise. Another way to solve Eq. (5.49) is to transform this Riccati-type equation to the second order Schroedinger-type one and to integrate the latter in quasi-classical approximation.

Let us apply the obtained results for the calculation of the frozen number density of the lightest supersymmetric particle (LSP), which must be stable if R-parity is conserved and is a popular candidate for dark matter (the topic will be briefly introduced in Sect. 9.2.1). The annihilation cross section is estimated as

$$\sigma_{ann}v \sim \alpha^2/m_{LSP}^2. \quad (5.60)$$

Correspondingly, the energy density of the LSP would be

$$\rho_{LSP} = m_{LSP}n_{LSP} \approx \frac{n_\gamma\,m_{LSP}^2}{M_{Pl}}\ln\left(\frac{\alpha^2 M_{Pl}}{m_{LSP}}\right). \quad (5.61)$$

For $m_{LSP} = 100$ GeV, which is a reasonable value in a minimal supersymmetric model, we find

$$\Omega_{LSP} \approx 0.05. \quad (5.62)$$

This value is very close to the observed 0.25 and makes the LSP a natural candidate for dark matter.

Another interesting example is the frozen number density of magnetic monopoles, which may exist in spontaneously broken gauge theories containing $O(3)$ subgroup (Polyakov 1974; 't Hooft 1974). The cross section of the monopole-antimonopole annihilation can be estimated as

$$\sigma_{ann} v \sim g^2/M_M^2, \qquad (5.63)$$

where M_M is the monopole mass. Correspondingly, the present day energy density of magnetic monopoles would be (Zeldovich and Khlopov 1978; Preskill 1979)

$$\rho_M = \frac{n_\gamma M_M^2}{g^2 M_{Pl}}. \qquad (5.64)$$

The slow diffusion of monopoles in the cosmic plasma due to the mutual attraction would slightly diminish the result, but not too much. If $M_M \sim 10^{17}$ GeV, as predicted by GUT models, monopoles would overclose the Universe by about 24 orders of magnitude assuming that their initial abundance was close to that of thermal equilibrium. This problem played a driving role for the suggestion of inflationary cosmology.

5.4 Neutrino Spectrum and Effective Number of Neutrino Species

As it is shown in Eqs. (5.29) and (5.30), massless particles keep their equilibrium distribution even after their interaction is switched off. This means that if the thermal equilibrium was initially established due to sufficiently strong interactions, the spectrum would remain that of equilibrium even after all the interactions have been switched off. An impressive example of such a situation is represented by the CMB photons: their spectrum is measured to be of the black body equilibrium form with a precision better than 10^{-4}, though these photons stopped interacting with the cosmic plasma and between themselves after the hydrogen recombination, which took place at the redshift $z_{rec} \approx 1100$ (see Sect. 10.1).

However, this is not true for neutrinos. Their spectrum started to deviate noticeably from the equilibrium one at redshift $\sim 10^{10}$, i.e. near the neutrino decoupling temperature, despite the fact that the non-vanishing mass of neutrinos is absolutely unessential at such temperatures. The point is that at that period there were two weakly interacting components in the primeval plasma with different temperatures: the neutrino part, consisting of ν_e, ν_μ, and ν_τ, and the electromagnetic part, consisting of photons and e^+e^- pairs, which were in the process of annihilation, heating up the electromagnetic part of the plasma, as is described in Sect. 5.2.3. The neutrino

5.4 Neutrino Spectrum and Effective Number of Neutrino Species

decoupling from e^+e^- pairs was not instantaneous and the residual energy exchange between the hotter electron/positron component and the cooler neutrino gas was going on. The probability of the energy transfer depends upon the particle energy and, as a result, the neutrino spectrum became distorted. This phenomenon was discovered in Dolgov and Fukugita (1992a, b), where the shape of the spectrum distortion was analytically calculated

$$\delta f_{\nu_e}/f_{\nu_e}^{eq} \approx 3 \times 10^{-4} \frac{E}{T} \left(\frac{11}{4} \frac{E}{T} - 3 \right). \qquad (5.65)$$

This analytical estimate was confirmed by the precise numerical solution of the integro-differential kinetic equation (Dolgov et al. 1997, 1999) and by subsequent work (Mangano et al. 2002), which confirmed this result. For discussion, history, and the list of references, see Dolgov (2002).

Due to such energy influx from the hotter electron-positron pairs to cooler neutrinos, the real neutrino energy density increased with respect to the would-be equilibrium one with $T_\nu = 0.714 \, T_\gamma$. This effect can be described by an increase of the effective number of the equilibrium neutrino species, $\Delta N_\nu = 0.035$. There is an additional contribution to the effective neutrino number, $\Delta N_\nu = 0.011$, which comes from a decrease of the photon number density due to the deviation of the $\gamma e^- e^+$ plasma from the ideal gas (e.g. from a decrease of the photon energy density due to non-zero plasma frequency) (Heckler 1994; Lopez et al. 1999). Because of this effect, the neutrino energy density relative to that of ideal photons rises by 1.011. Hence the total number of the effective neutrino species in the standard model is

$$N_\nu = 3.046, \qquad (5.66)$$

instead of the usual three. Physically we have exactly three standard neutrinos, but their energy density is slightly above the equilibrium value and is normalized to photons with slightly smaller energy density than the usual black body radiation. An increase of the number of neutrino species has negligible effect on the BBN (Dolgov and Fukugita 1992a, b), but may be noticeable in future CMB measurements.

In conclusion to this section, let us note that though we mentioned above that the neutrino temperature is approximately 1.4 times smaller than the temperature of photons, i.e. today it should be 1.95 K, would neutrino be massless, the distribution of neutrinos has the non-equilibrium form

$$f_\nu \approx [\exp(p/T_\nu) + 1]^{-1}, \qquad (5.67)$$

i.e. the magnitude of neutrino momentum enters instead of the neutrino energy and so the parameter T_ν does not have meaning of temperature. The correction (5.65) is neglected here.

Problems

5.1 Derive Eq. (5.9).

5.2 Derive Eq. (5.10).

5.3 Derive Eq. (5.16). [Hint: the calculations of integrals with the equilibrium distribution functions are discussed in detail in Landau and Lifshitz (1980).]

5.4 Check that the equilibrium Bose and Fermi distribution functions (5.4) annihilate the collision integral (5.19) if the T-invariance is not broken.

5.5 Show that the Bose condensed distribution functions (5.25) annihilate the collision integral (5.19) if and only if $\mu = m$.

5.6 Why in the distribution function p and t are taken as independent variables, while in Eq. (5.30) we treated the momentum as a function of the time, namely $p = p(t)$?

5.7 Find the frozen number densities of protons and electrons in a charge-symmetric universe. [Answer: $n_p/n_\gamma \approx 10^{-19}$, $n_e/n_\gamma \approx 10^{-16}$.]

5.8 What number density would have antiprotons if $(n_p - n_{\bar{p}})/n_\gamma = 10^{-9}$?

References

R. Cowsik, J. McClelland, Phys. Rev. Lett. **29**, 669 (1972)
A.D. Dolgov, Pisma Zh. Eksp. Teor. Fiz. **29**, 254 (1979)
A.D. Dolgov, Phys. Rept. **370**, 333 (2002) [hep-ph/0202122]
A.D. Dolgov, M. Fukugita, JETP Lett. **56**, 123 (1992a) [Pisma Zh. Eksp. Teor. Fiz. **56**, 129 (1992)]
A.D. Dolgov, M. Fukugita, Phys. Rev. D **46**, 5378 (1992b)
A.D. Dolgov, S.H. Hansen, D.V. Semikoz, Nucl. Phys. B **503**, 426 (1997) [hep-ph/9703315]
A.D. Dolgov, S.H. Hansen, D.V. Semikoz, Nucl. Phys. B **543**, 269 (1999) [hep-ph/9805467]
A.D. Dolgov, Y.B. Zeldovich, Rev. Mod. Phys. **53**, 1 (1981) [Usp. Fiz. Nauk **139**, 559 (1980)]
S.S. Gershtein, Y.B. Zeldovich, JETP Lett. **4**, 120 (1966) [Pisma Zh. Eksp. Teor. Fiz. **4**, 174 (1966)]
P. Gondolo, G. Gelmini, Nucl. Phys. B **360**, 145 (1991)
A.F. Heckler, Phys. Rev. D **49**, 611 (1994)
G. 't Hooft, Nucl. Phys. B **79**, 276 (1974)
L.D. Landau, E.M. Lifshitz, *Statistical Physics*, 3rd edn. (Butterworth-Heinemann, Oxford, 1980)
B.W. Lee, S. Weinberg, Phys. Rev. Lett. **39**, 165 (1977)
R.E. Lopez, S. Dodelson, A. Heckler, M.S. Turner, Phys. Rev. Lett. **82**, 3952 (1999) [astro-ph/9803095]
G. Mangano, G. Miele, S. Pastor, M. Peloso, Phys. Lett. B **534**, 8 (2002) [astro-ph/0111408]
K.A. Olive et al., Particle data group collaboration. Chin. Phys. C **38**, 090001 (2014)
A.M. Polyakov, JETP Lett. **20**, 194 (1974) [Pisma Zh. Eksp. Teor. Fiz. **20**, 430 (1974)]
J. Preskill, Phys. Rev. Lett. **43**, 1365 (1979)
M.I. Vysotsky, A.D. Dolgov, Y.B. Zeldovich, JETP Lett. **26**, 188 (1977) [Pisma Zh. Eksp. Teor. Fiz. **26**, 200 (1977)]
Y.B. Zeldovich, *Advances in Astronomy and Astrophysics*, vol. 3 (Academic Press, New York, 1965)
Y.B. Zeldovich, M.Y. Khlopov, Phys. Lett. B **79**, 239 (1978)
Y.B. Zeldovich, L.B. Okun, S.B. Pikelner, Usp. Fiz. Nauk **87**, 113 (1965)

Chapter 6
Inflation

6.1 Introduction and History

The idea of inflation was probably the most important breakthrough in cosmology of the XX century after that of the big bang. Historically, the first papers in which an exponential expansion was invoked for solving some problems of the FRW cosmology were those by Starobinsky (1980), where it was mentioned that an exponential expansion leads to a flat geometry of the Universe, and by Kazanas (1980), who found that a similar expansion could make the Universe isotropic. A few months later, the famous paper by Guth "Inflationary universe: A possible solution to the horizon and flatness problems" was published (Guth 1981). This work initiated a stream of papers that remains unabated to the present day. In the Starobinsky model, the initial de Sitter-like stage was created by R^2 corrections to the Einstein-Hilbert action, while in the scenarios proposed by Kazanas and Guth the vacuum-like energy, which might dominate during a first order phase transition, was suggested as a driving force of the exponential expansion. It was soon understood that the latter mechanism was not satisfactory because it would have created an inhomogeneous Universe consisting of many relatively small bubbles in an exponentially expanding vacuum-like background. The first workable mechanism of inflation based on a slowly evolving scalar field was suggested by Linde (1982) and, independently, by Albrecht and Steinhardt (1982). The most appealing inflationary scenario is probably the so-called chaotic inflation, proposed by Linde (1983). For a review on inflationary models and the associated main issues, see e.g. Linde (1990), Kinney (2015), Dolgov (2010), Baumann (2015).

There is a significant "pre-inflationary" literature directly related to the subject. The idea that the Universe avoided an initial singularity and underwent an exponential period during which the mass of the cosmological matter rose by many orders of magnitude was discussed by Gliner (1966) and Gliner and Dymnikova (1975). A de Sitter-like (exponentially expanding) non-singular cosmology was considered by Gurovich and Starobinsly (1979) and by Starobinsky (1979). In the latter paper, an important result was obtained, namely that during the "initial" exponentially

---

The original version of this chapter was revised: The errors in this chapter have been corrected. The correction to this chapter can be found at https://doi.org/10.1007/978-3-662-48078-6_13.

expanding stage gravitational waves were produced and that they may be observable at the present time. If observed, it would be one of the strongest "experimental" evidence for the existence of a primordial inflation. However, an absence of primordial gravitational waves would not kill the idea of inflation because their predicted intensity is model dependent and may be quite low.

Another prediction of inflation is the spectrum of primordial density perturbations, which is already verified by available data. Pioneering calculations of the spectrum were done by Mukhanov and Chibisov (1981) and confirmed later by many other studies (Linde 1990; Kinney 2015; Dolgov 2010; Baumann 2015).

It was shown in the paper by Sato (1981) that an exponential expansion induced by a first order phase transition would never be terminated for certain under-critical values of the parameters. This happened to be a serious shortcoming of the suggested later first inflationary scenarios. It was also noticed by Sato (1981) that an exponential expansion might permit astronomically interesting antimatter domains.

6.2 Problems of Pre-inflationary Cosmology

Despite the great success in the description of the Universe on the basis of General Relativity, the FRW cosmology suffered from quite a few serious problems, which were initially considered as virtually unsolvable. The only option in the market was the anthropic principle: the conditions in the Universe must be such that they allow observers to exist and to ask the question why the Universe is suitable for life.

First of all, the origin of the cosmological expansion was a mystery. Gravity was believed to be universally attractive and the sudden repelling force which acted briefly at the beginning and then disappeared without trace was difficult to digest, to say the least.

Second, the temperature of the CMB coming to us from different patches in the sky is almost exactly the same, though celestial points separated by more than one degree never knew about each other in the FRW cosmology. This is called the *horizon problem* or the *causality problem*.

A similar problem is related to the fact that the observed Universe is almost homogeneous at large scales, while no mechanism to make it the same everywhere was known.

The cosmological energy density is not much different from the critical one and so the geometry of the 3-dimensional space is close to be Euclidean. To achieve this state at the present time, the Universe had to be extremely well fine tuned at its early stage. The geometry should have been flat with a precision of about 10^{-15} at the BBN epoch and at the level of $\sim 10^{-60}$ at the Planck time. This is called the *flatness problem*.

Last but not least, for the creation of cosmic large scale structures (galaxies, clusters, to say nothing about stars and planets) the presence of primordial density perturbations at astronomically large scales is necessary. However, no single reasonable mechanism of generation of density perturbations at such large scales was known.

6.2 Problems of Pre-inflationary Cosmology

All these cosmic mysteries can be uniquely and beautifully solved if initially (say, at some very early time) the Universe exponentially expanded with the scale factor rising as

$$a(t) \sim \exp(H_I t), \qquad (6.1)$$

where the Hubble parameter H_I was approximately constant during at least about 60–70 e-folding times, i.e. the duration of inflation should satisfy the condition $H_I \Delta t > 60$.

6.2.1 Kinematics and Main Features of Inflation

Before discussing the above mentioned problems in more detail, we briefly present some mechanisms that could lead to an exponential cosmological expansion. An important condition for the (quasi-)exponential expansion is that the Hubble parameter should be (quasi-)constant. For simplicity, in this section where we discuss the kinematics, we assume that H is strictly constant, namely the exponential expansion is created by a cosmological constant (see Sect. 4.3.4) or, which is the same, by a vacuum energy-momentum tensor, which has the form

$$T_{\mu\nu}^{(vac)} = \rho^{(vac)} g_{\mu\nu}, \qquad (6.2)$$

where $\rho^{(vac)} \equiv \Lambda M_{Pl}^2/(8\pi)$. The vacuum has thus the "equation of state"

$$P^{(vac)} = -\rho^{(vac)}. \qquad (6.3)$$

So for the vacuum the parameter w introduced in Eq. (4.13) is -1. As follows from Eq. (4.12), the vacuum energy does not change with time, $\rho^{(vac)} = const$. Of course, for an inflationary scenario this cannot be exactly true, because it would mean that the exponential expansion would exist forever. In realistic inflationary models, the expansion could be governed, for instance, by a scalar field ϕ called inflaton with energy density only *approximately* constant, see Sect. 6.3.1. Inflationary models are described in Sect. 6.3. In the course of the cosmological expansion, the energy density of the inflaton field dropped down, first very slowly when $\phi \approx const$, and later, when ϕ begun to oscillate, the stored vacuum-like energy of ϕ turned into the energy of a hot "soup" of elementary particles. At the first stage, the Universe looked as a dark expanding empty place. The second stage was the big bang, when the primeval plasma was created. It is impressive that the total mass/energy of matter inside the observed Universe volume is by far larger than the initial mass/energy inside the microscopic volume from which the Universe originated. Still it agrees with the energy conservation law (4.12).

6.2.2 Flatness Problem

The cosmological 3-geometry is determined by the ratio of the total cosmological energy density to the critical energy density, $\Omega = \rho/\rho_c$, where $\rho_c = 3M_{\rm Pl}^2 H^2/(8\pi)$ as given by Eq. (4.14). Using Eq. (4.17), we find that Ω evolves as a function of the scale factor as

$$\Omega(a) = \left[1 - \left(1 - \frac{1}{\Omega_0}\right)\frac{\rho_0 a_0^2}{\rho a^2}\right]^{-1}, \tag{6.4}$$

where the subindex 0 is again used to denote the present day values of the corresponding quantities. The cosmological constant is not explicitly included in Eq. (6.4), but, as it is common in the literature, it can be conveniently taken into account by adding the proper vacuum energy density to the total energy density ρ.

If we assume that ρ is the energy density of some kind of normal matter, it drops down as $\rho \sim 1/a^n$ with $n = 3$ or 4, respectively for non-relativistic or relativistic matter. In this case, the product ρa^2 tends to infinity when $a \to 0$. This implies that in the past Ω had to be very precisely tuned to 1 to be still close to 1 today. For example, the fine-tuning of $|\Omega - 1|$ should be 10^{-15} at the BBN and 10^{-60} at the Planck era. Otherwise, the Universe would have recollapsed in a much shorter time than its present age, 10^{10} yrs, or would have expanded too fast to allow any structure formation. However, if at some stage ρa^2 rose with rising a, then the necessary fine tuning can be automatically realized. For example, an inflationary period with $\rho a^2 \sim \exp(Ht)$ and $Ht > 65$ would be sufficient.

The evolution of $\Omega(a)$ is schematically presented in Fig. 6.1. For small values of a, the energy density is approximately constant and Ω tends to 1. For large values of a, the product ρa^2 drops down and Ω starts deviating from 1. The upper and lower curves correspond, respectively, to the cases in which $\Omega > 1$ and $\Omega < 1$. The line $\Omega = 1$ separating the two previous scenarios does not change with a. For normal matter with ρa^2 going to zero at large a, the upper curve tends to infinity, while the lower one goes to zero.

6.2.3 Horizon Problem

To create the same temperature of the CMB over the whole celestial sphere, photons should have exchanged energy among themselves along the whole sphere. The distance that a photon, propagating with the speed of light, could pass during the all history of the Universe is determined by the equation of motion for a massless particle

$$ds^2 = dt^2 - a^2(t)dr^2 = 0. \tag{6.5}$$

6.2 Problems of Pre-inflationary Cosmology

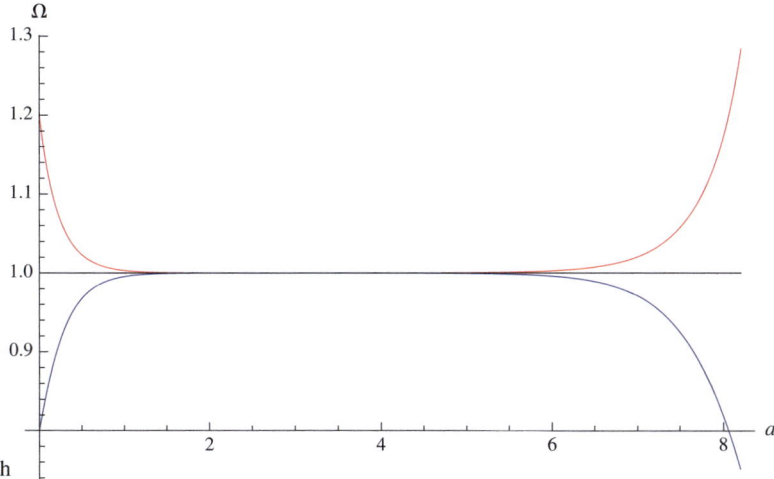

Fig. 6.1 Evolution of Ω in terms of a rising scale factor a. The *upper* and *lower curves* correspond to the cases $\Omega > 1$ and $\Omega < 1$, respectively. The special value $\Omega = 1$ does not change with a—the *middle straight line*. The scale factor a is presented in logarithmic scale of arbitrary units

So the distance $dl = a(t)dr$ passed by a photon during the time t would be

$$l(t) = a(t) \int_0^t \frac{dt'}{a(t')} = \begin{cases} 2t \text{ radiation dominated expansion, } a(t) \sim t^{1/2}, \\ 3t \text{ matter dominated expansion, } a(t) \sim t^{2/3}. \end{cases} \quad (6.6)$$

In reality, individual photons propagated to much shorter distances, because they slowly diffuse due to their interactions with the cosmological plasma till the hydrogen recombination at $z_{rec} \approx 1100$. Still the interactions between different parts of the plasma can be realized by macroscopic physics, e.g. by sound waves with a speed comparable to the speed of light $c_s = c/\sqrt{3}$. After recombination, interactions among CMB photons can be neglected. The maximum distance at which the temperature might be equilibrated is about the Universe age at recombination, $d_{causal} \sim t_{rec} \approx 10^{13}$ s. After recombination, d_{causal} rises due to the cosmological expansion by $z_{rec} \approx 10^3$ and today reaches the value $\sim 10^{16}$ s. The angular size of this path on the sky is

$$\theta_{max} = \frac{10^{16} \text{ s}}{2\pi t_U} \approx 1°, \quad (6.7)$$

where $t_U \approx 10^{10}$ years is the Universe age. θ_{max} is the maximum angle for which an exchange of information and energy in the FRW cosmology is possible. However, if the FRW expansion regime was preceded by an exponential expansion, the causally connected region would be extended by the factor $l_{infl} = H_I^{-1}[\exp(H_I t) - 1]$ and with $H_i t_i > 70$, the same as in Sect. 6.2.2, the whole observed Universe would be causally connected.

6.2.4 Origin of the Cosmological Expansion

The equation of state (6.3) for the inflaton field and the second Friedmann equation (4.11) explain the origin of the cosmological expansion. Indeed, for $P = -\rho$ the acceleration becomes positive: $\ddot{a}/a = +8\pi\rho/(3M_{Pl}^2)$. So when the quasi-constant inflaton field dominated the cosmological energy density, it created gravitational repulsion at cosmological scales, and the Universe started expanding. The acceleration later changed to normal deceleration, but the expansion continued, becoming in a sense the motion by inertia with diminishing speed. It was believed that the accelerated expansion was present only in the early Universe, during inflation. However, it was established during the last two decades that the Universe started expanding with acceleration again at a relatively recent cosmological epoch, at redshift of order 1, see Sect. 4.6 and Chap. 11.

It is worth noting that antigravity, which may apparently be induced by matter with negative pressure such that $|P| > \rho/3$, cannot be created by any finite body, since it is well known that the gravitational field of such bodies is created by their mass, which is an integral from the pressure density over the volume of the body, and it is always positive in non-pathological theories. The impact of the negative pressure is cancelled by surface effects. So cosmic ships cannot fly using negative pressure, at least if standard General Relativity is valid.

6.2.5 Smoothing Down the Universe and Creation of Primordial Density Perturbations

Inflation is able to fulfill two apparently opposite tasks. Firstly, it smoothed down density perturbations at very large scales, which would not be causally connected in a FRW universe. The cosmological energy density in regions where matter was never in causal contact might have very different values, while we observe that the Universe looks pretty homogeneous at very large scales, comparable to the present day horizon. The flattening of the density contrast at a given length is forced by an exponential rise of the length with practically fixed height of the contrast.

Simultaneously, inflation could generate small density perturbations inside the present day horizon by the amplification and the exponential expansion of quantum fluctuations during the quasi-de Sitter (inflationary) stage. This mechanism is discussed in more detail in Sect. 6.6. These density perturbations were amplified during the matter dominated stage in the course of the cosmological history and became seeds for large scale structure formation, explaining clumpiness of the Universe at galactic, galaxy cluster, and supercluster scales. The spectrum of density perturbations predicted by inflation, the so-called Harrison-Zeldovich spectrum (Harrison 1970; Zeldovich 1972), well agrees with astronomical observations.

6.2.6 Magnetic Monopole Problem

This is an example of probably non-existing problem, which greatly stimulated work on inflationary models. A magnetic monopole is an object with an elementary magnetic charge (e.g. a single South pole without the North one or vice versa) and was proposed by Dirac (1931). The theory was not really satisfactory because it demanded an unphysical string going from the monopole to infinity. However, with the famous Dirac quantization condition

$$q_e q_m = \frac{n}{2}, \tag{6.8}$$

where q_e and q_m are, respectively, the electric and the magnetic charges, such a string becomes unobservable.

The situation became very different after the papers by Polyakov (1974) and t'Hooft (1974), who independently discovered that in some spontaneously broken gauge theories with unbroken electromagnetic subgroup $U(1)$ there exist topologically stable classical solutions, which are not elementary particles and possess a magnetic charge satisfying the condition (6.8). By definition, a classical localized solution has a size much larger than its Compton wavelength, $d \gg 1/m$. The monopole solution has the form

$$\phi^a = \frac{r^a}{r} v f(r), \quad A_j^a = \frac{r^j}{q_e r^2} \varepsilon_{aij} F(r), \quad A_t^a = 0, \tag{6.9}$$

where ϕ^a is a Higgs-like scalar field, A_j^a is a vector gauge field, a is the $O(3)$ group index, i and j are space indices, $f(r)$ and $F(r)$ are functions of the radial coordinate r only and have boundary conditions $f(0) = F(0) = 0$ and $f(+\infty) = F(+\infty) = 1$, and v is the vacuum expectation value of the Higgs field. By assumption, $O(3)$ is a subgroup of some symmetry group of the GUT under consideration. It is interesting that the space vector r^a bears the index of the group of rotation in internal space, i.e. a here does not numerate the space coordinates, x, y, or z, but runs over the group indices. For example, if the internal symmetry group would be the $SU(3)$ of color, then a runs over the three color indices. ε_{aij} has mixed space and group indices. The characteristic size of the monopole is equal to the inverse mass of the Higgs or gauge boson, $d \sim 1/m_X$, while the monopole mass, M, is of the order $m_X/q_e^2 \sim v/q_e$. For a GUT with an energy scale $M_{GUT} \sim 10^{14}$ GeV, the mass of the monopole would be about $M \sim 10^{16}$ GeV. Properties of classical topologically stable objects appearing in spontaneously broken gauge theories are reviewed in Vilenkin (1985), Vilenkin and Shellard (1994), Dolgov (1992).

If one believes that GUTs are the correct way of unification of the strong and the electroweak interactions and that in the early Universe the temperature reached a value of the order the GUT scale, then magnetic monopoles had to be abundant in the early Universe and their present mass density should be much larger than the observed one (Zeldovich and Khlopov 1978; Preskill 1979). Magnetic monopoles would have thus overclosed the Universe. We can prove this by using the same approach as we applied to calculate the frozen density of massive stable particles in the Universe. The only difference in the calculations is that, in contrast to usual dark matter particles, monopoles and antimonopoles are mutually attracted, which somewhat enhances the probability of their annihilation. We can use the result of Sect. 5.3.2, according to which the energy density of GUT monopoles is 24 orders of magnitude larger than that allowed by data, Eq. (5.64). An enhancement of the annihilation due to the mutual attraction could somewhat change this result, but it still remain extremely large. More detailed calculations of monopole-antimonopole annihilation can be found in Dolgov and Zeldovich (1980).

The calculations of frozen densities of massive particles performed in Sect. 5.3.2 have been done under the assumption that the initial density of these particles was thermal, i.e. it was determined by thermal equilibrium. If the initial temperature of the Universe was smaller than the monopole mass, their density would be suppressed by the factor $\exp(-M/T)$. Though this assumption is probably not correct, it does not help to solve the magnetic monopole problem. Strictly speaking, we do not know the probability of production of classical objects (such as monopoles) in elementary particle collisions, but most probably it is strongly suppressed. Colliding particles must produce a certain highly coherent state of vector (gauge) and scalar fields with some non-trivial topology. The phase space of such a state is extremely small, probably at the level of $\exp(-CMd)$, where M is the mass of the object, d is its size, and C is a constant which is probably large. For classical objects, $Md \gg 1$. Thus the monopole production should be strongly suppressed even at high T. However, as we have already said, it does not solve the overabundance problem of magnetic monopoles. The point is that there is another mechanism to produce monopoles, the so-called topological mechanism (Kibble 1976). Such a mechanism can be visualized with the example of the production of cosmic strings: in causally non-connected regions in the Universe, the variation of the phase of a complex scalar field, ϕ, along a closed loop is not necessarily zero but could be $2\pi n$ and, if there is a singular state of ϕ inside this loop such that the loop radius cannot be shrunk down to zero, a cosmic string would be created. With this mechanism, one would expect on average one string per cosmological horizon. Detailed calculations can be found in Vilenkin (1985), Vilenkin and Shellard (1994), Dolgov (1992). A magnetic monopole is, in particular, a state of a vector field directed out of the center of a sphere surrounding the monopole, like the needles of a hedgehog. Such a configuration could be accidentally formed in the process of cosmological cooling when a gauge symmetry was spontaneously broken. Inside such a sphere, a magnetic monopole would be certainly created. The probability of this configuration is quite large and so monopoles would destroy the Universe. Inflation saved us from this gloomy destiny.

In conclusion, let us mention a striking phenomenon discovered by Rubakov (1981, 1982, 1982): in the vicinity of a magnetic monopole, protons would quickly decay. In other words, monopoles catalyse proton decay. Such a process could be a cheap energy source. Though it has no direct relation to the subject of this chapter, it might contribute to the generation of the baryon asymmetry of the Universe if the amount of monopoles were not negligibly small.

6.3 Mechanisms of Inflation

6.3.1 Canonical Scalar Inflaton with Power Law Potential

In the simplest case, a quasi-exponential cosmological expansion is assumed to be created by a real scalar field with the action

$$S[\phi] = \int d^4x \sqrt{-g} \left[\frac{1}{2} g^{\mu\nu} \partial_\mu \phi \partial_\mu \phi - U(\phi) \right], \quad (6.10)$$

where $U(\phi)$ is the potential of ϕ usually taken as a polynomial

$$U = \frac{1}{2} m^2 \phi^2 + \frac{\lambda_\phi}{4} \phi^4. \quad (6.11)$$

Such potential leads to a renormalizable theory for the self-interacting field ϕ.

In the FRW background metric, the field ϕ satisfies the equation of motion

$$\ddot\phi + 3H\dot\phi - \frac{\nabla^2 \phi}{a^2} + U' = 0, \quad (6.12)$$

where $U' = dU/d\phi$. The energy-momentum tensor of ϕ is

$$T^{\mu\nu} = \frac{2}{\sqrt{-g}} \frac{\delta S}{\delta g_{\mu\nu}} = \partial^\mu \phi \partial^\nu \phi - g^{\mu\nu} \left[\frac{1}{2}(\partial \phi)^2 - U(\phi) \right]. \quad (6.13)$$

If ϕ slowly changes as a function of the space and time coordinates, then the derivatives in the above expression can be neglected and $T^{\mu\nu} \approx g^{\mu\nu} U(\phi)$, i.e. this energy-momentum tensor is approximately equal to that of the vacuum (6.2). An important difference is that $\rho^{(vac)}$ is strictly a constant, while $U(\phi)$ slowly drops down because ϕ also slowly moves to the equilibrium point, where $U' = 0$. Usually the subtraction constant is taken in such a way that at the same point where $U' = 0$, the potential U vanishes, the real vacuum energy disappears, and the exponential expansion terminates.

A slow variation of ϕ can be achieved due to a large value of the Hubble parameter or, as it is usually said, due to a large Hubble friction. If we neglect the space derivative term, Eq. (6.12) coincides with the equation of motion of a point-like body in Newtonian mechanics with the liquid friction term $3H\dot\phi$. Evidently, the motion under such conditions proceeds with an almost constant and small velocity, $\dot\phi \approx const$. To see this, let us neglect the higher derivative term, $\ddot\phi$ in Eq. (6.12), and explicitly solve it. This approximation is called *slow roll approximation*. In this case, the equation of motion is reduced to the first order equation

$$\dot\phi = -\frac{U'}{3H}. \quad (6.14)$$

If the cosmological energy density ρ is dominated by the slowly varying inflaton field ϕ, then the Hubble parameter is equal to

$$H^2 = \frac{8\pi U}{3 M_{\mathrm{Pl}}^2}. \tag{6.15}$$

From Eq. (6.14), $dt = -U' d\phi/(3H)$, and thus the number of e-folding during the slow roll regime can be estimated as

$$N = \int H dt = \frac{8\pi}{M_{\mathrm{Pl}}^2} \int \frac{d\phi U(\phi)}{U'(\phi)}. \tag{6.16}$$

We take for simplicity a power law potential, $U(\phi) = g\phi^n/n$, though more complicated forms are possible and easy to analyze in the assumed slow roll approximation. For a power law potential, we find

$$N = \frac{4\pi}{n M_{\mathrm{Pl}}^2} \left(\phi_{in}^2 - \phi_{fin}^2 \right) \approx \frac{4\pi}{n M_{\mathrm{Pl}}^2} \phi_{in}^2, \tag{6.17}$$

where ϕ_{in} and ϕ_{fin} are, respectively, the initial and the final values of ϕ and it is assumed that $\phi_{fin} \ll \phi_{in}$. For a successful inflation, it is necessary that $N > 65\text{--}70$, which implies $\phi_{in}^2 \geq 5.6 n\, M_{\mathrm{Pl}}^2$. The final value of ϕ is determined by the condition that H becomes comparable or smaller than the frequency ω_ϕ of the oscillations of ϕ near the bottom of the potential. After that, the exponential expansion turns into a power law expansion. For the harmonic potential $U = m_\phi^2 \phi^2$ ($n = 2$), the initial value of ϕ^2 should be larger than $11\, M_{\mathrm{Pl}}^2$. The frequency of the oscillations near the minimum of U is $\omega_2 = m_\phi$, so, from the condition $H = \omega_2$ with $H^2 \approx 4\pi U/(3 M_{\mathrm{Pl}}^2)$, the minimum amplitude of the field would be $\phi_{fin}^2 = (3/4\pi)\, M_{\mathrm{Pl}}^2 \approx 0.24\, M_{\mathrm{Pl}}^2$. For quartic potential ($n = 4$), $U = \lambda \phi^4/4$, the frequency of the oscillation is $\omega_4 = \sqrt{\lambda}\phi$, and inflation terminates at $\phi_{fin}^2 = 0.48\, M_{\mathrm{Pl}}^2$.

For a successful inflation, it is thus necessary that the field amplitude is larger than M_{Pl}. At first sight, this looks disturbing. However, there is no reason to worry about, because the observable quantity is the energy density of ϕ and it remains much smaller than M_{Pl}^4 because the conditions $m_\phi \ll M_{\mathrm{Pl}}$ or $\lambda \ll 1$ must be imposed to avoid too large density perturbations, see Sect. 6.6.

The slow roll approximation is valid if the following two conditions are satisfied

$$\ddot{\phi} \ll 3 H \dot{\phi}, \tag{6.18}$$

and

$$\dot{\phi}^2 \ll 2 U(\phi). \tag{6.19}$$

6.3 Mechanisms of Inflation

To this end, we need

$$\left|\frac{U''}{U}\right| \ll \frac{8\pi}{3M_{\text{Pl}}^2}. \tag{6.20}$$

For instance, in the case of a non-self-interacting massive field with $U = m^2\phi^2/2$, the slow roll approximation would be valid if

$$\phi^2 > \frac{4\pi}{3} M_{\text{Pl}}^2. \tag{6.21}$$

With ϕ exactly at the lower limit, the number of e-foldings is not enough, but a slightly larger ϕ can do the job. The harmonic potential does not exceed the Planck value if

$$\phi^2 < \frac{M_{\text{Pl}}^4}{m_\phi^2}. \tag{6.22}$$

If we take ϕ equal to the upper bound, namely $\phi_{in} = M_{\text{Pl}}^2/m_\phi$, and $m_\phi \sim 10^{-6} M_{\text{Pl}}$, which is demanded by the condition of sufficiently small density perturbations, the number of e-foldings would be huge, $N = 10^{13}$. On the other hand, the characteristic time when all this happens is tiny, $t_{inf} \sim 10^{-31}$ s. The duration of inflation is thus very short, but during this very short period our huge Universe was inflated out of an extremely tiny initial volume.

6.3.2 Other Mechanisms of Inflation

In addition to inflationary models based on slowly evolving scalar fields with different forms of the potential, which are essentially described in the previous subsection, there are quite a few more exotic suggestions made later. The discussion here is by necessity very brief. Details can be found in Linde (1990), Kinney (2015), Dolgov (2010), Baumann (2015).

Among them, there is the scenario of double field or hybrid inflation (Adams and Freese 1991; Linde 1991, 1994). It is realized by two scalar fields with an interaction potential that can be taken, for instance, as

$$U(\phi, \chi) = \frac{1}{2}(\lambda_1\phi^2 - v^2)\chi^2 + \frac{\lambda_2}{4}\chi^4 + U(\phi). \tag{6.23}$$

Here $U(\phi)$ is a slow-rolling inflaton potential and $\lambda_{1,2}$ are positive coupling constants. The inflaton field ϕ is initially large and the system evolves along the valley $\chi = 0$. When ϕ^2 drops below v^2/λ_1, the field χ acquires a non-vanishing vacuum expectation value, $\chi^2 \sim v^2/\lambda_2$, so the effective mass of ϕ becomes large, $m_{eff}^2 \sim (\lambda_1/\lambda_2)v^2$, and ϕ quickly evolves down to the equilibrium point $\phi = 0$, efficiently producing particles and heating the Universe (see the discussion in Sect. 6.4).

Inflation can be naturally realized by a pseudo-Goldstone field, that is why the name "natural" (Freese et al. 1990). Probably some explanation is necessary here. As a result of the spontaneous breaking of a global symmetry, there appears a massless scalar boson θ. If, however, in addition to the spontaneous breaking, the symmetry is also broken explicitly, such a boson would acquire a small mass, which would lead to a slow motion of θ to the minimum of the potential and its vacuum-like energy may create sufficient inflation. In this sense, the model is similar to that considered in the previous subsection. Note that, before the symmetry breaking, since θ is massless, it may have an arbitrary value from 0 to 2π, so initially it could be naturally displaced from the would-be equilibrium point, which appears after θ became massive.

A quite unusual model, called k-inflation, was suggested in Armendariz-Picon et al. (1999). It is a scalar field model that does not contain a potential but the kinetic term has a non-canonical form. The Lagrangian of this model is

$$\mathscr{L} = p(\phi, X), \qquad (6.24)$$

where $X = (\partial \phi)^2/2$. In the FRW metric with a homogeneous field $\phi = \phi(t)$, $X = (\dot{\phi})^2/2$. It is assumed that p vanishes when $\phi \to 0$, so it can be expanded as

$$p = K(\phi)X + L(\phi)X^2. \qquad (6.25)$$

The energy-momentum tensor in this theory has the form

$$T_{\mu\nu} = \frac{\partial p}{\partial X} \partial_\nu \phi \, \partial_\nu \phi - p g_{\mu\nu}, \qquad (6.26)$$

and therefore the function p can be interpreted as the pressure density of the field ϕ. Its energy density, according to Eq. (6.26), is equal to

$$\rho = \dot{\phi} \frac{\partial p}{\partial \dot{\phi}} - p. \qquad (6.27)$$

The extremal points of $p(X)$, where its derivative over X vanishes, has the vacuum-like equation of state $p = -\rho$, and when the system is close to it the Universe expands quasi-exponentially. As it is argued in Armendariz-Picon et al. (1999), all the intersection points of the energy density with the line $p = \rho$ are attractors of the (future) evolution, if the function $p(\phi, X)$ is properly selected.

There is a model of inflationary expansion based on a modification of gravity at large curvatures. It is called Starobinsky or R^2 inflation (Starobinsky 1979). At first sight, it has nothing in common with scalar inflaton models, though it is not exactly so. In this model, we introduce an R^2 term into the Einstein-Hilbert action

6.3 Mechanisms of Inflation

$$S = -\frac{M_{\text{Pl}}^2}{16\pi} \int d^4x \sqrt{-g} R \to S = -\frac{M_{\text{Pl}}^2}{16\pi} \int d^4x \sqrt{-g} R \left(1 - \frac{R}{6m^2}\right). \quad (6.28)$$

This essentially changes the dynamic of the curvature scalar, R. In standard General Relativity, R is not a dynamical quantity. It is algebraically determined by the trace of the energy-momentum tensor

$$R = -\frac{8\pi T_\mu^\mu}{M_{\text{Pl}}^2}. \quad (6.29)$$

After the addition of R^2, the equation of motion for the metric tensor becomes of higher (4th) order and the curvature becomes a dynamical quantity satisfying (in the homogeneous case) the Klein-Gordon equation for a scalar field with mass m, which is often called "scalaron",

$$\ddot{R} + 3H\dot{R} + m^2 \left(R + \frac{8\pi}{M_{\text{Pl}}^2} T_\mu^\mu \right) = 0. \quad (6.30)$$

The Hubble parameter is expressed through R as

$$R = -6\dot{H} - 12H^2. \quad (6.31)$$

One can check that in the absence of matter these equations have a solution with an almost constant H describing an exponential cosmological expansion in the very early Universe. The scalaron can later decay, producing elementary particles and terminating inflation in the same way as it happens in other inflationary models, see the next section.

6.4 Universe Heating

The evolution of the inflaton ϕ can be divided into two quite different regimes. Initially, ϕ changes very slowly and its energy-momentum tensor has approximately the vacuum form, as one can see from Eq. (6.13), where the derivatives of ϕ are neglected. This is the period of inflation when the cosmological scale factor rose by many orders of magnitude. It took place when the potential term U' in the equation of motion can be nearly neglected. To this end, it is necessary that

$$U''(\phi) < H_I^2, \quad (6.32)$$

where H_I is the Hubble parameter during inflation. In particular, for the harmonic potential $U = m^2\phi^2/2$, this condition means that $H_I > m$. Still ϕ is not exactly constant because the weak force induced by U' slowly pushes ϕ to the minimum of the potential, which is usually taken at $\phi = 0$. Correspondingly, the Hubble parameter $H \sim \sqrt{U}/M_{\text{Pl}}$ gradually drops down and the condition (6.32) ultimately fails to fulfill. As it is shown in the previous section, for $U = m^2\phi^2/2$ this happens when $\phi = \sqrt{3/4\pi}\, M_{\text{Pl}}$, and for $U = \lambda\phi^4/4$ the boundary value is $\phi = (3/\sqrt{2\pi})\, M_{\text{Pl}}$. After it, the role of the expansion in the evolution of ϕ becomes less essential and the field started oscillating with adiabatically decreasing amplitude. The decrease is induced by the redshift due to the cosmological expansion (in other words, by the Hubble friction) and due to the back reaction of particle production. This rather short period would be proper to call big bang, because during that brief time interval an empty, cold Universe filled only with the field ϕ exploded, creating hot relativistic particles.

The simplest way to describe the particle production is with a perturbation approach, which is valid when the coupling of the inflaton to the other fields/particles is sufficiently weak. Perturbative calculations were first done in Albrecht et al. (1982), Dolgov and Linde (1982), Abbott et al. (1982). Later, it was understood (Dolgov and Kirilova 1990; Traschen and Brandenberger 1990) that non-perturbative effects could be quite essential. In the first of these papers, the possibility of a parametric resonance excitation (see below) in the process of boson production was mentioned. However, the conclusion was that in the model considered there the parametric resonance was not efficient because of the fast redshift of the produced particles out of the resonance region and also due to their mutual scattering. However, sufficiently wide resonances might strongly enhance the efficiency of particle production (Traschen and Brandenberger 1990; Kofman et al. 1994, 1997).

6.4.1 Perturbative Production

Let us consider an example in which the inflaton is coupled to some other lighter field through the trilinear coupling

$$L_{int} = f\phi(t)\chi^\dagger \chi, \tag{6.33}$$

where $\phi(t)$ is supposed to be a homogeneous classical field satisfying Eq. (6.12) with the harmonic potential $U = m^2\phi^2/2$. At this stage, we neglect the contribution of the interaction with the field χ (6.33) into the inflation equation of motion. The back reaction of the particle production to the evolution of ϕ is taken into account below, Eqs. (6.42) and (6.43). In Eq. (6.33), χ is a fermionic field and f is the dimensionless coupling constant of the interaction. In the case χ is a scalar field, the interaction term would be $g\phi|\chi|^2$ with g the coupling constant with the dimension of energy.

When the Hubble parameter becomes smaller than the mass of ϕ, the solution of Eq. (6.12) is

$$\phi(t) = \frac{M_{\text{Pl}}}{\sqrt{3\pi}\, m_\phi} \frac{\sin\left[m_\phi (t + t_0)\right]}{t + t_0} \tag{6.34}$$

6.4 Universe Heating

At the lowest order, the amplitude of the production of pairs of, for simplicity massless, scalars with momenta k_1 and k_2 is given by the matrix element of the interaction Lagrangian between the initial vacuum and final χ-particle-antiparticle states integrated over the 4-dimensional volume (similar to perturbative calculations in non-relativistic quantum mechanics)

$$A(k_1, k_2) = g \int d^4x \phi(t) \langle k_1, k_2 | \chi^\dagger \chi | 0 \rangle, \tag{6.35}$$

where $|0\rangle$ is the initial vacuum state and $\langle k_1, k_2 |$ is the final particle-antiparticle state with momenta k_1 and k_2.

Using the expansion of the field operator χ in terms of its creation-annihilation operators, see appendix C, Eqs. (C.2), (C.4), (C.8), (C.9), we find

$$A(k_1, k_2) = (2\pi)^3 g\, \delta(\mathbf{k_1} + \mathbf{k_2}) \tilde{\phi}(\omega_1 + \omega_2), \tag{6.36}$$

where $\omega_j = |\mathbf{k}_j|$ is the particle energy, and

$$\tilde{\phi}(\omega) = \int dt\, e^{i\omega t} \phi(t). \tag{6.37}$$

In what follows, we will interchangingly use the notations E or ω for the particle energy.

The probability of boson production per unit volume is

$$N_b \equiv \frac{W}{V} = \frac{1}{V} \int \frac{d^3k_1\, d^3k_2}{(2\pi)^6\, 4E^2} |A|^2 = \frac{g^2}{8\pi^2} \int_{E>0} dE |\tilde{\phi}(2E)|^2. \tag{6.38}$$

The volume factor V, as usually, comes from the square of the δ-function

$$[\delta(\mathbf{k_1} + \mathbf{k_2})]^2 = \frac{V}{(2\pi)^3} \delta(\mathbf{k_1} + \mathbf{k_2}). \tag{6.39}$$

If $\omega \gg t^{-1}$, then the integration over time in the interval $\Delta t \gg \omega^{-1}$ in Eq. (6.37) gives approximately $\pi\, \phi_0 \delta(\omega - 2E)$, where ϕ_0 is the slowly changing amplitude of the oscillating ϕ, as given by Eq. (6.34). The square of the δ-function, as we know, is $\Delta t\, \delta(\omega - 2E)/(2\pi)$. So the result for the rate of boson production per unit time and unit volume is

$$\dot{N}_b = \frac{N_b}{\Delta t} = \frac{g^2 \phi_0^2}{32\pi} = \frac{g^2 M_{\text{Pl}}^2}{96\pi^2 m_\phi^2 (t + t_0)^2}. \tag{6.40}$$

This corresponds to the following decay rate of the field ϕ

$$\Gamma_\phi = \frac{\dot{N}_b}{N_\phi} = \frac{g^2}{32\pi m_\phi}. \tag{6.41}$$

This, as one can expect, is the decay width of the ϕ-scalar into a pair of $\chi\bar{\chi}$ particles.

Because of the energy loss due to particle production, $\phi(t)$ should decrease faster than what is given by Eq. (6.34). For $\Gamma_\phi \ll \omega$, this can be taken into account by the substitution $\phi(t) \to \phi(t)\exp(-\Gamma_\phi t/2)$. However, this approximation is applicable only for the harmonic potential. In this case, the back reaction from particle production can be described by an additional term proportional to Γ_ϕ in the equation of motion

$$\ddot{\phi} + (3H + \Gamma_\phi/2)\dot{\phi} + U' = 0. \tag{6.42}$$

In the general case of arbitrary potential $U(\phi)$, the equation of motion with an account of back reaction of particle production becomes an integro-differential equation, non-local in time. In the one loop approximation, it was derived in Dolgov and Hansen (1999). For example, in flat spacetime the equation has the form

$$\ddot{\phi} + U'(\phi) = \frac{g^2}{4\pi^2}\int_0^{t-t_{in}} \frac{d\tau}{\tau}\phi(t-\tau). \tag{6.43}$$

Remind that up to this point we assumed that ϕ is coupled to a scalar field with the interaction term $g\phi\chi^2$. It can be generalized to the FRW metric, as it is done in Dolgov and Freese (1995) and Arbuzova et al. (2012).

Returning to the simple harmonic case, we note that the thermalization of the produced bosons is usually faster than the expansion rate. In this case, the temperature of the plasma can be simply evaluated as follows. We assume that ϕ completely decayed, that the particles were produced instantly at the moment when the Hubble parameter $H = 2/(3t)$ became equal to the decay rate Γ_ϕ, and that $t \gg t_0$, see Eq. (6.34).

Using Eq. (6.40) we can estimate the energy density of the produced fermions changing $\dot{N}_b$ to $\dot{N}_f$ as

$$\rho_f = \frac{\Gamma_\phi^2 M_{\text{Pl}}^2}{6\pi} = \frac{f^4}{96\pi^3} M_{\text{Pl}}^2 m_\phi^2 \tag{6.44}$$

and, correspondingly, the temperature of the Universe heating would be

$$T_h = \left(\frac{30\rho_f}{\pi^2 g_*}\right)^{1/4} = \left(\frac{30}{96\pi^5 g_*}\right)^{1/4} f\sqrt{M_{\text{Pl}} m_\phi}. \tag{6.45}$$

For a more accurate evaluation of T_h, let us take into account the non-instantaneous character of particle production and the decrease of the amplitude of the oscillations of $\phi_0(t)$, caused not only by the Universe expansion but also by the back reaction of

6.4 Universe Heating

particle production. With these factors taken into account, the energy densities of ϕ and of the produced fermions satisfy the equation

$$\dot\rho_\phi = -\Gamma_\phi \rho_\phi - 3H\rho_\phi, \quad \dot\rho_f = \Gamma_\phi \rho_\phi - 4H\rho_f, \tag{6.46}$$

where Γ_ϕ is given by Eq. (6.41). The equations can be solved as

$$\rho_\phi(t) = \frac{\rho_\phi^{(in)}}{a^3(t)} e^{-\Gamma_\phi t}, \quad \rho_f(t) = \frac{\rho_\phi^{(in)}}{a^4(t)} \int_0^t dt'\, a(t') e^{-\Gamma_\phi t'}. \tag{6.47}$$

We assumed here that $t_{in} = 0$ and $a(t_{in}) = 1$. The time t_h when the particle production was accomplished can be estimated from the condition $\rho_\phi(t_h) = \rho_f(t_h)$. When ρ_ϕ was larger than ρ_f, the expansion regime in a good approximation can be taken as a non-relativistic one, i.e. $a(t) \sim t^{2/3}$. Calculating numerically the integral, we find $\Gamma_\phi t_h = 1.073$. So for the energy density of relativistic matter we obtain essentially the same result as (6.44) with the extra suppression factor $1/\exp(1.073) = 0.34$. Correspondingly, the heating temperature would be approximately 0.76 smaller than (6.45).

The mass of the inflaton is bounded by $m < 10^{-6} M_{Pl}$ to avoid too large density perturbations, and the coupling constant to fermions is natural to expect to be bounded by $f < 10^{-3}$. This bound follows from the condition that the quartic coupling $\lambda \phi^4$ radiatively induced through fermionic loop gives $\lambda \sim f^4$ and λ is also bounded by the density perturbations as $\lambda < 10^{-13}$. Hence the heating temperature after inflation would be rather low, $T_h \leq 5 \cdot 10^{-8} M_{Pl} \approx 5 \cdot 10^{11}$ GeV. However, there are more efficient mechanisms of the Universe heating, which are discussed in the next subsection.

6.4.2 Non-perturbative Phenomena

A rigorous theory of particle production by an external field is based on the Bogolyubov transformation (Bogoliubov 1947, 1958). Let us consider again a classical scalar field $\phi(t)$ coupled to a quantum complex scalar χ, as described by an interaction Lagrangian of the form $g\phi|\chi|^2$, where g is the coupling constant with dimension of mass. In this subsection, we consider a flat spacetime and we postpone considerations of General Relativity effects to the next subsections. The equation of motion for the Fourier modes of χ is

$$\left[\partial_t^2 + \mathbf{k}^2 + m^2 - g\phi(t)\right] f_k(t) = 0. \tag{6.48}$$

This equation is similar to Eq. (C.6), we have only added the term describing the interaction with the external field ϕ.

It is usually assumed, though it is not true in cosmology, that at early and late times, $t \to \pm\infty$, the interaction disappears, $\phi(t) \to 0$, so χ satisfies the free equation of motion and its Fourier mode has the form $f_k(t) \sim \exp[\pm i(\omega_k t - i\mathbf{k}\mathbf{x})]$ and the complex conjugate one for the annihilation term. This particular choice of the solution ensures positive energy of the quanta. At later times, when $\phi(t) \neq 0$, the mode functions $f_k(t)$ are determined by Eq. (6.48) and can be found analytically within a perturbation approach, in quasi-classical approximation (the so-called WKB approximation), or the equation can be solved numerically. At asymptotically large times, when according to our assumption the field ϕ also disappeared, $\phi \to 0$, the form of the solution is evident

$$f_k(t \to +\infty) \to \alpha_k e^{-i\omega t} + \beta_k e^{i\omega t}, \tag{6.49}$$

so the mode expansion (C.2) evolves as

$$\chi(t \to +\infty) = \int \widetilde{dk}\, \left[e^{(-i\omega t + i\mathbf{k}\cdot\mathbf{x})}(\alpha_k a_\mathbf{k} + \beta_k^* b_{-\mathbf{k}}^\dagger) \right. $$
$$\left. + e^{(i\omega t - i\mathbf{k}\cdot\mathbf{x})}(\alpha_k^* b_{\mathbf{k}}^\dagger + \beta_k a_{-\mathbf{k}}) \right], \tag{6.50}$$

where $\widetilde{dk}$ is defined in (C.10). The problem is analogous to the well known quantum mechanical problem of calculation or reflection/transition coefficients in an external potential. These coefficients satisfy the flux conservation condition $|R|^2 + |T|^2 = 1$. There is a similar relation here

$$|\alpha_k|^2 - |\beta_k|^2 = 1. \tag{6.51}$$

Now one can define new creation and annihilation operators for particles and antiparticles

$$\tilde{a}_\mathbf{k} = \alpha_k a_\mathbf{k} + \beta_k^* b_{-\mathbf{k}}^\dagger, \quad \tilde{b}_\mathbf{k} = \alpha_k b_\mathbf{k} + \beta_k^* a_{-\mathbf{k}}^\dagger. \tag{6.52}$$

It is important that, due to the relation (6.51), the commutators between the new operators remain the canonical ones (C.4).

The operator of the final particle number is given by $\tilde{N}_\mathbf{k} = \tilde{a}_\mathbf{k}^\dagger \tilde{a}_\mathbf{k}/[2k^0 V]$. The number of particles in the final state of momentum $\mathbf{k}$ is equal to $N_\mathbf{k} = \langle 0|\tilde{N}_\mathbf{k}|0\rangle = |\beta_k|^2$. The total number density of produced particles is

$$n = \frac{1}{V}\frac{V}{(2\pi)^3}\int d^3k\, N_\mathbf{k} = \int \frac{d^3k}{(2\pi)^3}|\beta_k|^2. \tag{6.53}$$

Let us calculate β_k in perturbation theory. Expanding $f = f_0 + f_1$, we have $f_0 = \exp(-i\omega t)$ and the equation of motion (6.48) becomes

$$(\partial_t^2 + \mathbf{k}^2 + m^2)f_1 = g\phi(t)\exp(-i\omega t). \tag{6.54}$$

6.4 Universe Heating

Using the Green's function method, we find

$$f_1(t) = -g \int \frac{d\omega'}{2\pi} \frac{\tilde{\phi}(\omega' - \omega)}{\omega'^2 - \mathbf{k}^2 - m^2} e^{-i\omega' t} . \tag{6.55}$$

Taking the residue at the pole $\omega' = -\sqrt{\mathbf{k}^2 + m^2} = -\omega$, we find that the coefficient of $\exp(+i\omega t)$ is $\beta_k = ig[\tilde{\phi}(2\omega)]^*/2\omega$, in agreement with the results of Sect. 6.4.1. Details on quasi-classical calculations can be found, for instance, in Dolgov and Kirilova (1990), Dolgov (2002).

The Bogolyubov transformation technique allows the calculation of the number density of the produced particles for an arbitrary external time-dependent field, but only in the case when the external field disappears at $t \to \pm\infty$. A more appropriate quantity, especially in cosmology, is the energy density of the produced particles. The point is that the particle number operator is not a local operator, as it is known from the basics of quantum field theory, and hence the notion of particle depends upon the definition of the vacuum state. It creates serious ambiguities in the interpretation. For example, the well known Unruh effect (Fulling 1973; Davies 1975; Unruh 1976) predicts that an accelerated observer moving in the vacuum observes a black body radiation, while an inertial observer detects nothing. However, the total energy density of the system remains zero. Indeed, the energy-momentum tensor of the vacuum is assumed to be zero and so it would remain zero in any coordinate frame. The observation of the thermal bath of particles by an accelerated detector means that the vacuum energy (defined in the accelerated frame) is non-zero and such that it exactly compensates the non-vanishing energy-momentum tensor of the bath, so the gravitational action of the system in the inertial and accelerated frames equally vanish.

On the other hand, the consideration of the evolution of the energy density of the system of interacting $\phi - \chi$ fields allows to describe the particle production for any value of $\phi(t)$ and not only for $\phi = const$. The operator of energy density in the FRW metric is equal to the time-time component of the energy-momentum tensor

$$T_{\mu\nu}(\phi, \chi) = \partial_\mu \phi \partial_\nu \phi - g_{\mu\nu} \left[\frac{1}{2} \partial_\alpha \phi \partial^\alpha \phi - U_\phi(\phi)\right]$$
$$+ 2\partial_\mu \chi^\dagger \partial_\nu \chi - g_{\mu\nu} \left[\partial_\alpha \chi^\dagger \partial^\alpha \chi - U_\chi(\chi)\right] - g_{\mu\nu} g \phi \chi^\dagger \chi . \tag{6.56}$$

We assume, for simplicity, that $\phi(t)$ is a classical field and neglect quantum corrections to its energy. Though they can be easily included, they are not important for our results, while they are essential for χ quantum effects. We take the simplest harmonic potentials $U_\phi(\phi) = m_\phi^2 \phi^2/2$ and $U_\chi(\chi) = m_\chi^2 \chi^2/2$. Let us calculate the vacuum expectation value of $\rho = T_{00}$ and study its evolution with time. The operator of the total energy density consists of the following three parts. The energy density of the field $\phi(t)$ is

$$\rho_\phi = \frac{1}{2}(\dot\phi^2 + m_\phi^2 \phi^2) . \tag{6.57}$$

The energy density of the interaction is

$$\rho_{int} = g\phi\langle 0|\chi^\dagger \chi|0\rangle = g\phi(t) \int \widetilde{dk}\, |f_k(t)|^2, \qquad (6.58)$$

where the condition that the annihilation operator kills the vacuum $a_k|0\rangle = 0$ and the commutator (C.4) (for b_k) were used. This expression is quadratically divergent at high k, but the difference between the initial and the final energy densities should be finite, if the interaction is softly switched on. The last term is the most interesting. In contrast to the other two, it remains non-zero in the limit of $\phi = 0$. Following the same procedure as above, we find

$$\rho_\chi = \int \widetilde{dk}\, \left[\left(m_\chi^2 + k^2\right)|f_k(t)|^2 + |\dot{f}_k(t)|^2\right]. \qquad (6.59)$$

This expression is quadratically ultraviolet divergent. It is a known result for the vacuum energy even for non-interacting fields.

It is instructive to check if the above result agrees with the perturbative calculations based on the Bogolyubov transformation. The solution of Eq. (6.48) is given by the expression in (6.49). Substituting it into Eq. (6.59), we find

$$\rho_\chi = \int \widetilde{dk}\, \left[\left(m_\chi^2 + k^2\right)\left(|\alpha_k|^2 + |\beta_k|^2 + \alpha_k^*\beta_k\, e^{-i\omega_k t} + \alpha_k\beta_k^*\, e^{i\omega_k t}\right) \right.$$
$$\left. + \omega_k^2 \left(|\alpha_k|^2 + |\beta_k|^2 - \alpha_k^*\beta_k\, e^{-i\omega_k t} - \alpha_k\beta_k^*\, e^{i\omega_k t}\right)\right]. \qquad (6.60)$$

The oscillating interference terms mutually cancel and, using the relation (6.51), we find

$$\Delta\rho_\chi = \rho_\chi(t \to +\infty) - \rho_\chi(t \to -\infty) = 2\int \widetilde{dk}\, \omega_k^2 |\beta_k|^2, \qquad (6.61)$$

in agreement with (6.53). The factor two here comes from the sum of the contributions of particles and antiparticles. Note that for a $|\beta_k|^2$ decreasing faster than $1/k^2$, which is usually the case if the field ϕ is switched on adiabatically, the result is not ultraviolet divergent.

6.4.3 Parametric Resonance

As it was noted at the beginning of this section, a parametric resonance can be excited in particle production processes and could strongly enhance the efficiency of the Universe heating. Parametric resonance is a well known phenomenon of an exponential rise of the amplitude of oscillations when the parameters of the oscillator are periodically changed with the frequency, which is an integer fraction of the

6.4 Universe Heating

eigenfrequency of the oscillator (see below). This effect is well known to children all over the world, as they use it to increase the swing amplitude by periodic squatting up and down.

Let us now consider the Fourier mode of a real scalar field χ, which satisfies the equation of motion (6.48), and assume that the field ϕ changes with time as $\phi(t) = \phi_0 \cos(m_\phi t)$. Such a behavior of ϕ is typical at the end of inflation. Substituting this expression into Eq. (6.48), we come to the well-known Mathieu equation

$$\ddot{\chi} + \omega_0^2 (1 + h \cos mt) \chi = 0, \tag{6.62}$$

where $\omega_0^2 = m_\chi^2 + k^2$ and $h = g\phi_0/\omega_0^2$. When $h \ll 1$ and the value of m is close to $2\omega_0/n$ (where n is an integer), Eq. (6.62) describes a parametric resonance, which leads to oscillations of χ with an exponentially growing amplitude. For $h \ll 1$, the solution of Eq. (6.62) can be presented as a product of a slowly (but exponentially) rising amplitude by a quickly oscillating function with frequency ω_0

$$\chi = \chi_0(t) \cos(\omega_0 t + \alpha). \tag{6.63}$$

The amplitude χ_0 satisfies the equation

$$-\ddot{\chi}_0 \cos(\omega_0 t + \alpha) + 2\omega_0 \dot{\chi}_0 \sin(\omega_0 t + \alpha) = h\omega_0^2 \chi_0 \cos mt \cos(\omega_0 t + \alpha). \tag{6.64}$$

Let us multiply Eq. (6.64) by $\sin(\omega_0 t + \alpha)$ and average over the period of oscillation. The right hand side would not vanish on the average if $m = 2\omega_0$. In this case, χ_0 would exponentially rise if $\alpha = \pi/4$

$$\chi_0 \sim \exp\left(\frac{1}{4} h\omega_0 t\right). \tag{6.65}$$

In this way we recovered the standard result of the parametric resonance theory for the lowest frequency mode. There are higher frequency modes of χ-oscillations and they also resonate. Their existence can be established in a similar way. A more detailed discussion of parametric resonances can be found, for instance, in Landau and Lifshitz (1976) or in any mathematical book on the Mathieu equation.

In the quantum language, parametric resonances can be understood as an enhancement of the particle production into the plasma already populated by such particles. Clearly this enhancement exists only for bosons—this is the mechanism of laser amplification. On the opposite, fermions are reluctant to be produced due to the Fermi exclusion principle. Indeed, the equation of motion of quantum fermions does not show a resonance behavior (Dolgov and Kirilova 1990).

In the cosmological case, there are two effects that are potentially dangerous for a parametric resonance. First, there is the redshift of the momenta of the produced particles, which pushed them out of the resonance mode. The effects of the cosmological expansion are taken into account by a proper modification of the equation of motion for the Fourier modes of χ

$$\ddot{f}_k + 3H \dot{f}_k + \left(\frac{k^2}{a^2(t)} + m_\chi^2 + g\phi \right) f_k = 0 \,. \tag{6.66}$$

This equation can be easily solved numerically and the solution indeed demonstrates the resonance behavior, if the resonance is sufficiently wide. The second phenomenon that could inhibit the resonance is the possible scattering of the produced particles off other particles in the cosmological plasma. This effect has not been properly studied (to the best of our knowledge). However, it was evidently absent at the initial stage of the production when the Universe was almost empty. More detail about the Universe heating can be found, for instance, in Allahverdi et al. (2010). It is shown that broad parametric resonances can convert a substantial fraction of the inflaton energy density into matter in a time interval small compared to the Hubble time. The inflaton field started oscillating and creating particles when ϕ dropped down to $M_{\rm Pl}$. Correspondingly, the energy density of the inflaton at the onset of particle production was $\rho_\phi \sim m_\phi^2 \phi^2 \sim 10^{-12} M_{\rm Pl}^2$ and the temperature of the Universe at the beginning of the big bang could be as high as $T \sim 10^{-3} M_{\rm Pl} \sim 10^{16}$ GeV or maybe an order of magnitude below.

It has been recently found (Dolgov et al. 2015) that the efficiency of heating may be enhanced up to an order of magnitude if the inflaton oscillates not as pure cosine but closer to a periodic succession of θ-functions.

If $g\phi$ in Eq. (6.66) always remains smaller than $(m_\chi^2 + k^2/a^2)$, the square of the effective eigenfrequency of χ-oscillations keeps to be positive. However, if $g\phi$ is large, the square of the frequency would be negative and there appears a new type of instability, generically leading to a faster rise of the χ amplitude than it happens due to a parametric resonance. This is the case of the so-called tachyonic heating (Felder et al. 2001a,b). Note that for another type of coupling $\lambda_{\phi\chi} \phi^2 \chi^2$, the tachyonic heating evidently does not appear. This mechanism is called tachyonic because of the similarity of Eq. (C.5) to the equation of motion of tachyons, i.e. of particles with negative mass squared

$$(\partial_t^2 - \Delta + m_\zeta^2)\zeta = 0 \,. \tag{6.67}$$

Tachyons are particles that are supposed to propagate faster than light. However, this is not the case. The group velocity of tachyonic waves can indeed be larger than c, but the velocity of the wave front, which is determined by the asymptotics of the refraction index at infinite energy, does not exceed the speed of light. The vacuum state of tachyons is unstable, similar to the Higgs boson vacuum state, and presumably an additional quartic self-interaction would stabilize the field. A similar tachyonic instability was suggested earlier in Greene et al. (1997). It was assumed that the coupling $\lambda_{\phi\chi}$ was negative, so the field χ rose exponentially, as the usual tachyon and stabilization is achieved by some additional quartic coupling.

6.4.4 Particle Production in a Gravitational Field

The production of particles by the inflaton field is accompanied by the simultaneous production of particles by the cosmological gravitational field. This process could be quite efficient at the end of inflation. Creation of particles by the isotropic FRW metric was pioneered by Parker (1968, 1969) and by Bronnikov and Tagirov (1968, 2004) in an earlier, but unknown, paper in Russian, where the special case of de Sitter spacetime was considered. It was shown (Parker 1968, 1969) that in general in a conformally flat metric, including the particular case of de Sitter spacetime (Bronnikov and Tagirov 1968, 2004), massless conformally invariant particles are not created (see below). A study of particle production was further developed in a series of papers and books (Grib and Mamaev 1967, 1971; Chernikov and Shavokhina 1973; Birrell and Davies 1982; Zeldovich and Novikov 1983; Grib et al. 1995). Massless particles can be produced in a non-isotropic space (Zeldovich 1970; Zeldovich and Starobinsky 1972; Hu et al. 1973; Hu 1974; Berger 1974; Lukash et al. 1974), leading to a rapid isotropization of the space.

The FRW metric belongs to a special class of so-called conformally flat metrics, which, after a redefinition of coordinates, can be presented as the product of the Minkowsky type metric by a conformal factor

$$ds^2 = a^2(r, \eta)(d\eta^2 - d\mathbf{r}^2). \tag{6.68}$$

For the spatially flat FRW metric, where $ds^2 = dt^2 - a^2(t)\mathbf{r}^2$, one only needs to redefine the time as $dt/a(t) = d\eta$ and to express the scale factor in terms of the *conformal time* η. The corresponding expressions are presented in Table 6.1 for the matter dominated (MD), radiation dominated (RD), and de Sitter (dS) expansion regimes.

As we can see in what follows, it is convenient to rescale the metric as well as scalar, spinor, and vector fields in the following way

$$g_{\mu\nu} = a^2 \tilde{g}_{\mu\nu}, \quad \chi = \tilde{\chi}/a, \quad \psi = \tilde{\psi}/a^{3/2}, \quad A_\mu = \tilde{A}_\mu. \tag{6.69}$$

Table 6.1 Scale factors as functions of the conformal time for different expansion regimes

Expansion regime	Cosmological scale factor $a(t)$	Conformal time $\eta(t)$	Conformal scale factor $a(\eta)$
MD	$a \sim t^{2/3}$	$\eta \sim t^{1/3}$	$a(\eta) \sim \eta^2$
RD	$a \sim t^{1/2}$	$\eta \sim t^{1/2}$	$a(\eta) \sim \eta$
dS	$a \sim \exp(Ht)$	$\eta \sim -1/He^{Ht}$	$a(\eta) \sim -1/(H\eta)$

The matter action for complex scalars, spinors, and photons has the form

$$S_{tot} = \int d^4x \sqrt{-g} \Big[g^{\mu\nu} \partial_\mu \chi^* \partial_\nu \chi - m_\chi^2 |\chi|^2 - \lambda_\chi |\chi|^4$$
$$+ \bar\psi \left(i g^{\mu\nu} \Gamma_\mu \nabla_\nu - m_\psi \right) \psi - g^{\mu\alpha} g^{\nu\beta} F_{\mu\nu} F_{\alpha\beta} \Big]. \quad (6.70)$$

where Γ_μ is a generalization of the Dirac γ_μ matrices for curved spacetime. In the FRW metric, they have the form $\Gamma_\mu = a\gamma_\mu$, where γ_μ commutes as $[\gamma_\mu, \gamma_\nu] = \eta_{\mu\nu}$, $[\Gamma_\mu, \Gamma_\nu] = g_{\mu\nu}$, and ∇_μ is the covariant derivative for the spin-1/2 field. In the FRW metric, it has the form $\nabla_\mu = \partial_\mu + 3 \partial_\mu \ln a / 2$. To derive it, one has to use the so-called tetrad (vierbein) formalism, by which spinors are described in General Relativity.

One can easily check that the action (6.70) is invariant with respect to the transformation (6.69) (i.e. it has the same form in terms of the new fields) if $m_\chi = m_\psi = 0$ and $a = const$. Under this transformation, all the masses go into ma. However, in what follows we should take $a = a(t) \neq const$ to transform the FRW metric into the MInkowsky one: we need to redefine the metric in such a way that $g_{\mu\nu}$ goes into $\eta_{\mu\nu}$. This transformation can be done only if the rescaling factor depends upon time, i.e. $a = a(t)$. In this case, the action (6.70) for the scalar field is not invariant, while massless spinors and electromagnetic fields remain scale invariant. The Parker theorem, according to which massless particles are not produced in a FRW gravitational field, is fulfilled for the latter, while massless scalars can be produced. Indeed, the transformed electromagnetic and spinor fields satisfy free equations of motion, so evidently the initial solution $f_k = \exp(i\omega t)$ always remains such and the Bogolyubov coefficients vanish, $\beta_k \equiv 0$.

On the other hand, the equation of motion for the Fourier modes of $\tilde\chi$, derived from the action (6.70), has the form

$$f_k'' + \left(k^2 + m_\chi^2 a^2 - \frac{a''}{a} \right) f_k = 0, \quad (6.71)$$

where the prime means the derivative with respect to the conformal time. Clearly this equation is not conformally invariant, even for $m = 0$, due to the presence of the last term, with an exception for the radiation dominated regime when $a'' = 0$.

A scalar field is conformally invariant if the non-minimal coupling to the curvature scalar $\xi R |\chi|^2$, with $\xi = 1/6$, is added to the action. In this case, the term a''/a disappears from Eq. (6.71) and massless scalars are not produced in the cosmological background, but live there as free non-interacting fields. For arbitrary ξ, Eq. (6.71) turns into

$$f_k'' + \left[k^2 + m_\chi^2 a^2 + (6\xi - 1) \frac{a''}{a} \right] f_k = 0. \quad (6.72)$$

In the case of radiation dominated expansion, $a'' = 0$, so massless particles do not feel the cosmological expansion. For the matter dominated and de Sitter cases,

6.4 Universe Heating

$a''/a = 2/\eta^2$. Moreover, at the de Sitter stage $a^2 \sim 1/\eta^2$, so the equation can be solved also for massive fields in terms of the Hankel functions as

$$f_k = C_1 \sqrt{y}\, H_\nu^{(1)}(y) + C_2 \sqrt{y}\, H_\nu^{(2)}(y), \tag{6.73}$$

where $\nu = \left(9/4 - m_\chi^2/H^2 - 12\xi\right)^{1/2}$ and $y = -k\eta$. The coefficients $C_{1,2}$ can be determined from the condition that for short waves, i.e. $k \gg H$, the curvature effects are not essential and the solution should approach that in flat spacetime. In the limit $y \gg 1$ but finite ν, the Hankel functions have the following asymptotic behavior

$$H_\nu^{(1,2)}(y) \approx \sqrt{\frac{2}{\pi y}} \exp\left[\pm i\left(y - \pi\nu/2 - \pi/4\right)\right]. \tag{6.74}$$

Since for small Ht the conformal time is $\eta \approx -H^{-1} + t$, the correct positive energy mode is $H_\nu^{(1)}(k\eta) \sim \exp(-ikt)$ and so we shall take $C_1 = \sqrt{\pi/2}$ and $C_2 = 0$, see Chap. 5 in Birrell and Davies (1982). We should keep in mind, however, that this result is justified for short waves only, which at the beginning of inflation were much shorter than the horizon, $1/H$, i.e. $k \gg H$. In other words, we consider only waves with a length that was initially very small but in the course of the expansion they went outside of the horizon, i.e. $k^{-1} \exp(Ht) \geq H^{-1}$.

It follows from Eq. (6.73) that in the course of the evolution, the positive energy mode does not acquire an additional negative frequency mode and so during a pure de Sitter expansion particles are not produced. Gravitational production of heavy particles at the end of inflation was considered in Chung et al. (1998, 1999) and a more general case, which includes post inflationary radiation dominated or matter dominated stages, was studied in Kuzmin and Tkachev (1998, 1999). It was found there that particles were predominantly created when the Hubble parameter was close to their mass.

In conclusion, let us consider the production of photons in the FRW cosmology. The electromagnetic field is conformally invariant and according to the Parker theorem cannot be generated in a conformally flat gravitational field. This, however, is only true in classical electrodynamics. Quantum corrections are known to break conformal invariance and the classical Maxwell equations acquire an additional term (Dolgov 1981), leading to the generation of electromagnetic waves in the early Universe. At the end of inflation, this mechanism might create large scale cosmological magnetic fields.

6.5 Generation of Gravitational Waves

The generation of gravitational waves at the end of inflation is essentially the same process as the production of massless particle. However, this subject is very important and deserves special attention. In particular, the possible detection of very long gravitational waves would be an unambiguous proof of inflation. Though gravitons are massless, their equation of motion is not conformally invariant (Grishchuk 1975) and so they are produced in the FRW spacetime. The generation of gravitational waves at the de Sitter (inflationary) stage was first studied by Starobinsky (1979) (see also Rubakov et al. (1982), where the intensity of gravitational waves is calculated in newer inflationary models). A review on gravitational waves produced in the early Universe can be found in Maggiore (2000), Buonanno (2015).

The derivation of the equation of motion for gravitational waves is straightforward but quite tedious. Gravitational waves are considered as tensor perturbations of the metric

$$g_{\mu\nu} = g_{\mu\nu}^{(b)} + h_{\mu\nu}, \tag{6.75}$$

where $g_{\mu\nu}^{(b)}$ is the background metric and $|h_{\mu\nu}| \ll 1$ is the amplitude of the gravitational wave. In the case under scrutiny, we introduce tensor perturbations to the FRW metric in conformal time in the usual way as

$$ds^2 = a^2(\eta)\left(\eta_{\mu\nu} + h_{\mu\nu}\right) dx^\mu dx^\nu. \tag{6.76}$$

Instead of $h_{\mu\nu}$, it is common to consider the quantity

$$\psi_{\mu\nu} = h_{\mu\nu} - \frac{1}{2} g_{\mu\nu}^{(b)} h, \tag{6.77}$$

where $h = h_\mu^\mu$. Here and below, indices are raised with the background metric. By an appropriate choice of coordinates, the following conditions can be imposed

$$h = 0, \quad \text{and} \quad \nabla_\mu h_\nu^\mu = 0. \tag{6.78}$$

These conditions ensure that vector and scalar components are excluded. Due to the vanishing mass of gravitons, three more conditions on plane gravitational waves can be imposed, so only two components orthogonal to the wave propagation remain independent. With such a choice of the gauge conditions, $\psi_{\mu\nu} = h_{\mu\nu}$. After some algebra, we find the wave equation for the propagation of the k-mode

$$h''_{\mu\nu} + 2\frac{a'}{a} h'_{\mu\nu} + k^2 h_{\mu\nu} = 0. \tag{6.79}$$

Rescaling the metric as $h_{\mu\nu} = \tilde{h}_{\mu\nu}/a$, we find that all the components of $\tilde{h}_{\mu\nu}$ satisfy the same equation as a massless scalar field with the minimal coupling ($\xi = 0$) to gravity, see Eqs. (6.71) and (6.72). Here and in what follows, we omit the subindex k in h_k.

We consider gravitational wave production in a simple model when the initial de Sitter stage is instantly changed to a radiation domination expansion. It may be

6.5 Generation of Gravitational Waves

close to the realistic situation if inflation finished with a very fast Universe heating through inflaton production of relativistic particles. At the de Sitter stage, Eq. (6.71) has the solution (6.73) with $\nu = 3/2$

$$\tilde{h} = e^{-iz}\left(1 - \frac{i}{z}\right), \quad (6.80)$$

where $z = k\eta$ and the second solution is not taken according to the arguments presented above, after Eq. (6.74). In a radiation dominated universe, $a'' = 0$, and the solution has the simple form

$$\tilde{h} = \alpha e^{-iz} + \beta e^{iz}. \quad (6.81)$$

It is instructive to present the relation between the physical and the conformal times more accurately than what is done in Table 6.1, though some of these relations are not necessary for the calculations of the intensity of the generated gravitational waves. The conformal time is expressed through the physical time as

$$\eta - \eta_0 = \int_{t_0}^{t} \frac{dt'}{a(t')}. \quad (6.82)$$

At the de Sitter epoch, which is assumed to last from $t_I^{(in)}$ till $t_I^{(fin)}$, the scale factor is equal to $a(t) = a_I^{(in)} \exp[H_I(t - t_I^{(in)})]$. So for $t < t_I^{(fin)}$ we have

$$\eta - \eta_0 = \frac{1}{t_I^{(in)} H_I}\left[1 - e^{-H_I(t - t_I^{(in)})}\right], \quad (6.83)$$

where we took $t_0 = t_I^{(in)}$. We also chose $\eta_0 = 1/H_I$, $t_I^{(in)} = 0$, and $a_I^{(in)} = 1$, and obtain the same relations as in Table 6.1.

For $t \geq t_I^{(fin)}$, the scale factor evolves as $a_R(t) = a_R^{(in)}[(t + t_1)/t_2]^{1/2}$, where evidently $a_R^{(in)} = a_I^{(fin)} = \exp(H_I t_I^{(fin)})$. t_2 can be determined from the condition $a(t_I^{(fin)}) = a_I^{(fin)}$, so $t_2 = t_1 + t_I^{(fin)}$. At last, t_1 can be determined from the continuity of the cosmological energy density at the moment of the change of regime $t = t_I^{(fin)}$ from the de Sitter to the radiation dominated phase

$$\rho_I = \frac{3H_I^2 M_{\text{Pl}}^2}{8\pi} = \rho_R = \frac{3M_{\text{Pl}}^2}{32(t + t_1)^2}. \quad (6.84)$$

Hence $t_1 = 1/(2H_I) - t_I^{(fin)}$ and

$$a_R(t) = a_I^{(fin)} (2H_I)^{1/2} \left(t - t_I^{(fin)} + \frac{1}{2H_I}\right)^{1/2}. \quad (6.85)$$

Correspondingly, for the radiation dominated regime the conformal time is

$$\eta = -\frac{2}{H_I a_I^{(fin)}} + \frac{2}{\sqrt{2H_I} \, a_I^{(fin)}} \left(t - t_I^{(fin)} + \frac{1}{2H_I} \right)^{1/2}, \quad (6.86)$$

and the scale factor as a function of the conformal time evolves as

$$a_R(\eta) = (a_I^{(fin)})^2 H_I \eta + 2 a_I^{(fin)}. \quad (6.87)$$

For the matching, we demand the continuity of the solutions (6.80) and (6.81), as well as that of their first derivatives at $\eta = \eta_I^{(fin)} = -1/(H_i a_I^{(fin)})$. To simplify the notation, we denote $\eta_I^{(fin)} \equiv \eta_1$. The coefficients α and β are determined by the equations

$$\alpha e^{-iz_1} + \beta e^{iz_1} = e^{-iz_1} \left(1 - \frac{i}{z_1} \right)$$

$$\alpha e^{-iz_1} - \beta e^{iz_1} = e^{-iz_1} \left(1 - \frac{i}{z_1} - \frac{1}{z_1^2} \right), \quad (6.88)$$

where $z_1 = k\eta_1$. We thus find

$$\alpha = 1 - \frac{i}{z_1} - \frac{1}{2z_1^2}, \quad \beta = \frac{e^{-2iz_1}}{2z_1^2}. \quad (6.89)$$

Clearly $|\alpha|^2 - |\beta|^2 = 1$, as expected.

We can calculate the energy density of the gravitational radiation with frequency k using properly modified Eq. (6.61). However, it is necessary to take into account the effects of the cosmological expansion. The energy density of the field ϕ is equal to the time-time component of its energy-momentum tensor

$$\rho = T_{tt} = \frac{T_{\eta\eta}}{a^2} = \frac{(\phi')^2 + (\partial_j \phi)^2}{2a^2} = \frac{(f_k' - a' f_k/a)^2 + f_k^2}{2a^4}, \quad (6.90)$$

where the prime means the derivative over η and at the last step we made the transition to the conformally rescaled Fourier mode of $\phi = \tilde{\phi}/a$ with the comoving wave number k. Since $a'/a = Ha$, the energy density in the frequency interval dk can be written as

$$\rho_k = \frac{1}{4\pi^2} \frac{f_k^2 + (\partial_z f - Haf/k)^2}{a^4}, \quad (6.91)$$

where the cosmological scale factor is taken at the moment of matching $\eta_1 = -1/(H_I a_I^{(fin)})$. Note that $k/a = p$, where p is the physical wave number or frequency, which redshifts in the course of the cosmological expansion.

Using the arguments leading to Eq. (6.61), we eventually find

$$\rho_k = \frac{1}{4\pi^2} k^3 dk |\beta_k|^2 = \frac{1}{16\pi^2} \frac{dk}{k} H_I^4. \tag{6.92}$$

The spectrum is frequency independent in logarithmic interval. The cosmological energy fraction of inflationary gravitational waves in this simple model is

$$\Omega_{GW} = \frac{H_I^2}{6\pi M_{\text{Pl}}^2}. \tag{6.93}$$

This result is true for the waves with physical momenta that were stretched beyond the Hubble horizon during inflation, i.e. the maximum frequency should be about $k \sim H_I$, which becomes today $H_I/(z_I + 1)$, where z_I is the redshift of the end of inflation. If we take $H_I = 10^{-5} M_{\text{Pl}}$ and the Universe heating temperature $T_{heat} \sim 10^{-3} M_{\text{Pl}}$, then $z_I = T_{heat}/2.7$ K, so the maximum frequency today would be about $10^{-2} T_{CMB}$. This leads to an abrupt cutoff in the frequency spectrum above approximately 10^8 Hz. The minimum frequency corresponds to the wave with the present day length of the order of the contemporary Hubble horizon.

If the length of a gravitational wave were shorter than the cosmological horizon at redshift $z_{eq} \approx 10^4$, when the radiation dominated regime changed to the matter dominated one, then the fraction of its cosmological energy density would drop by 4 orders of magnitude. So for $H_I = 10^{-5} M_{\text{Pl}}$ we would expect $\Omega_{GW} \sim 10^{-15}$. Longer waves would not be so much redshifted and their energy fraction could be up to 4 orders of magnitude higher. Today such waves should be longer than 10^8 yrs and their frequency smaller than 10^{-16} Hz. Let us note that in more realistic scenarios, which include, in particular, possible matter dominated regimes after the end of inflation, the spectrum of inflationary gravitational waves is not exactly flat but depends upon the inflaton potential. The fraction of the cosmological energy of gravitational waves also depends upon the inflationary model and may considerably vary. Moreover, it could be strongly suppressed if there existed an early stage of dominance of primordial black holes (Dolgov et al. 2015; Dolgov and Ejlli 2011), which later evaporated restoring a radiation dominated phase. If this regime was realized, gravitational waves of much higher frequencies could be generated by primordial black hole interactions.

6.6 Generation of Density Perturbations

The absence of a mechanism responsible for the creation of density perturbations on cosmological scales was a fundamental unsolved puzzle in the old FRW cosmology. Quantum or thermal fluctuations in the cosmological plasma might have sufficiently large amplitudes, but only on very small scales. Now we know that this problem can be brilliantly solved by inflation. In short, the mechanism of the creation of primordial density perturbations works as follows. The wavelength of quantum fluctuations is exponentially stretched by the inflationary expansion from micro-scales up to galactic, galaxy cluster, and even the present day horizon scales. Moreover, the amplitude of these fluctuations is amplified by the process of expansion analogously to the amplification of tensor perturbations considered in Sect. 6.5. The inflationary mechanism for the creation of density perturbations and the prediction of its spectrum (Mukhanov and Chibisov 1981), which turns out to be in good agreement with the data, is a great successes of the Standard Model of cosmology.

Let us start with a real quantum field ϕ in a FRW background. We assume that the field satisfies the equation of motion (6.12). It is convenient to quantize the conformally rescaled field, $\tilde{\phi} = a\phi$, in conformal time, see Eqs. (6.68) and (6.69). We expand the field, as usually, in terms of creation/annihilation operators, see Eq. (C.2)

$$\tilde{\phi} = \int \widetilde{dk} \left[a_\mathbf{k} e^{i\mathbf{k}\cdot\mathbf{x}} f_k(\eta) + a_\mathbf{k}^\dagger e^{-i\mathbf{k}\cdot\mathbf{x}} f_k^*(\eta) \right], \qquad (6.94)$$

where the Fourier amplitudes satisfy the equation

$$f_k'' + \left[k^2 - \frac{a''}{a} \right] f_k + a^3 U'(f_k/a) = 0, \qquad (6.95)$$

which is similar to Eq. (6.71). Here $U' = dU/d\phi$, so:

1. $a^3 U'(f_k/a) = m^2 a^2 f_k$ for a free massive field with $U(\phi) = m^2 \phi^2/2$.
2. $a^3 U'(f_k/a) = \lambda f_k^3$ for a self-interacting field ϕ with $U(\phi) = \lambda \phi^4/4$. This result is true if only one k-mode dominates.

We now consider the inflationary regime with $H^2 \gg m^2$ and/or $H^2 \gg \lambda f^2$. Equation (6.95) can be solved analytically if $\lambda = 0$. The solution is given by Eq. (6.73). For vacuum quantum fluctuations, $C_2 = 0$. In the limit of small mass, when $\nu = 3/2$, the solution simplifies to (6.80). The spectrum of vacuum quantum fluctuations of ϕ can be calculated as

$$\langle |\phi^2| \rangle_{vac} = \frac{1}{a^2} \int \widetilde{dk}\, \widetilde{dk'} \langle [a_\mathbf{k} e^{i\mathbf{k}\cdot\mathbf{x}} f_k(\eta) + a_\mathbf{k}^\dagger e^{-i\mathbf{k}\cdot\mathbf{x}} f_k^*(\eta)]$$
$$[a_{\mathbf{k}'} e^{i\mathbf{k}'\cdot\mathbf{x}} f_{k'}(\eta) + a_{\mathbf{k}'}^\dagger e^{-i\mathbf{k}'\cdot\mathbf{x}} f_{k'}^*(\eta)] \rangle_{vac}$$
$$= \frac{1}{4\pi^2} \int dp\, p \left(1 + \frac{H^2}{p^2} \right), \qquad (6.96)$$

where $p = k e^{-Ht}$ is the physical momentum and we use the condition $a_k|vac\rangle = 0$, the commutation relation (C.4), the solution (6.80) for f_k, and $y = -k\eta = ke^{-Ht}/H$. The first quadratically divergent term in this expression corresponds to the infinitely large value of the quantum operators in coinciding space points.

6.6 Generation of Density Perturbations

It is the same as that in flat spacetime and should be subtracted. The second term describes the effect of the cosmological expansion and this is what we need. Note that initially we assumed that $k \gg H$ and the second term was much smaller than 1. This describes quantum fluctuations in de Sitter spacetime with wavelengths smaller than the cosmological horizon. However, the exponential cosmological expansion pushes these waves beyond the horizon, where $He^{Ht} \gg 1$, and the second term rises to be much larger than 1. It is noteworthy that $H \sim 1/t$, with $a \sim t^{1/2}$ and $t^{2/3}$ in the radiation dominated and matter dominated regimes, respectively, so the ratio Ha/k drops down. In other words, quantum fluctuations strongly rise at inflation but decrease in the radiation dominated or matter dominated regimes.

Despite the rising quantum fluctuations of the inflaton, its energy density does not change (the Bogolyubov coefficients remain zero). Nevertheless, the fluctuations create stochastic density perturbations in the following indirect way. We can see this if we separate the inflaton field $\phi(x,t)$ into a classical homogeneous part $\phi_0(t)$ and a small quantum fluctuation $\delta\phi(x,t)$

$$\phi(x,t) = \phi_0(t) + \delta\phi(x,t). \tag{6.97}$$

During the inflationary epoch, the quantum part satisfies the equation

$$\delta\ddot{\phi} + 3H\delta\dot{\phi} - e^{-2Ht}\partial_i^2\delta\phi - \frac{\partial^2 V(\phi_0)}{\partial\phi^2}\delta\phi = 0, \tag{6.98}$$

obtained from Eq. (6.12) by a first order expansion in $\delta\phi$. For large Ht, the third term in the equation is redshifted away and $\delta\phi$ satisfies the same equation as $\dot{\phi}_0(t)$, as one can see differentiating Eq. (6.12) over time with constant H. Equation (6.98) has two solutions. One of them decreases as $\exp(-3Ht)$ for $\partial^2 V/\partial\phi^2 \ll H^2$. The second solution varies relatively slowly. At large t, the first solution can be neglected and we can write

$$\delta\phi(x,t) = -\delta\tau(x)\dot{\phi}_0(t). \tag{6.99}$$

If $\delta\phi$ is small, this is equivalent to an x-dependent retardation of the classical field motion to the equilibrium point

$$\phi(x,t) = \phi_0(t - \delta\tau(x)). \tag{6.100}$$

Correspondingly, inflation ends at different moments in different space points. This is the physical reason for the generation of density perturbations. Since the energy density in the Universe during inflation is dominated by the inflaton field ϕ, one can write $\rho(x,t) = \rho(t - \delta\tau(x))$, forgetting possible subtleties connected with the freedom in the choice of coordinates. The problem of gauge freedom and fixation of a convenient gauge is considered in Chap. 12, where the evolution of the density perturbations in the FRW background is discussed. Thus we come to

$$\frac{\delta\rho}{\rho} = -\delta\tau\frac{\dot{\rho}}{\rho} = 4H\delta\tau(x). \tag{6.101}$$

At the last step, the relation $\dot{\rho}/\rho = -4H$ has been used. It is valid in the radiation dominated stage, which, by assumption, was formed in the heated Universe when inflation was over. The spectrum of the density perturbations is thus determined by the spectrum of $\delta t = -\delta\phi/\dot{\phi}_0$. The spectrum of $\delta\phi$ can be read-off from Eq. (6.96)

$$\langle \delta\phi(x,t)^2 \rangle = \left(\frac{H}{2\pi}\right)^2 \int \frac{dk}{k}, \tag{6.102}$$

and the power spectrum of density perturbations is given by

$$\langle \left(\frac{\delta\rho}{\rho}\right)^2 \rangle = \frac{4H^4}{\pi^2 \dot{\phi}_0^2} \int \frac{dk}{k}. \tag{6.103}$$

If inflation is realized with a slow roll regime, as it is discussed in Sect. 6.3.1, then $\dot{\phi} = U'(\phi)/(3H)$, $H^2 = 8\pi U/(3M_{\text{Pl}}^2)$, and we can estimate the magnitude of the density perturbations as

$$\frac{\delta\rho}{\rho} = 16 \left(\frac{8\pi}{3}\right)^{1/2} \frac{U^{3/2}}{U' M_{\text{Pl}}^3}. \tag{6.104}$$

For $U = m^2\phi^2/2$, we obtain

$$\left(\frac{\delta\rho}{\rho}\right)_m = 16 \left(\frac{\pi}{3}\right)^{1/2} \left(\frac{m\phi^2}{M_{\text{Pl}}^3}\right), \tag{6.105}$$

while for $U = \lambda\phi^4/4$ we find

$$\left(\frac{\delta\rho}{\rho}\right)_\lambda = 2 \left(\frac{8\pi\lambda}{3}\right)^{1/2} \left(\frac{\phi^3}{M_{\text{Pl}}^3}\right). \tag{6.106}$$

Since, roughly speaking, $\delta\rho/\rho < 10^{-5}$ and at the end of inflation $\phi \sim M_{\text{Pl}}$, we conclude that $m \leq 10^{-6} M_{\text{Pl}}$ and $\lambda \leq 10^{-12}$ to agree with observational data.

In the limit of constant H and $\dot{\phi}$, the flat Harrison-Zeldovich spectrum of perturbations is obtained. However, we need to take into account the slow variations of $\dot{\phi}$ and H and to estimate these quantities when the wavelength of the perturbations became equal to the cosmological horizon, because after that the corresponding modes remained constant. This leads to some small corrections to the flat spectrum. The discussion of this and a few more corrections can be found, for instance, in Mukhanov (2005), Weinberg (2008), Gorbunov and Rubakov (2011).

References

L.F. Abbott, E. Farhi, M.B. Wise, Phys. Lett. B **117**, 29 (1982)
F.C. Adams, K. Freese, Phys. Rev. D **43**, 353 (1991) [hep-ph/0504135]
A. Albrecht, P.J. Steinhardt, Phys. Rev. Lett. **48**, 1220 (1982)
A. Albrecht, P.J. Steinhardt, M.S. Turner, F. Wilczek, Phys. Rev. Lett. **48**, 1437 (1982)
R. Allahverdi, R. Brandenberger, F.Y. Cyr-Racine, A. Mazumdar, Ann. Rev. Nucl. Part. Sci. **60**, 27 (2010). arXiv:1001.2600 [hep-th]
E.V. Arbuzova, A.D. Dolgov, L. Reverberi, JCAP **1202**, 049 (2012). arXiv:1112.4995 [gr-qc]
C. Armendariz-Picon, T. Damour, V.F. Mukhanov, Phys. Lett. B **458**, 209 (1999) [hep-th/9904075]
D. Baumann. arXiv:0907.5424 [hep-th]
B.K. Berger, Ann. Phys. **83**, 458 (1974)
N.D. Birrell, P.C.W. Davies, *Quantum Fields in Curved Space* (Cambridge University Press, Cambridge, 1982)
N.N. Bogoliubov, Izv. AN SSSR Fiz. **11**, 77 (1947)
N.N. Bogoliubov, JETP **34**, 58 (1958)
K.A. Bronnikov, E.A. Tagirov, Preprint R2-4151, JINR (1968)
K.A. Bronnikov, E.A. Tagirov, Grav. Cosmol. **10**, 249 (2004) [gr-qc/0412138]
A. Buonanno. gr-qc/0303085
C.G. Callan Jr, Phys. Rev. D **26**, 2058 (1982)
N.A. Chernikov, N.S. Shavokhina, Teor. Mat. Fiz. **16**, 77 (1973)
D.J.H. Chung, E.W. Kolb, A. Riotto, Phys. Rev. Lett. **81**, 4048 (1998) [hep-ph/9805473]
D.J.H. Chung, E.W. Kolb, A. Riotto, Phys. Rev. D **59**, 023501 (1999) [hep-ph/9802238]
P.C.W. Davies, J. Phys. A **8**, 609 (1975)
P.A.M. Dirac, Proc. Roy. Soc. Lond. A **133**, 60 (1931)
A.D. Dolgov, Sov. Phys. JETP **54**, 223 (1981) [Zh. Eksp. Teor. Fiz. **81**, 417 (1981)]
A.D. Dolgov, Phys. Rept. **222**, 309 (1992)
A.D. Dolgov, in *Multiple facets of quantization and supersymmetry*, eds by. M. Olshanetsky, A. Vainshtein (World Scientific, 2002) [hep-ph/0112253]
A.D. Dolgov, Phys. Atom. Nucl. **73**, 815 (2010). arXiv:0907.0668 [hep-ph]
A.D. Dolgov, D. Ejlli, Phys. Rev. D **84**, 024028 (2011). arXiv:1105.2303 [astro-ph.CO]
A. Dolgov, K. Freese, Phys. Rev. D **51**, 2693 (1995) [hep-ph/9410346]
A.D. Dolgov, S.H. Hansen, Nucl. Phys. B **548**, 408 (1999) [hep-ph/9810428]
A.D. Dolgov, D.P. Kirilova, Sov. J. Nucl. Phys. **51**, 172 (1990) [Yad. Fiz. **51**, 273 (1990)]
A.D. Dolgov, A.D. Linde, Phys. Lett. B **116**, 329 (1982)
A.D. Dolgov, P.D. Naselsky, I.D. Novikov. astro-ph/0009407
A.D. Dolgov, A.V. Popov, A.S. Rudenko. arXiv:1412.0112 [astro-ph.CO]
A.D. Dolgov, B. Ya Zeldovich, Rev. Mod. Phys. **53**, 1 (1981) [Usp. Fiz. Nauk **139**, 559 (1980)]
G.N. Felder, J. Garcia-Bellido, P.B. Greene, L. Kofman, A.D. Linde, I. Tkachev, Phys. Rev. Lett. **87**, 011601 (2001a) [hep-ph/0012142]
G.N. Felder, L. Kofman, A.D. Linde, Phys. Rev. D **64**, 123517 (2001b) [hep-th/0106179]
K. Freese, J.A. Frieman, A.V. Olinto, Phys. Rev. Lett. **65**, 3233 (1990)
S.A. Fulling, Phys. Rev. D **7**, 2850 (1973)
E.B. Gliner, Sov. Phys. JETP **22**, 378 (1966) [Zh. Eksp. Teor. Fiz. **49**, 542 (1965)]
E.B. Gliner, I.G. Dymnikova, Pisma Astron. Zh. **1**, 7 (1975)
D.S. Gorbunov, V.A. Rubakov, *Introduction to the Theory of the Early Universe: Cosmological Perturbations and Inflationary Theory* (World Scientific, Hackensack, 2011)
B.R. Greene, T. Prokopec, T.G. Roos, Phys. Rev. D **56**, 6484 (1997) [hep-ph/9705357]
A.A. Grib, S.G. Mamaev, Yad. Fiz. **10**, 1276 (1969) [Sov. J. Nucl. Phys. **10**, 722 (1970)]
A.A. Grib, S.G. Mamaev, Yad. Fiz. **14**, 800 (1971)
A.A. Grib, S.G. Mamayev, V.M. Mostepanenko, *Vacuum Quantum Effects in Strong Fields* (Friedman Laboratory Publishing, St. Petersburg, 1995)
L.P. Grishchuk, Sov. Phys. JETP **40**, 409 (1975) [Zh. Eksp. Teor. Fiz. **67**, 825 (1974)]

V.T. Gurovich, A.A. Starobinsky, Sov. Phys. JETP **50**, 844 (1979) [Zh. Eksp. Teor. Fiz. **77**, 1683 (1979)]
A.H. Guth, Phys. Rev. D **23**, 347 (1981)
E.R. Harrison, Phys. Rev. D **1**, 2726 (1970)
G. 't Hooft, Nucl. Phys. B **79**, 276 (1974)
B.L. Hu, Phys. Rev. D **9**, 3263 (1974)
B.L. Hu, S.A. Fulling, L. Parker, Phys. Rev. D **8**, 2377 (1973)
D. Kazanas, Astrophys. J. **241**, L59 (1980)
T.W.B. Kibble, J. Phys. A **9**, 1387 (1976)
W.H. Kinney. arXiv:0902.1529 [astro-ph.CO]
L. Kofman, A.D. Linde, A.A. Starobinsky, Phys. Rev. Lett. **73**, 3195 (1994) [hep-th/9405187]
L. Kofman, A.D. Linde, A.A. Starobinsky, Phys. Rev. D **56**, 3258 (1997) [hep-ph/9704452]
V. Kuzmin, I. Tkachev, JETP Lett. **68**, 271 (1998) [Pisma Zh. Eksp. Teor. Fiz. **68**, 255 (1998)] [hep-ph/9802304]
V. Kuzmin, I. Tkachev, Phys. Rev. D **59**, 123006 (1999) [hep-ph/9809547]
L.D. Landau, E.M. Lifshitz, *Mechanics*, 3rd edn. (Butterworth Heinemann, Amsterdam, 1976)
A.D. Linde, Phys. Lett. B **108**, 389 (1982)
A.D. Linde, Phys. Lett. B **129**, 177 (1983)
A.D. Linde, Contemp. Concepts Phys. **5**, 1 (1990) [hep-th/0503203]
A.D. Linde, Phys. Lett. B **259**, 38 (1991)
A.D. Linde, Phys. Rev. D **49**, 748 (1994) [astro-ph/9307002]
V.N. Lukash, A.A. Starobinsky, Zh Eksp, Teor. Fiz. **66**, 1515 (1974)
M. Maggiore, Phys. Rept. 331, 283 (2000) [gr-qc/9909001]
V. Mukhanov, *Physical Foundations of Cosmology* (Cambridge University Press, Cambridge, 2005)
V.F. Mukhanov, G.V. Chibisov, JETP Lett. **33**, 532 (1981) [Pisma Zh. Eksp. Teor. Fiz. **33**, 549 (1981)]
L. Parker, Phys. Rev. Lett. **21**, 562 (1968)
L. Parker, Phys. Rev. **183**, 1057 (1969)
A.M. Polyakov, JETP Lett. **20**, 194 (1974) [Pisma Zh. Eksp. Teor. Fiz. **20**, 430 (1974)]
J. Preskill, Phys. Rev. Lett. **43**, 1365 (1979)
V.A. Rubakov, JETP Lett. **33**, 644 (1981) [Pisma Zh. Eksp. Teor. Fiz. **33**, 658 (1981)]
V.A. Rubakov, Nucl. Phys. B **203**, 311 (1982)
V.A. Rubakov, M.V. Sazhin, A.V. Veryaskin, Phys. Lett. B **115**, 189 (1982)
K. Sato, Mon. Not. Roy. Astron. Soc. **195**, 467 (1981)
K. Sato, Phys. Lett. B **99**, 66 (1981)
A.A. Starobinsky, JETP Lett. **30**, 682 (1979) [Pisma Zh. Eksp. Teor. Fiz. **30**, 719 (1979)]
A.A. Starobinsky, Phys. Lett. B **91**, 99 (1980)
J.H. Traschen, R.H. Brandenberger, Phys. Rev. D **42**, 2491 (1990)
W.G. Unruh, Phys. Rev. D **14**, 870 (1976)
A. Vilenkin, Phys. Rept. **121**, 263 (1985)
A. Vilenkin, E.P.S. Shellard, *Cosmic Strings and Other Topological Defects* (Cambridge University Press, Cambridge, 1994)
S. Weinberg, *Cosmology* (Oxford University Press, Oxford, 2008)
Y.B. Zeldovich, Pisma. Zh. Eksp. Teor. Fiz. **12**, 443 (1970)
Y.B. Zeldovich, Mon. Not. Roy. Astron. Soc. **160**, 1P (1972)
Y.B. Zeldovich, M.Y. Khlopov, Phys. Lett. B **79**, 239 (1978)
Y.B. Zeldovich, I.D. Novikov, *Structure and Evolution of the Universe* (University of Chicago Press, Chicago, Illinois, 1983) [*Structura i Evolyutsiya Vselennoi* (Nauka, Moscow, Russia, 1975)]
Y.B. Zeldovich, A.A. Starobinsky, Sov. Phys. JETP **34**, 1159 (1972) [Zh. Eksp. Teor. Fiz. **61**, 2161 (1971)]

Chapter 7
Baryogenesis

7.1 Observational Data

The observed part of the Universe is practically 100% populated by particles, despite almost identical properties of particles and antiparticles. A small fraction of antiprotons in cosmic rays, at the level of 10^{-4} with respect to protons, can be explained by their secondary origin in collisions of energetic cosmic particles. A similar situation is with cosmic positrons, though there are some exciting data on a positron excess, both at high and low energies (see below).

Quite strong bounds on the possible existence of cosmic antimatter arise from observations of 100 MeV cosmic γ-rays, which may presumably be created by $\bar{p}p$ annihilations into pions and subsequent decays of π^0 into photons (there could be also energetic photons from the annihilation of positrons produced by pion decay). The absence of an excessive γ radiation allows to conclude that the nearest antigalaxy cannot be closer than ~ 10 Mpc (Steigman 1976). However, we cannot say much about galaxies outside of the Virgo Supercluster. The observed colliding galaxies at any distance or the galaxies in common intergalactic gas clouds are of the same kind of matter (or antimatter). In particular, the fraction of antimatter in the two colliding galaxies in the Bullet Cluster is bounded by $n_{\bar{B}}/n_B < 3 \times 10^{-6}$ (Steigman 2008). Very restrictive bounds are found for a baryon symmetric universe, namely a universe consisting of an equal amount of (large) matter-antimatter domains (Cohen et al. 1998). In this case, the annihilation would be so efficient that the nearest antimatter domain should practically be at the cosmological horizon, namely at a few Gpc distance (Cohen et al. 1998). Probably this bound could be relaxed in some modifications of the standard scenario of spontaneous CP breaking.

According to the analysis of cosmic electromagnetic radiation, in particular of ~ 100 MeV photons from $\bar{p}p$ annihilation and of the 0.511 MeV line from e^+e^- annihilation at low energies, the fraction of antistars in a galaxy should be generally below $10^{-5} - 10^{-6}$ of the total amount of stars there. In particular, for our Galaxy, as it is shown in Ballmoos (2014), the amount of antistars is bounded by $N_{\bar{*}}/N_* < 4 \cdot 10^{-5}$ within 150 pc from the Sun.

An unambiguous proof for the existence of primordial antimatter would be an observation of sufficiently heavy antinuclei, starting from $^4\overline{\text{He}}$. According to theoretical estimates (Duperray et al. 2005), antideuterium could be created in energetic cosmic ray reactions of $\bar{p}\,p$ or $\bar{p}\,\text{He}$ collisions with a flux of $\sim 10^{-7}$ m^2/s^{-1}/sr/(GeV/n), i.e. 5 orders of magnitude lower than the observed flux of antiprotons. The fluxes of the secondary-produced $^3\overline{\text{He}}$ and $^4\overline{\text{He}}$ are predicted to be much smaller, respectively 4 and 8 orders of magnitude below that of antideuterium (Duperray et al. 2005). On the other hand, the production of antinuclei was measured at LHC by the Alice group (Martin 2013) and the results were reported at a seminar (Kalweit 2014). Though the production rate looks significant, with a suppression factor of about 1/300 per each extra antinucleon added to a produced antinucleus, such events are quite rare in cosmology and their contribution to the total cosmological production is very small. At the present time, there is only an upper bound on the flux of cosmic antihelium (Sasaki 2008):

$$\overline{\text{He}}/\text{He} < 3 \cdot 10^{-7}. \tag{7.1}$$

In the near future, this bound is expected to be improved to $\overline{\text{He}}/\text{He} < 3 \cdot 10^{-8}$ (Boezio 2008; Picozza and Morselli 2008) and $\overline{\text{He}}/\text{He} < 10^{-9}$ (Alcaraz 1999).[1] To summarize, the current situation is roughly the following. We have the observations $\bar{p}/p \sim 10^{-4}$ and He$/p \sim 0.1$, and the upper limit $\overline{\text{He}}/\text{He} < 3 \cdot 10^{-7}$. Theoretical predictions for secondary production are $\bar{d} \sim 10^{-5}\bar{p}$, $^3\overline{\text{He}} \sim 10^{-9}\bar{p}$, and $^4\overline{\text{He}} \sim 10^{-13}\bar{p}$.

There are also other types of limits emerging from considerations of the BBN (Chap. 8), which exclude large fluctuations of the baryonic number density at distances larger than about 1 Mpc, and from the study of the angular fluctuations of the CMB (Chap. 10), which forbids noticeable isocurvature fluctuations on scales larger than about 10 Mpc.

The total amount of baryonic matter in the Universe can be determined from BBN and CMB data under the assumption of negligible amount of antimatter. Before the precise data on the angular fluctuations of the CMB became available, the only measurements of Ω_B and of the effective number of new light particle species, N_{eff}, came from the data on light element abundances, see Chap. 8. Now, both types of measurements give closely coinciding results, though the BBN measures the amount of baryons when the Universe was a hundred seconds old, while the CMB measures the amount at $t_U \sim 370{,}000$ yrs. The agreement between the two different methods is a very strong argument in favor of the correctness of the general cosmological picture. It is interesting that the data on the amount of the directly visible baryons in the contemporary Universe show a smaller baryonic density by a factor of a few. The current CMB measurement (Ade 2014) gives the value

$$\Omega_B h^2 = 0.02205 \pm 0.00028. \tag{7.2}$$

---

[1] http://ams-02project.jsc.nasa.gov.

Very similar results follow from the analysis of the BBN. The observation of ^{4}He implies the baryonic density (Izotov et al. 2013)

$$\Omega_B h^2 = 0.0234 \pm 0.0019 \ (68\% \ \text{C.L.}). \tag{7.3}$$

Primordial deuterium is much more sensitive to the amount of baryons. The recent measurement $D/H = (2.53 \pm 0.04) \cdot 10^{-5}$ gives (Cooke et al. 2014):

$$\Omega_B h^2 = 0.02202 \pm 0.00045, \tag{7.4}$$

As we have already mentioned, before the accurate measurements of the spectrum of the angular fluctuations of the CMB, Ω_B was determined by the abundance of the primordial deuterium, which by this reason was called "baryometer". According to the above presented data, the density of the baryonic number with respect to the number density of CMB photons is

$$\eta = (n_B - n_{\bar{B}})/n_\gamma = (6.1 \pm 0.3) \cdot 10^{-10}. \tag{7.5}$$

7.2 General Features of Baryogenesis Models

7.2.1 Sakharov Principles

The predominance of matter over antimatter was beautifully explained by Sakharov (1967) as dynamically generated in the early Universe due to three conditions, today called *Sakharov principles*:

1. Non-conservation of baryon number.
2. Breaking of C and CP invariance.
3. Deviation from thermal equilibrium.

As we show below, none of these conditions is really obligatory, but baryogenesis models without them require some exotic mechanism. We first discuss normal baryogenesis scenarios, in which the three Sakharov principles hold.

7.2.1.1 Non-conservation of Baryon Number

The non-conservation of baryons is theoretically justified. GUT scenarios, supersymmetric models, and even the electroweak theory predict the violation of the baryon number, namely processes with $\Delta B \neq 0$. However, at the present time this prediction is not confirmed by direct experiments. Despite an extensive search, only lower bounds on the proton lifetime and on the period of neutron-antineutron oscillations are established. The only "experimental piece of data" in favor of the non-conservation of baryons is our Universe: we exist, ergo baryons are not conserved. Half a century ago, from the same experimental fact, our existence, an opposite conclusion of baryon conservation was deduced. Theory is an important input in understanding of what we see.

Probably we should clarify the last statement. Inflation seems to be a necessary ingredient to create a universe that is suitable for life. Moreover, the predicted inflationary spectrum of the density perturbations well agrees with data. In this sense, inflation may be seen as an experimental fact. On the other hand, inflation would be impossible if the baryon number were conserved. For a successful solution of the FRW cosmology problems, inflation should last at least 65 e-foldings, $a \sim e^{65}$. Let us assume that the baryonic number is conserved. At the present time, the baryon energy density is about 10^4 times the energy density of CMB photons. When we travel backward in time, the baryon to photon energy density ratio drops as the cosmological redshift. At the moment of the QCD phase transition, which took place at $z \sim 10^{12}$, the ratio ρ_B/ρ_γ was about 10^{-8}. Prior to the QCD phase transition, baryon number carriers were relativistic quarks and at that early period ρ_B/ρ_γ remained constant. Hence we should conclude that at the end of inflation this ratio would be approximately the same as at the QCD phase transition, namely $\rho_B/\rho_\gamma \sim 10^{-8}$. During inflation, all the matter of the Universe was in the form of the inflaton field with constant energy density. If ρ_{tot} dropped down with time, inflation would have been impossible. However, the energy density of matter with a conserved quantum number cannot stay constant but drops as $1/a^3$ or $1/a^4$. Traveling backward in time, it means that ρ_B rises and in less than 6 Hubble times becomes comparable to the inflaton energy density. However, 6 Hubble times are not enough to create our good old Universe.

7.2.1.2 Breaking of C and CP Invariance

C and CP violations were discovered and confirmed in direct experiments. In the first part of the XX century, the common belief was that physics was invariant with respect to the separate action of all the three transformations, namely mirror reflection, P, charge conjugation, C, and time reversal, T. The weakest link in this chain of discrete symmetries was P, found to be broken in 1956 (Lee and Yang 1956; Wu et al. 1957).

It was immediately assumed that the world was symmetric with respect to the combined transformation from particles to mirror reflected antiparticles, CP (Landau 1957). Both P and C are 100% broken in weak interactions but still some symmetry between particles and antiparticles was saved by this assumption. This symmetry crashed down pretty soon, in 1964 (Christenson et al. 1964). After this discovery, life in the Universe became possible.

At the present time, only the CPT symmetry has survived. It is the only one with a rigorous theoretical justification, the CPT theorem (Luders 1954, 1957; Schwinger 1951; Pauli 1955), based on the solid grounds of Lorenz-invariance, canonical spin-statistics relation, and positive definite energy. Still models without CPT are considered, e.g. for the explanation of some neutrino anomalies and for baryogenesis.

7.2.1.3 Deviation from Thermal Equilibrium

Thermal equilibrium is always broken for massive particles, but usually very weakly. To estimate the effect, let us approximate the collisional integral in the kinetic equation as $I_{coll} = \Gamma(f_{eq} - f)$, see Eq. (5.26), where Γ is the interaction rate. Let us assume that Γ is large, so that deviations from equilibrium are small, $\delta f / f_{eq} \ll 1$. Substituting f_{eq} into the left hand side of the kinetic equation, as it is done in Eq. (5.29), with $T \sim m$ and $\Gamma \sim \alpha m$ we find

$$\frac{\delta f}{f_{eq}} \approx \frac{Hm^2}{\Gamma T E} \sim \frac{Tm^2}{\Gamma E M_{Pl}} \sim \frac{m}{\alpha M_{Pl}}. \tag{7.6}$$

Since the Planck mass is very large, deviations from equilibrium might be significant only at large temperatures or tiny Γ. However, if the fundamental gravity scale were in the TeV range (Arkani-Hamed et al. 1998), equilibrium could be strongly broken even at the electroweak scale.

Another source of deviation from equilibrium in the cosmological plasma could be a first order phase transition from, say, an unbroken to broken symmetry phase in a non-abelian gauge theory with a spontaneous symmetry violation. There might be a rather long non-equilibrium period of the coexisting two phases.

7.2.2 CP Breaking in Cosmology

There are many scenarios of baryogenesis, each performs a rather modest task to explain only one number, the observed asymmetry (7.5). It is a great challenge for astronomers to check if η is constant or it may vary at different space points, i.e. $\eta = \eta(x)$. A few questions in this connection deserve attention. What is the characteristic scale l_B of possible variations of the baryon number density? May there be astronomically large domains of antimatter nearby or only very far away? Answers to these questions depend, in particular, upon the mechanisms of CP violation realized in cosmology, which are described below. For more details, see Dolgov (2005).

There are three possibilities to violate CP in cosmology:

1. *Explicit CP violation.* It is realized by complex coupling constants in the Lagrangian of the theory, in particular by complex Yukawa couplings transformed by the vacuum expectation value of the Higgs field $\langle\phi\rangle \neq 0$ into a non-vanishing phase in the Cabibbo-Kobayashi-Maskawa (CKM) mixing matrix. However, in the MSM of particle physics based on $U_Y(1) \times SU_L(2) \times SU(3)$, CP violation at $T \sim$ TeV is too weak, at least by 10 orders of magnitude, to allow for the generation of the observed baryon asymmetry. Indeed, CP violation in the MSM is absent for two quark families because the phase in a 2×2 quark mass matrix can be rotated away. At least three families are necessary. It could be an anthropic principle explanation to say that for this reason there are three generations.

If the masses of different up- or down-type quarks were equal, CP violation could also be rotated away, because the unit matrix is invariant with respect to a unitary transformation. If the mass matrix were diagonal, in the same representation as the flavor matrix, CP violation could also be rotated away. Thus CP breaking is proportional to the product of the mixing angles and to the mass differences of all the up- and down-type quarks

$$A_- \sim \sin\theta_{12} \sin\theta_{23} \sin\theta_{31} \sin\delta\, (m_t^2 - m_u^2)(m_t^2 - m_c^2)(m_c^2 - m_u^2) \quad (7.7)$$
$$(m_b^2 - m_s^2)(m_b^2 - m_d^2)(m_s^2 - m_d^2)/\, M^{12}.$$

At high temperatures, namely $T \geq$ TeV, where the electroweak baryon non-conservation is operative, the characteristic mass is $M \sim 100$ GeV and $A_- \sim 10^{-19}$. For a successful baryogenesis, an extension of the Standard Model is definitively necessary, because A_- is too small.

2. *Spontaneous CP violation* (Lee 1974). It could be realized by a complex scalar field Φ with a CP symmetric potential having two separated minima at $\langle \Phi \rangle = \pm f$. The Lagrangian is supposed to be CP invariant, but these two vacuum states have opposite signs of CP violation. Such CP breaking is locally indistinguishable from the explicit one, but globally leads to a charge symmetric universe with an equal amount of matter and antimatter. As we mentioned at the beginning of this section, antimatter domains, if they exist, should be very far from us, at $l_B \geq$ Gpc. Moreover, there is another problem with this mechanism, namely walls between matter and antimatter domains could destroy the observed homogeneity and isotropy of the Universe (Zeldovich et al. 1974). To avoid this problem, a mechanism of wall destruction is necessary.
3. *Stochastic or dynamical CP violation* (Dolgov 1992; Balaji et al. 2004, 2005). If a complex scalar field χ were displaced from its equilibrium point in the potential, e.g. by quantum fluctuations at inflation, and did not relax down to the equilibrium at baryogenesis, it would create a CP violation proportional to the amplitude of the field but without the problems of a spontaneous CP violation. Later, after baryogenesis is over, χ can relax down to zero. In this way, domain walls do not appear. Inhomogeneous $\eta(x)$ with domains of matter and antimatter can be created with such a CP violation. The domain size depends upon the details of the scenario.

7.3 Models of Baryogenesis

This is a long, but probably incomplete, list of baryogenesis scenarios:

1. Heavy particle decays (Sakharov 1967).
2. Electroweak baryogenesis (Kuzmin et al. 1985). It is too weak in the (minimal) Standard Model of particle physics, but may work with TeV gravity.
3. Baryo-through-leptogenesis (Fukugita and Yanagita 1986).

7.3 Models of Baryogenesis

4. Supersymmetric condensate baryogenesis (Affleck and Dine 1985).
5. Spontaneous baryogenesis (Cohen and Kaplan 1987, 1988; Cohen et al. 1993).
6. Baryogenesis by primordial black hole evaporation (Hawking 1974; Zeldovich 1976a).
7. Space separation of B and $\bar{B}$ at astronomically large distances (Omnes 1969, 1970), which is probably not effective. However, antibaryons might be moved to higher dimensions (Dvali and Gabadadze 1999) and in this case the separation would be microscopically small and might be even realistic.
8. Baryogenesis due to CPT violation (Dolgov 2010).

In all these scenarios, new physics beyond the Standard Model of particle physics is necessary. In what follows, we will very briefly describe some of them. More details can be found in Dolgov (1992, 1997), Rubakov and Shaposhnikov (1996), Riotto and Trodden (1999), Dine and Kusenko (2003).

7.3.1 Baryogenesis by Heavy Particle Decays

This is the earliest scenario of baryogenesis proposed in the pioneering paper by Sakharov (1967). Later it was understood that baryogenesis by heavy particle decay is naturally realized in GUT scenarios through decays of heavy gauge or Higgs-like bosons, X, with mass around 10^{15} GeV. These bosons can decay, for instance, into qq and $\bar{q}\bar{l}$ pairs, so the baryon number is evidently not conserved. In this particular example, the difference between the baryon and the lepton numbers, $(B - L)$, is conserved, which is true in GUT models based on the $SU(5)$ symmetry group as well as in the electroweak theory, see the next section.

Due to the large mass of X-bosons, deviations from equilibrium can be significant. CP violation might also be sufficiently large: since we know nothing about it, we are free to allow maximum CP violation. This mechanism could thus be efficient enough to generate the observed asymmetry. The problem with GUT models is that the temperature of the GUT scale might not be reachable after inflation. On the other hand, baryogenesis might proceed with under-abundant X-bosons created out of equilibrium.

Particles and antiparticles can have different decay rates into charge conjugated channels if C and CP are broken, while the total widths are equal due to CPT invariance. If only C is broken, but CP is not, then partial widths, summed over spins, are the same because CP invariance implies the equality of the reaction rates between particles and antiparticles with mirror reflected particle helicities, $\sigma = \mathbf{s} \cdot \mathbf{p}/p$:

$$\Gamma(X \to f, \sigma) = \Gamma(\bar{X} \to \bar{f}, -\sigma). \tag{7.8}$$

If both C and CP are broken, partial widths would be different, but the difference appears only at higher orders in perturbation theory. At the lowest order, the amplitudes of charged conjugated processes must be equal, $A = \bar{A}^*$, because of the hermicity of the Lagrangian. The same would also be true for higher order contributions if they were real. An imaginary part is generated by re-scattering in the final state (with non-conservation of B or L), as can be seen from the S-matrix unitarity condition, $SS^\dagger = I$. Written in terms of the T-matrix defined as $S = I + iT$, it reads

$$i(T_{if} - T_{if}^\dagger) = -\sum_n T_{in} T_{nf}^\dagger = -\sum_n T_{in}^\dagger T_{nf}. \tag{7.9}$$

Here the summation is done over all the open reaction channels from the state i to the state f, and integration of the intermediate states over the phase space is assumed.

Thus, at least at the second order in perturbation theory, the amplitudes acquire non-vanishing imaginary parts, which are not reduced to the common but opposite phase of the charge conjugated amplitudes. So the amplitudes of charge conjugated processes stopped being complex conjugate and might have different absolute values. To achieve this, there should not be less than three different particle states. Let us assume that there are only two states, i and f, so the following reactions are possible: elastic $i \leftrightarrow i$ and $f \leftrightarrow f$, and inelastic $i \leftrightarrow f$. In this case, the unitarity condition (7.9) is reduced to

$$2 \mathscr{I}m\, T_{ii}[\lambda] = -\int d\tau_i |T_{if}|^2 - \int d\tau_f |T_{ii}|^2, \tag{7.10}$$

where $[\lambda]$ denotes the set of polarization states of the participating particles and $d\tau$ is the infinitesimal phase space volume.

CPT invariance demands equality of amplitudes of charge conjugated reactions with opposite signs of particle helicities

$$T_{ii}[\lambda] = T_{\bar{i}\bar{i}}[-\lambda], \tag{7.11}$$

so, after summing over the polarization, we find $\Gamma_{if} = \Gamma_{\bar{i}\bar{f}}$. Hence to destroy the equality of partial widths of charge conjugated processes, $\Gamma_{if} = \Gamma_{\bar{i}\bar{f}}$, at least three interacting states are necessary with the following reaction channels open

$$i \leftrightarrow f, \quad i \leftrightarrow k, \quad k \leftrightarrow f. \tag{7.12}$$

Let us consider an example of X-boson decaying into the channels

$$\begin{aligned} X &\to qq, \quad X \to \bar{q}\bar{l}, \\ \bar{X} &\to \bar{q}\bar{q}, \quad \bar{X} \to ql \end{aligned} \tag{7.13}$$

and assume that the partial widths are different due to C and CP violation

$$\begin{aligned} \Gamma_{X \to qq} &= (1 + \Delta_q)\Gamma_q, \quad \Gamma_{X \to \bar{q}\bar{l}} = (1 - \Delta_l)\Gamma_l, \\ \Gamma_{\bar{X} \to \bar{q}\bar{q}} &= (1 - \Delta_q)\Gamma_q, \quad \Gamma_{\bar{X} \to ql} = (1 + \Delta_l)\Gamma_l. \end{aligned} \tag{7.14}$$

7.3 Models of Baryogenesis

The parameters Δ would be non-zero due to re-scattering in the final state $qq \leftrightarrow \bar{q}\bar{l}$ and their charge conjugated ones.

If X is a gauge boson, then $\Gamma \sim \alpha$ and $\Delta \sim \alpha$, where $\alpha \sim 1/50$ is the fine structure constant at the GUT scale. The asymmetry is proportional to $\eta \sim (2/3)(2\Delta_q - \Delta_l)$. Assuming that CP violation is not suppressed at all ($\sin \delta_- \sim 1$), we can roughly estimate the magnitude of the cosmological baryon asymmetry as

$$\eta \sim \frac{\delta f}{f} \frac{\Delta \Gamma}{\Gamma} \sim \frac{m}{M_{\text{Pl}}}. \tag{7.15}$$

Small numerical coefficients omitted here would somewhat diminish the result. For example, the subsequent entropy dilution by about 1/100 is not included. For a successful leptogenesis/baryogenesis, the mass of the decaying particles should be larger than 10^{10} GeV, or the fundamental gravity force at short distances should be noticeably stronger than in standard General Relativity, namely $M_{\text{Pl}}^{(fund)} \ll 10^{19}$ GeV.

Thus we see that GUT scenarios could naturally lead to the observed baryon asymmetry of the Universe. The baryon number conservation is broken because quarks and leptons belong to the same multiplet of the symmetry group. The mass of the gauge bosons of GUT models, $m_X \sim 10^{15}$ GeV, is high enough to ensure a sufficiently large deviation from equilibrium, see Eq. (5.29) or (7.15). CP violation can be easily unsuppressed at the GUT scale, $T \sim m_X$. So far so good, but the problem is that such high temperatures may have never been reached in the Universe after inflation. However, even if the Universe was not hot enough, always having $T < m_X$, X-bosons could be produced out of equilibrium at the end of inflation by gravity, as is discussed in Sect. 6.4.4. Still one should take care of an overabundant production of gravitinos (Khlopov and Linde 1984; Ellis et al. 1984) if Supergravity is realized.

7.3.2 Electroweak Baryogenesis

It is remarkable that the Standard Model of particle physics has all the necessary ingredients for a successful baryogenesis. It is known from experiments that the C and the CP symmetries are broken. CP violation in the Standard Model is easily introduced either by a complex quark mass matrix with at least three generations (Kobayashi and Maskawa 1973) or, and it is essentially the same, by complex coupling constants of the Higgs field. Even more surprisingly, the baryon number is not conserved by electroweak interactions (Hooft 1976a, b). This is a rather complicated phenomenon and is related to the so-called quantum chiral anomaly (Adler 1969; Bell and Jackiw 1969). The classical electroweak Lagrangian conserves baryonic charge. Quarks always enter in bilinear combinations $\bar{q}q$, so a quark can disappear only in a collision with an antiquark. The classical baryonic current is thus conserved

$$\partial_\mu J_B^\mu = \sum_j \partial_\mu (\bar{q}_j \gamma^\mu q_j) = 0. \tag{7.16}$$

However, quantum corrections destroy this conservation law, and, instead of zero on the right hand side, one gets

$$\partial_\mu J_B^\mu = \frac{g^2}{16\pi^2} C G_{\mu\nu} \tilde{G}^{\mu\nu}, \tag{7.17}$$

where C is a numerical constant, $\tilde{G}^{\mu\nu} = G_{\alpha\beta}\varepsilon^{\mu\nu\alpha\beta}/2$, and the gauge field strength $G_{\mu\nu}$ in a non-abelian gauge theory is given by the expression

$$G_{\mu\nu} = \partial_\mu A_\nu - \partial_\nu A_\mu + g[A_\mu, A_\nu]. \tag{7.18}$$

Here $A_\mu \equiv A_\mu^a$ is a "vector" in the group space, which is indicated by the upper index a, which we omitted for simplicity.

An important fact is that the anomalous current non-conservation is proportional to the total derivative of a vector operator: $G_{\mu\nu}\tilde{G}^{\mu\nu} = \partial_\mu K^\mu$, where the anomalous current K^μ is

$$K^\mu = 2\varepsilon^{\mu\nu\alpha\beta} \left(A_\nu \partial_\alpha A_\beta + \frac{2}{3}ig A_\nu A_\alpha A_\beta \right). \tag{7.19}$$

The last term in this expression does not vanish for non-abelian gauge theories only, because the antisymmetric product of three vector potentials A_ν can be non-zero only due to different group indices (e.g. for the electroweak group it should contain the product of W^+, W^- and the isospin one part of Z^0).

A total derivative is usually unobservable because one can get rid of it integrating by parts. However, this may not be true for K^μ in (7.19). The gauge field strength $G_{\mu\nu}$ should indeed vanish at infinity, but the potential A_μ does not necessarily vanish. It turns out that different vacuum states, which all have $G_{\mu\nu} = 0$, differ by the value of K^0. Since the difference $J_B^\mu - K^\mu$ is conserved, transitions from one vacuum state to another leads to a change in the baryonic charge. The path from one vacuum to another is separated by a potential barrier, where $G_{\mu\nu} \neq 0$. As we know from quantum mechanics, the barrier penetration at small energies is exponentially suppressed, and indeed the probability of processes with $\Delta B \neq 0$ contains the extremely small factor $\exp(-16\pi^2/g^2) \sim 10^{-160}$ (Hooft 1976a,b). However, at high energies or high temperatures (comparable or above the barrier height), transitions between different vacua can be achieved by classical motion over the barrier. The height of the barrier, as calculated in Manton (1983), Klinkhamer and Manton (1984), is about a few TeV. In fact the barrier disappears at high temperatures together with the W- or Z-boson masses according to the law $m_W^2(T) = m_W^2(0)(1 - T^2/T_c^2)$. This also occurs in the same TeV region. So one may expect that at high temperatures the non-conservation of the baryon number is not suppressed. It has been argued that above the electroweak phase transition the processes with $\Delta B \neq 0$ are much faster than the Universe expansion rate, so that any preexisting baryon asymmetry would be washed out. To be more precise, electroweak interactions (even with the chiral anomaly) conserve the difference between the baryonic and leptonic charges,

7.3 Models of Baryogenesis

$(B - L)$. At high temperatures, only $(B + L)$ could be erased, while a preexisting $(B - L)$ would survive.

To generate a baryon asymmetry, a deviation from thermal equilibrium is necessary (below we demonstrate an example where it is not so, but generally this must be fulfilled). Deviations from thermal equilibrium due to non-vanishing masses of the intermediate W- and Z-bosons and of the Higgs boson are not sufficiently strong, as we can be seen from Eq. (7.6). However, in gauge theories with a spontaneously broken symmetry, there is another source of equilibrium breaking. The symmetry is restored at high temperatures (Kirzhnits 1972; Kirzhnits and Linde 1972) and it breaks in the course of the Universe cooling. The situation is very similar to the rotational symmetry restoration in ferromagnets and its breaking at low temperatures when spontaneously magnetized domains are created. The phase transition from the unbroken to broken phase could be either first or second order. In the course of a (delayed) first order phase transition, the two phases coexist in the cosmological plasma: broken and unbroken ones. This is surely not an equilibrium state, so it could be favorable for a baryon asymmetry generation. If the electroweak phase transition is second order, then everything goes smoothly, thermal equilibrium is not disturbed, and a charge asymmetry is not generated even below the phase transition. The type of phase transition depends upon the mass of the Higgs boson (or bosons, in extended models). For high masses, the transition is second order, while for low masses it is first order. The boundary value of the mass is not well known even in the MSM and different estimates give the critical value somewhere between 50 and 100 GeV. Now the Higgs boson mass is known, $m_H = 125.7 \pm 0.4$ GeV (Olive 2014), which excludes a first order phase transition at least in the minimal version of the Standard Model of particle physics. If so, the electroweak interactions at high temperatures could play the role of terminator of asymmetry and not be its creator at lower temperatures.

However, the above picture may be incorrect for the following reason. Processes of quark and lepton transformations with a non-vanishing change of baryonic (and leptonic) charge at high temperatures are accompanied by a change in the structure of the gauge and Higgs fields. Roughly speaking, classical field configurations, the so-called sphalerons (Manton 1983; Klinkhamer and Manton 1984), should be present in the course of the transition

$$A_k^{sph} = \frac{i\varepsilon_{klm}x^l\tau^m}{r^2} f_A(\xi),$$

$$\phi^{sph} = \frac{iv}{\sqrt{2}} \frac{\tau^i x_i}{r} (0, 1) f_\phi(\xi), \qquad (7.20)$$

where $\xi = gvr$, v is the vacuum expectation value of the Higgs field, and the functions f have the properties $f(0) = 1$ and $f(+\infty) = 0$. The size of these objects is much larger than their Compton wavelength, and for this reason they are called classical field configurations. It is assumed that sphalerons are in thermal equilibrium, so their number density is determined by the Boltzmann exponent, $\exp(-F/T)$, where

F is their free energy. In the broken symmetry phase, $F = O(\text{TeV})$, while in the symmetric phase $F \sim T$. If this is true, the processes with violation of the baryon number are not suppressed at high temperatures. However, the rate of production of classical field states in the collision of elementary particles is not known and, strictly speaking, we cannot say if they are in equilibrium or not. The analogy with magnetic monopoles in non-abelian gauge theories, which are also classical states (see Sect. 6.2.6), suggests that the production of similar states in a two-body or a few-body collision is exponentially suppressed. Nothing is known about the probability of production of monopoles or sphalerons in, say, a hundred-particle collision. To create a pair of monopole-antimonopole or a sphaleron, one presumably needs to create a special coherent field configuration which is quite unlikely in the primeval plasma. If this is true, then electroweak processes do not produce or destroy baryons in a significant amount. No analytical way to solve this problem is known at the present stage. These are non-perturbative and multi-particle processes. The only available approach to the calculation of the sphaleron transition rates is via numerical lattice simulations. Results of different groups show that the probability of the production per unit volume and unit time is of the order $\alpha^n T^4$, where $\alpha \approx 0.01$ is the fine structure constant and $n = 4$ or 5, depending upon the concrete simulations (Arnold and McLerran 1987; Ambjorn et al. 1991). Such a probability is high enough to ensure an abundant sphaleron formation. However, a similar probability is expected if sphalerons have a finite number of degrees of freedom equal to the number of the lattice cubics inside their volume, but not an infinitely large number as it is in reality.

7.3.3 Baryo-Through-Leptogenesis

The Baryo-through-leptogenesis is probably the most popular mechanism nowadays. The process of creation of a baryon asymmetry proceeds in two steps. First, we have the creation of a lepton asymmetry by L non-conserving decays of heavy ($m \sim 10^{10}$ GeV) Majorana neutrinos, N. Their existence was originally postulated for the realization of the so-called see-saw mechanism (Minkowski 1977; Gell-Mann et al. 1979; Yanagida 1979; Glashow 1979), which was suggested to explain the small value of neutrino masses as a result of the mixing with very heavy Majorana neutrinos with off-diagonal components of the mixing matrix. The decays of heavy Majorana neutrinos do not conserve the leptonic number, so one of the Sakharov conditions is naturally fulfilled. The leptogenesis part proceeds in a way very similar to the GUT baryogenesis considered in Sect. 7.3.1.

CP violation is in a sense simpler to introduce into a Majorana fermion sector than for Dirac fermions. As we have shown in Sect. 7.2.2, for the CP breaking in the mass matrix of Dirac fermions, at least three generations of fermions with unequal masses are necessary and there is only one CP-odd phase. For Majorana fermions, there is much more freedom. The Majorana mass term is written as

$$\mathscr{L}_M = M_{ij} v_i C v_j + h.c., \qquad (7.21)$$

where C is the operator of charge conjugation and $h.c.$ means the hermitian conjugate. Now all the elements of M_{ij}, including the diagonal ones, may be complex. One can kill three phases in M_{ii} by three phase rotations of ν_i. No freedom is left after that, and the three phases of M_{12}, M_{23}, and M_{31} remain arbitrary. In the Majorana case there can thus be three independent CP-odd phases. If we add three more heavy Majorana neutrinos, three more CP-odd phases would emerge. So finally the mass matrix of three flavor light and heavy Majorana neutrinos has six independent phases: three in the sector of light neutrinos and three in the sector of heavy neutrinos. They are currently unknown and therefore allowed to be of order unity. The phases measured in neutrino oscillations are not directly related to the phases in heavy neutrino decays, and thus low energy measurements cannot teach us anything about the CP violation in leptogenesis scenarios in a model independent way.

The next step can take place during the electroweak stage. The generated lepton asymmetry is transformed into a baryon asymmetry by C and CP conserving sphaleron processes in thermal equilibrium. Sphalerons do not individually conserve the baryon, B, and the lepton, L, numbers, but they conserve $(B-L)$. The initial non-vanishing L may thus be redistributed in equilibrium in almost equal shares between B and L. For a review of this scenario, see Buchmuller et al. (2004, 2005a, b), Paschos (2004), Chen (2007). The mechanism clearly require that sphalerons are abundantly created in the primeval plasma, see the discussion in Sect. 7.3.2.

The mechanism considered in this subsection looks very attractive. Leptonic and baryonic charges are naturally non-conserved. Heavy particles (Majorana neutrinos) to break thermal equilibrium are present. Three CP-odd phases of order unity might be there. However, the magnitude of the asymmetry η is just of the right size in the most favorable situation. Any deviation from the most favorable case would destroy the successful prediction of the model. Maybe such a rigid framework is an attractive feature of the model. On the other hand, an extra dilution of the asymmetry through the entropy production by new massive particles or by phase transitions in the primeval plasma would destroy the successful prediction of the model and might demand a significantly larger mass for the heavy Majorana neutrinos.

7.3.4 Evaporation of Primordial Black Holes

This model does not require any violation of the baryonic charge in the particle physics sector. However, in a sense, black hole evaporation violates the baryon number conservation. There is even a general statement that black holes break all global symmetries. Indeed, if a conserved charge does not create any long-range field, as the electric charge does, then such a charge could disappear inside a black hole without trace. If, for instance, a black hole was formed solely out of baryons, nevertheless it would evaporate into an (almost) equal number of baryons and antibaryons. For an external observer, the baryonic number is not conserved. Such a process might in principle create a cosmological baryon asymmetry if, for some reason, black holes in the early Universe predominantly captured antibaryons with respect to baryons.

An excessive capture of antibaryons might be induced, e.g., by a larger mobility of antibaryons in the primordial plasma due to breaking of C and CP invariance. In this way, it would be possible to generate an excess of baryons over antibaryons in the space outside the black holes.

Another possibility discussed in Hawking (1974), Zeldovich (1976a) is that the process of black hole evaporation (Hawking 1975; Carr and Hawking 1974; Page 1976) could be baryon asymmetric and small black holes in the process of their evaporation would enrich the Universe with baryons. These black holes could either completely disappear or evolve down to stable Planck mass remnants, but in both cases the Universe outside black holes would have a non-vanishing baryon charge and an equal amount of antibaryon charge would be buried inside the black holes, which either completely disappeared or survived and became cosmological dark matter.

At first sight, thermal evaporation of black holes cannot create any charge asymmetry by the same reason as no charge asymmetry can be generated in thermal equilibrium. However, the spectrum of particles radiated by black holes is not black but gray due to the distortion of the spectrum by the propagation of the produced particles in the gravitational field of the black hole (Page 1976). Moreover, interactions among the produced particles are also essential. These two facts allow black holes to create an excess of matter over antimatter in the external space. As a possible "realistic" model, let us consider the following case (Zeldovich 1976a; Dolgov 1980, 1981). Let us assume that there exists a heavy A-meson decaying into two charge conjugated channels with unequal probabilities (due to C and CP violation):

$$A \to H + \bar{L} \text{ and } A \to \bar{H} + L, \tag{7.22}$$

where H and L are, respectively, heavy and light baryons, for instance t and u quarks. If the temperature of the black hole is larger or comparable with the mass of the A-boson, the latter would be abundantly produced at the horizon and decay while propagating in the gravitational field of the black hole. There is a non-zero probability of back capture of the decay products by the black hole and evidently the back-capture of the heavy baryons, H and $\bar{H}$, is larger than that of the light ones, L and $\bar{L}$. As a result, a net baryon asymmetry could be created outside the black hole. According to the calculations of Dolgov (1980, 1981), the baryon asymmetry may have the proper magnitude compatible with observations.

If at the moment of the black hole formation, which presumably took place at the radiation dominated stage, primordial black holes represented a very small fraction, say ε, of the total cosmological energy density, then at redshift $z = 1/\varepsilon$ after their formation, these black holes would dominate the cosmological energy density if their lifetime were larger than the time interval necessary to survive till that redshift, $t_{MD} = t_{in}/\varepsilon^2$. If τ_{evap} is the black hole lifetime with respect to the process of evaporation, then at $\tau_{evap} > t_{MD}$ the evaporation would be recreated a radiation dominated universe, but now with a non-vanishing baryon asymmetry.

For the convenience of the reader, we present some expressions for the quantities describing the process of black hole evaporation. Numerical coefficients of order unity are sometimes omitted. Precise expressions can be found in any modern

7.3 Models of Baryogenesis

textbook on black hole physics, e.g. Frolov and Novikov (1998). For a Schwarzschild black hole, the only dimensional parameter is its gravitational radius

$$r_g = \frac{2M_{BH}}{M_{Pl}^2}. \tag{7.23}$$

The black hole temperature, just on dimensional grounds, is the inverse of the gravitational radius. The exact value is

$$T_{BH} = \frac{1}{4\pi r_g} = \frac{M_{Pl}^2}{8\pi M_{BH}}. \tag{7.24}$$

The luminosity can be easily estimated as

$$L_{BH} \sim \sigma_{SB} T^4 r_g^2 \sim \sigma_{SB} \frac{M_{Pl}^4}{M_{BH}^2}, \tag{7.25}$$

where σ_{SB} is the Stefan-Boltzmann constant (in the natural units, discussed in appendix A, $\sigma_{SB} = \pi^2/60$) and $\sim r_g^2$ is the black hole surface. Knowing the black hole mass and its luminosity, it is straightforward to estimate its lifetime $\tau_{evap} \sim M_{BH}^3/M_{Pl}^4$. The precise calculations give (Page 1976)

$$\tau_{evap} = \frac{10240\,\pi}{N_{eff}} \frac{M^3}{M_{Pl}^4}, \tag{7.26}$$

where N_{eff} is the number of particle species with masses smaller than the black hole temperature (7.24). For example, a black hole with the mass $M_{BH} \sim 10^{15}$ g would have a radius approximately 10^{-13} cm and its temperature would be about 100 MeV. They could survive till the present time, $\tau_{evap} \approx t_U$.

According to the calculations of Dolgov (1980, 1981), the mass of the heavy decaying particles A should be in the interval $m \sim 10^6 - 10^{10}$ GeV to create the observed cosmological baryon asymmetry. Let us consider the following example. Assume that primordial black holes were created when the temperature of the Universe was about 10^{14} GeV. Such a value corresponds to a Universe age $t_U \approx 10^{-34}$ s. The mass inside the cosmological horizon at that moment was

$$M_h = M_{Pl}^2 t \approx 10^{38} \text{ g } (t/s) \approx 10^4 \text{ g}. \tag{7.27}$$

Primordial black holes with such masses might be in principle created. Their temperature would be $T_{BH} = 10^9$ GeV. For $N_{eff} \sim 100$, their lifetime would be

$$\tau_{evap} \sim 3 \cdot 10^{-15} \text{ s}. \tag{7.28}$$

During this time, the Universe would have cooled down to $T \sim 10^4$ GeV. The redshift from the moment of the primordial black hole creation would have been about 10^{10}. If the black hole production efficiency was such that only in one per 10^{10} horizon volumes a black hole was created, then their mass fraction at production was 10^{-10} and at the moment of their evaporation they would dominate the cosmological energy density and could create the observed baryon asymmetry. As we mentioned above, the Planck mass remnants of such primordial black holes, if stable, could be cosmological dark matter, see e.g. Dolgov et al. (2000).

If the today baryon asymmetry were generated by the evaporation process of classical (in contrast to quantum) black holes, then it would be natural to expect that the baryonic charge might be also non-conserved in the decay of small quantum black holes, since, as we have already mentioned, gravity breaks all global symmetries and at the Planck scale the effects should be unsuppressed. This observation was first done in Zeldovich (1976b, 1977), where it was stated that protons must decay due to transformation into virtual black holes which subsequently decay into, say, a positron and a meson. The probability of such a process can be estimated as follows. Let us consider the process inside a proton

$$q + q \to \bar{q} + l, \tag{7.29}$$

which could be meditated by a virtual black hole. Here q ($\bar{q}$) is a quark (antiquark) and l is a lepton. The rate of this reaction is equal to

$$\frac{\dot{n}}{n} = n\sigma_{BH} = \sigma_{BH} |\psi(0)|^2, \tag{7.30}$$

where $n \sim m_p^3$ is the number density of quarks inside the proton and σ_{BH} is their interaction cross section through the formation of a virtual black hole. Since the interaction arises from a dimension six operator, the amplitude has a factor $1/M_{\text{Pl}}^2$ and the cross section can be estimated as

$$\sigma_{BH} \sim \frac{m_p^2}{M_{\text{Pl}}^4}, \tag{7.31}$$

or, even simpler, the cross section is equal to the square of the gravitational radius. So we ultimately obtain that the proton lifetime with respect to the decay through the formation of a virtual quantum black hole is

$$\tau_p \sim \frac{M_{\text{Pl}}^4}{m_p^5}. \tag{7.32}$$

Inserting the usual Planck mass $M_{\text{Pl}} \sim 10^{19}$ GeV into Eq. (7.32), we predict that the proton lifetime is of the order 10^{45} yrs and there is no problem with the current experimental bound.

7.3 Models of Baryogenesis

In models with large extra dimensions, the fundamental gravity scale is $M_* \ll M_{Pl}$ and, replacing M_{Pl} with $M_* \sim 1$ TeV, leads to the quite short proton lifetime $\tau_p \sim 10^{-12}$ s. Hence, in order to avoid contradictions with the present experimental constraints, it was required (Adams et al. 2001) $M_* \geq 10^{16}$ GeV, much larger than the TeV scale. In our work (Bambi et al. 2007), the assumption was done that black holes lighter than the (effective) Planck mass must have zero electric and color charge and zero angular momentum—this statement is true in classical General Relativity and we made the conjecture that it holds in quantum gravity as well. If this is correct, the rates for proton-decay, neutron-antineutron oscillations, and lepton-violating rare decays would be suppressed to be below the existing experimental bounds even for large extra dimensions with TeV-scale gravity. So the disagreement with experimental limits could be avoided and, in principle, it is not excluded that a successful baryogenesis could proceed in the frameworks of TeV-scale gravity, and, in particular, in the minimal standard electroweak theory.

7.3.5 Spontaneous Baryogenesis

The spontaneous breaking of a $U(1)$-symmetry, which may be either related to the baryon number conservation or to mixture of the some quantum numbers that includes the baryonic one, would lead to the non-conservation of the baryonic number of physical particles. The baryon number of a Higgs-like field would be absorbed by the vacuum, so the total baryon number is formally conserved, but since only physical particles, which are excitations over the vacuum in the broken phase, are observed, reactions between these particles would proceed with a non-conservation of the number of real baryons. As a result, a baryon asymmetry may be generated in the broken symmetry phase. This kind of phenomenon may take place in some electroweak scenarios with several Higgs fields if their relative phase plays the role of Goldstone, or Nambu-Goldstone, boson, which appears after a spontaneous symmetry breaking (explained just below).

Generically, the model of spontaneous breaking of a global symmetry is described by a scalar field theory, with ϕ having the potential

$$U(\phi) = \lambda(|\phi|^2 - v^2)^2, \tag{7.33}$$

where v is a constant c-number. In the lowest energy state in this potential (vacuum), the field ϕ is non-vanishing, $\phi = v \exp(i\theta)$. A particular choice of the vacuum state among many degenerate ones, corresponding to different values of θ, results in a spontaneous symmetry breaking. The field $\theta(x)$ is called the Goldstone boson. If there is no explicit symmetry breaking but only the spontaneous one, the theory is invariant with respect to the transformation

$$\theta(x) \rightarrow \theta(x) + const. \tag{7.34}$$

This means that the field θ is massless. In other words, the curve where the potential $U(\phi)$ in Eq. (7.33) reaches its minimum is flat and θ can evolve along this curve without changing energy. If the bottom of the potential is tilted, so that the degeneracy in the potential energy of θ disappears, we speak about an explicit symmetry breaking (as, for instance, in the axion case). In this case, the θ-field typically acquires a non-vanishing mass and becomes a pseudo-Goldstone boson.

Let us consider the following toy model with the scalar field ϕ and two fermionic fields, "quarks" Q and "leptons" L. The theory is supposed to be invariant with respect to the "baryonic" $U(1)$-symmetry: $\phi \to \exp(i\alpha)\phi$, $Q \to \exp(i\alpha)Q$, and $L \to L$, where α is a constant phase. The corresponding Lagrangian has the form

$$\mathscr{L} = (\partial\phi)^2 - U(\phi) + i\bar{Q}\gamma^\mu\partial_\mu Q + i\bar{L}\gamma^\mu\partial_\mu L + (g\phi\bar{Q}L + h.c.), \quad (7.35)$$

where $U(\phi)$ is given in Eq. (7.33). In the spontaneously broken phase, when $\phi = v\exp(i\theta)$, the Lagrangian can be rewritten as

$$\mathscr{L} = v^2(\partial\theta)^2 - V(\theta) + i\bar{Q}\gamma^\mu\partial_\mu Q + i\bar{L}\gamma^\mu\partial_\mu L + \left[gv\exp(i\theta)\bar{Q}L + h.c.\right] + \cdots, \quad (7.36)$$

where the potential $V(\theta)$ describes a possible explicit symmetry breaking, which is not present in the original Lagrangian (7.35), and the radial degrees of freedom are supposed to be very heavy and are neglected. Indeed, one can study perturbations near the vacuum in the broken symmetry phase, introducing a new field ζ, the so-called radial excitations $\phi = (v + \zeta)\exp(i\theta)$. Evidently, the mass of ζ is $m_\zeta = 2\sqrt{\lambda}v$, so ζ is not excited at low energies.

Another representation of the Lagrangian (7.36) may be useful in consideration of the baryon asymmetry generation. Let us introduce the new quark field by the rotation $Q \to \exp(i\theta)Q$, so the Lagrangian turns into

$$\mathscr{L} = v^2(\partial\theta)^2 + \partial_\mu\theta J_B^\mu - V(\theta) + i\bar{Q}\gamma^\mu\partial_\mu Q + i\bar{L}\gamma^\mu\partial_\mu L + (gv\bar{Q}L + h.c.), \quad (7.37)$$

where $J_B^\mu = \bar{Q}\gamma^\mu Q$ is the baryonic currents of quarks. In this expression, the interaction of θ with the matter fields enters only linearly. It is imperative that the current J_B^μ is not conserved, otherwise the interaction term

$$\mathscr{L}_{int} = \partial_\mu\theta\, J_B^\mu \quad (7.38)$$

can be integrated away. This current is indeed non-conserved. Combining the equations of motion for Q and L, one sees that $\partial_\mu J_B^\mu = igv(\bar{L}Q - \bar{Q}L)$.

For the case of a homogeneous and only time-dependent field θ, the interaction Lagrangian (7.38) can be written as $\mathscr{L}_{int} = \dot{\theta}n_B$, where n_B is the density of the baryonic charge. One is thus tempted to identify $\dot{\theta}$ with the baryonic chemical potential, as it was done in Cohen and Kaplan (1987, 1988), Cohen et al. (1993). If this were so, the baryonic charge density would be non-zero even in thermal equilibrium, when the reaction rates are fast, while θ is not relaxed down to the dynamical equilibrium point at the minimum of the potential, where $\dot{\theta} = 0$. The charge density

7.3 Models of Baryogenesis

for small $\dot\theta$ would be equal to

$$n_B = \frac{1}{6} B_Q \dot\theta T^2, \qquad (7.39)$$

where B_Q is the baryonic charge of the quarks Q. However, this is not true, as can be seen immediately from the equation of motion for the θ-field (Dolgov and Freese 1995)

$$2v^2 \partial^2 \theta = -\partial_\mu J_B^\mu. \qquad (7.40)$$

In fact this equation is just the law of the total current conservation, $\partial_\mu J_{tot}^\mu = 0$, where J_{tot}^μ is the total baryonic current, which includes the contribution from the scalar field ϕ. Though the symmetry is spontaneously broken, the theory still "remembers" that it was symmetric. In the case of space-point independent $\theta = \theta(t)$, Eq. (7.40) is reduced to $2v^2 \ddot\theta = -\dot n_B$. It can be easily integrated, giving

$$\Delta n_B = -v^2 \Delta \dot\theta, \qquad (7.41)$$

which is evidently incompatible with Eq. (7.39). One should definitely trust Eq. (7.41) because this is simply the condition of the total current conservation, which is not disturbed by thermal corrections. Below, we will discuss in some detail why $\dot\theta$ cannot be interpreted as the baryonic chemical potential, and thus why Eq. (7.39) is incorrect, but let us first consider the generation of baryon asymmetry both in the pure Goldstone and pseudo-Goldstone cases. We have seen that in the Goldstone case the baryonic charge density is given by Eq. (7.41). The initial value of $\dot\theta$ is determined by inflation and depends on whether the symmetry was broken before the end of inflation or after that. We assume that the former is true, then the kinetic energy of the θ-field is given by $v^2 (\partial\theta)^2 \sim H_I^4$. This is the magnitude of quantum fluctuations in de Sitter spacetime, as it is described by the so-called Gibbons-Hawking temperature (Gibbons and Hawking 1977), $T_{GH} = H/(2\pi)$. So $\dot\theta \sim H_I^2/v$, where the Hubble parameter during inflation, H_I, can be found by matching the energy of the inflaton $\rho_{inf} \sim H_I^2 M_{Pl}^2$ and the thermal energy after reheating $\rho_{reh} \sim T_{reh}^4$. Comparing these expressions we find that

$$\eta \sim \frac{n_B}{T^3} \approx \frac{v T_{reh}}{M_{Pl}^2}. \qquad (7.42)$$

If the scale of the symmetry breaking v and the reheating temperature are not far from the Planck scale, the asymmetry would be large enough to explain the observed value, $\eta \approx 6 \cdot 10^{-10}$. However, a serious problem emerges in this scenario. It is known that all the regular classical motions during inflation are exponentially redshifted down to zero. The initial non-vanishing $\dot\theta$ came from quantum fluctuations at the inflationary stage. The characteristic size of the region with a definite sign of $\dot\theta$ is microscopically small, $l_B^{inf} \sim H^{-1}$, and even after the redshift $z_{reh} + 1 = T_{reh}/3$ K it remains much smaller than the size of baryonic domains now, $l_B > 10$ Mpc.

Let us now turn to the pseudo-Goldstone case, when θ has a non-vanishing potential $V(\theta) = \Lambda^4 \cos\theta$. If θ is close to the minimum of this potential, it can be approximated by the mass term, namely $V(\theta) \approx -1 + m^2 v^2 (\theta - \pi)^2/2$ with $m^2 = \Lambda^4/v^2$. The equation of motion for θ now acquires an extra term related to the potential force

$$v^2 \ddot{\theta} + 3H\dot{\theta} + V'(\theta) = \partial_\mu J_B^\mu . \tag{7.43}$$

Here we have also taken into account the Hubble friction term connected to the expansion of the Universe. We assume that initially θ is away from its equilibrium value $\theta_{eq} = \pi$. It is natural to assume that θ can be found anywhere in the interval $(0, 2\pi)$ with equal probability. During inflation, when $H \gg m$, the magnitude of θ remains practically constant due to the large friction term, $3H\dot{\theta}$. The region with a constant θ is exponentially inflated, $l_B \sim l_i \exp(Ht)$, and may be large enough to be bigger than the lower limit of the size of baryonic domains today. When inflation is over and the Hubble parameter falls below m, we can neglect the Hubble friction and the field θ starts oscillating in accordance with the equation

$$\ddot{\theta} + m^2 \theta = -\partial_\mu J_B^\mu / v^2 . \tag{7.44}$$

The oscillation of θ would produce both baryons and antibaryons, but with different number densities because the current j_B^μ is not conserved. To calculate this asymmetry, the following arguments have been used in the literature. The equation of motion for θ with the back reaction of the produced particles was assumed to be

$$\ddot{\theta} + m^2 \theta + \Gamma\dot{\theta} = 0 . \tag{7.45}$$

This equation has a solution correctly describing the decrease of the amplitude of θ due to production of particles, namely

$$\theta = \theta_i \exp(-\Gamma t/2) \cos(mt + \delta) . \tag{7.46}$$

Comparing Eqs. (7.44) and (7.45), one may conclude that

$$\partial_\mu J_B^\mu = v^2 \Gamma \dot{\theta} . \tag{7.47}$$

However, this identification is not correct (Dolgov and Freese 1995; Dolgov et al. 1997). It can easily be seen that, if Eq. (7.47) were true, then the energy of the produced particles would be larger than the energy of the parent field θ. This is impossible, of course. Indeed, if the expression (7.47) were correct, then the energy density of the produced baryons could be estimated as follows. The energy of each quark produced by the field oscillating with the frequency m is equal to $m/2$. The total number density of the produced quarks, $n_Q + n_{\bar{Q}}$, is larger than the density of the baryonic charge, $n_B = n_Q - n_{\bar{Q}}$. So the energy density of the produced baryons is larger than $m n_B$. From Eq. (7.47) it follows that n_B is linear in θ, while the energy density of the field θ is quadratic in θ. Thus in the limit of small θ the energy of

7.3 Models of Baryogenesis

the produced particles would be bigger than the energy of the field-creator. This contradicts the energy conservation and proves that the identification made above is wrong. In fact, the correct solution of the equation does not necessarily mean that the equation itself is correct. For example, one can describe the decaying field by the equation

$$\ddot{\theta} + (m - i\Gamma/2)^2\theta = 0. \tag{7.48}$$

This equation has the same solution (7.46), but does not permit to make the identification (7.47).

In Dolgov and Hansen (1999), Dolgov and Freese (1995), we have derived in the one loop approximation the equation of motion for θ with the account of the back reaction of the produced fermions. It is a non-local non-linear equation which, in the limit of small amplitudes of θ, has the same solution as Eqs. (7.45) and (7.48) but does not permit to make the wrong identification (7.47). The direct calculation of the particle production by the time-dependent field (7.46) gives the result (Dolgov et al. 1997)

$$n_B \sim \eta^2 \Gamma_{\Delta B}(\Delta\theta)^3, \tag{7.49}$$

where Γ is the width of the θ-decay with the non-conservation of the baryonic charge and $\Delta\theta$ is the difference between the initial and final values of θ. The asymmetry is proportional to the cube of the initial value of θ and not just to the first power, because the asymmetry oscillates as a function of time with alternating signs and thus the net effect appears due to a non-complete cancellation of the integral of the oscillating function with decreasing amplitude. The asymmetry in this case can roughly be estimated as

$$\eta = g^2(\Delta\theta)^3 \frac{v^2 m}{T^3}. \tag{7.50}$$

In this scenario, the size of the baryonic-antibaryonic domains, l_B, depends upon the model parameters and can be either larger than the present day horizon or much smaller, inside our visibility.

Let us now turn to the possibility of the interpretation of $\dot{\theta}$ as the baryonic chemical potential. It enters the Lagrangian as $\mathscr{L}_\theta = \dot{\theta} n_B$, exactly in the same way as a chemical potential should enter the Hamiltonian. However, from the relation between $\mathscr{L}$ and $\mathscr{H}$

$$\mathscr{H} = \frac{\partial \mathscr{L}}{\partial \dot{\phi}} \dot{\phi} - \mathscr{L} \tag{7.51}$$

follows that the contribution from $\mathscr{L}_\theta$ into the Hamiltonian formally vanishes. The Hamiltonian depends upon n_B through the canonical momentum, $P = \partial \mathscr{L}/\partial \dot{\theta} = 2v^2\dot{\theta} + n_B$. So from the kinetic term in the Lagrangian, $v^2(\partial\theta)^2$, one gets $\mathscr{H} = (P - n_B)^2/4v^2$. However, if the field θ is external (let us denote it now with the capital letter Θ), so that the Lagrangian does not contain its kinetic term, and Θ only comes there as $\dot{\Theta} n_B$, then we do not have any equation of motion for Θ: it is an external "constant" variable. In this case, the Hamiltonian would be $\mathscr{H} = -\dot{\Theta} n_B$ and this $\dot{\Theta}$ is the baryonic chemical potential. In this case, for sufficiently fast reactions, the baryonic charge density would be given by the expression (7.39).

However, for our dynamical field θ the equation of motion, which governs its behavior, does not permit θ to be an adiabatic variable that can change slowly with respect to the reactions with $\Delta B \neq 0$. A change of the baryonic charge implies a similar change in θ, so an equilibrium is never reached. In the pure Goldstone situation, this is seen of course from the equation of motion (7.40). For the pseudo-Goldstone case, the situation is slightly more complicated, but still the result is the same. Let us consider the Dirac equation for quarks in the presence of the θ-field

$$\left(i\gamma^\mu \partial_\mu - \dot{\theta}\right) Q = -gvL. \tag{7.52}$$

We neglected here a possible mass term, as it is not essential. In perturbation theory, one is tempted to neglect the right hand side of this equation because it is proportional to the small coupling constant g and to study the spectrum of the Dirac equation with vanishing right hand side. The dispersion relation for this equation is $E = p \pm \dot{\theta}$, where the signs $+$ and $-$ stand, respectively, for quarks and antiquarks. Thus energy levels of particles and antiparticles are shifted by $2\dot{\theta}$ and in equilibrium their number densities should be different. However, the point is that any change in the population numbers proceeds with the same speed as the change in θ or, in other words, the non-conservation of the current that can create a difference between Q and $\bar{Q}$ is proportional to the same coupling constant g entering the equation of motion (7.40) and governing the behavior of $\theta(t)$ in the Goldstone case. In the pseudo-Goldstone case, the variation of θ can be dominated by the potential term (7.43). Hence, it may change (oscillate) faster than just in the limit of vanishing potential (Goldstone limit) and one has even less ground to suppose that $\theta(t)$ is an adiabatic variable. In this case, the situation is worse than in the Goldstone case, because the rate of variation of the baryonic charge is much slower than the variation of θ and the system is even further from equilibrium.

It may be instructive to see how different fermion/antifermion levels are populated in the presence of the θ-field in the "rotated" fermion representation, $Q \to \exp(i\theta)Q$, when the Dirac equation has the form

$$i\gamma^\mu \partial_\mu Q = -gvL \exp(-i\theta). \tag{7.53}$$

This equation, in the limit of $g = 0$, has the same spectrum for particles and antiparticles, $E = p$, but the levels would be differently populated because the interaction term (in the right hand side) does not conserve energy. Assuming that $\theta(t)$ is a slowly varying function of time, we can write $\theta(t) \approx \dot{\theta}t$. Thus in the reactions with quarks their energy is increased by $\dot{\theta}$ in comparison with the energy of the participating particles, while the energy of antiquarks would decrease by the same amount. One sees from this example that the energies of particles and antiparticles are indeed getting different but the process of differentiation is proportional to the coupling g.

7.3.6 Baryogenesis by Condensed Scalar Baryons

Supersymmetric theories open new possibilities for baryogenesis. First, in high energy scale supersymmetric models the baryonic charge is not conserved. This usually happens at energies below the GUT scale. Second, there exist scalar partners of baryons with a non-zero baryonic charge, for example, superpartners of quarks (they are denoted χ in what follows). The potential for these fields generically has the so-called flat directions, along which the field can evolve without changing energy. This means, in particular, that the mass of χ is zero.

A massless scalar field is known to be infrared unstable in de Sitter background. Its vacuum expectation value is singular at $m = 0$ (Bunch and Davies 1978)

$$\langle \phi_m^2 \rangle = \frac{3H^4}{8\pi^2 m^2} . \tag{7.54}$$

If the field is strictly massless, then its fluctuations rise as $\langle \phi_0^2 \rangle = H^3 t / (2\pi)^2$, as it was argued in Vilenkin and Ford (1982), Linde (1982), (though see Dolgov and Pelliccia (2006) for a different result). If the mass is small but non-zero, the rise of the field terminates when its potential energy becomes equal to the kinetic one, $U(\phi) \sim H^4$. Wavelengths of quantum fluctuations are exponentially stretched up together with the expansion and during the inflationary stage "classical" condensates of light scalar fields can be developed. These condensates may store a baryonic charge (if the field, like e.g. χ, possesses it) and when inflation is over, the decay of χ would produce a baryon asymmetry. The picture is slightly more complicated by the following reasons. First, the field χ should not possess any conserved quantum number. The current conservation condition

$$D_\mu J^\mu = \partial_\mu J^\mu + 3H j_0 = 0 \tag{7.55}$$

makes any conserved current density vanish since $J^0 \sim \exp(-3Ht)$. Thus only colorless and electrically neutral combinations of the fields may condense. When, after some symmetry breaking, the flat directions become curved (e.g. the mass m_χ became non-zero) and the Universe expansion rate becomes smaller than the mass, $H < m_\chi$, the field can evolve down to the mechanical equilibrium point at $\chi = 0$. During this relaxation down to $\chi = 0$, the field χ can decay into quarks, most probably conserving the baryonic number B, and release the baryonic charge stored in the condensate into the baryonic charge of quarks. This is the basic idea of the Affleck-Dine scenario of baryogenesis (Affleck and Dine 1985). However, one should keep in mind that the baryonic charge is not accumulated in the amplitude of the scalar baryon field, but in its phase rotation analogous to a mechanical angular momentum, see below Eq. (7.59).

As a toy model possessing these properties, we consider the scalar baryon, χ, with the self-interaction potential

$$U_\lambda(\chi) = (\lambda/2)\left(2|\chi|^4 - \chi^4 - \chi^{4*}\right) = \lambda|\chi|^4\,(1 - \cos 4\theta)\,, \qquad (7.56)$$

where $\chi = |\chi|\exp(i\theta)$. There are four flat directions in this potential along the lines $\cos 4\theta = 1$ in the complex χ-plane. The potential breaks the symmetry with respect to the phase rotation $\chi \to \chi \exp(i\alpha)$. It means that the baryonic charge of χ is not conserved, as expected. In addition to the quartic potential (7.56), we add the following mass term

$$U_m(\chi) = m^2|\chi|^2\,[1 - \cos(2\theta + 2\alpha)]\,. \qquad (7.57)$$

Here α is some unknown phase. If $\alpha \neq 0$, C and CP are explicitly broken.

"Initially" (at inflation), χ is away from the origin, due to quantum fluctuations as discussed above, and when inflation is over, χ starts evolving down to the equilibrium point, $\chi = 0$, according to its equation of motion. For a homogeneous χ, the latter coincides with the equation of motion of a point-like particle in Newtonian mechanics

$$\ddot{\chi} + 3H\dot{\chi} + U'(\chi) = 0\,. \qquad (7.58)$$

The baryonic charge of χ is

$$B_\chi = \dot{\theta}|\chi|^2 \qquad (7.59)$$

and it is analogous to a mechanical angular momentum. When χ decays, its baryonic charge is transferred to that of quarks in a B-conserving process. Thus we can easily visualize the process without an explicit solution of the equations of motion.

For massless χ, the B-charge is accumulated in its "rotational" motion, induced by quantum fluctuations in a direction orthogonal to the valley. The space average value of the baryonic charge is evidently zero and, as a result, a globally charge symmetric universe is created. The size of the domain, l_B, with definite sign of the baryonic charge density, is determined by the size of the regions with a definite sign of $\dot{\theta}$. Normally, the size of the regions with definite $\dot{\theta}$ is microscopic and this leads to a very small l_B. However, if the Hubble parameter at inflation happens to be larger than the second derivative of the potential U_λ in the direction orthogonal to the valley, the field motion in the orthogonal direction is frozen during the exponential expansion and the size of the domains with a fixed value of B may be large enough.

The situation would be different if $m \neq 0$. In this case, the initial angular momentum or, which is the same, the initial baryonic charge of χ could be zero, but the rotational motion (or baryonic charge) may be created by a different direction of the valley at low χ. At large χ, the direction of the valley is determined by $U_\lambda(\chi)$, Eq. (7.56), while at small χ the quadratic part (7.57) dominates.

If the CP-odd phase in Eq. (7.57) is zero, namely $\alpha = 0$, but the flat direction of U_λ along which χ condenses is orthogonal to the flat directions of U_m, the field χ can rotate with 50 % probability clock-wise or anti-clockwise creating a baryonic or an antibaryonic universe. If inflation helps, such regions could be sufficiently large. This

7.3 Models of Baryogenesis

is an example of baryogenesis without an explicit C and CP violation and without the domain wall problem.

If the CP-odd phase α is small but non-vanishing, the rotation of χ when it approaches the m-valley proceeds with different probabilities in different directions. Hence, both baryonic and antibaryonic regions are possible with a dominance of one of them. Matter and antimatter domains may exist, but globally $B \neq 0$.

A very interesting picture appears if the field χ is coupled to the inflaton with the general renormalizable coupling (Dolgov and Silk 1993)

$$\mathscr{L}_{\chi\Phi} = \lambda |\chi|^2 (\Phi - \Phi_1)^2 \,. \tag{7.60}$$

In this case, the "gates" to the the valley can be open only for a short time, when the inflaton field Φ is close to Φ_1. The probability of penetration to the valley is thus small and χ can acquire a large baryonic charge condensate, giving large η, up to $\eta \sim 1$ only in a tiny fraction of space. This model would lead to a universe mostly having the normal small homogeneous baryon asymmetry $\eta = 6 \times 10^{-10}$, which could be created by one of the standard mechanisms described above, with relatively rare compact high-B regions. Depending upon the concrete model, the high-B regions may be symmetric with respect to baryons and antibaryons or dominated by one of them. This scenario is discussed in more detail in the next section.

7.4 Cosmological Antimatter

The prediction of antimatter is justly prescribed to Dirac (1928), who found on the basis of the Dirac equation the existence of a solution with an electric charge opposite to that of electron. He initially assumed that this "electron" with positive electric charge was the proton. At that time, physicists were rather reluctant to introduce new particles, in drastic contrast to the present days. However, Oppenheimer criticized such an interpretation, pointing out that in this case hydrogen would be very unstable. This led Dirac to conclude in 1931 that the "antielectron" was a *new* particle, the positron, with the same mass as the electron. Very soon after that, in 1933, Carl Anderson discovered positrons. Dirac received his Nobel prize immediately after this discovery, and Anderson got it three years later, in 1936.

In his Nobel lecture "Theory of electrons and positrons", on December 12, 1933, Dirac said about antimatter in the Universe: "It is quite possible that...these stars being built up mainly of positrons and negative protons. In fact, there may be half the stars of each kind. The two kinds of stars would both show exactly the same spectra, and there would be no way of distinguishing them by present astronomical methods". However, we see in what follows that there are such ways and we can conclude whether a star is made of antimatter making astronomical observations from Earth.

It is surprising that in 1898, 30 years before Dirac and 1 year after the discovery of electron by Thomson, Arthur Schuster (another British physicist) conjectured that there might be another sign of electricity. He called it antimatter and he supposed that there might be entire solar systems made of antimatter and indistinguishable from ours (Schuster 1898). Schuster made the wild guess that matter and antimatter were capable of annihilating and producing energy. It happened to be ingenious and true. He also believed that matter and antimatter were gravitationally repulsive, since, according to his assumption, antimatter particles had negative mass and gravitationally repelled from matter particles. Two such objects in close contact would have vanishing mass! As we know now, this is not the case and matter and antimatter have mutual gravitational attraction.

Presently, the common belief is that the Universe is populated only by matter and meager antimatter is of secondary origin. Nevertheless, despite quite strong observational restrictions on possible existence of antimatter domains and antimatter objects, as it is discussed at the beginning of this chapter, it is still not excluded that antimatter may be abundant in the Universe and even in the Galaxy, not too far from us. For this reason, there is an active search for cosmic antimatter by several instruments: BESS (Balloon Borne Experiment with Superconducting Solenoidal Spectrometer) (Sasaki 2008), PAMELA (Payload for Antimatter Matter Exploration and Light-nuclei Astrophysics) (Boezio 2008; Picozza and Morselli 2008), and AMS (AntiMatter Spectrometer or Alpha Magnetic Spectrometer) (Alcaraz 1999). Some new detectors are under discussion now. Existing and new missions could either eliminate or strongly diminish the remaining room for antimatter objects such as antistars. If we are lucky, we might discover antisolar systems, as was envisaged by Schuster and Dirac.

There are many theoretical models leading to an abundant creation of antimatter in the Universe. For example, if CP invariance is broken spontaneously (Lee 1974), the Universe would be equally populated by matter and antimatter. At least two scenarios of baryogenesis, discussed in Sects. 7.3.5 and 7.3.6, are quite favorable for the creation of antiworlds. Unfortunately, in their simple forms such models are strongly restricted either by astronomical observation or BBN and CMB data.

However, not all is that bad. We briefly describe below a scenario for which the existing bounds are not applicable or much weaker, and which allows for abundant antimatter even in the Galaxy, almost at hand. The scenario is based on the model described in the previous subsection. As it is mentioned there, we need to introduce the coupling (7.60) to close the gates to the flat directions for most of the time of the χ evolution. To make the size of antimatter domains or stellar like objects with high (anti)baryon density astronomically large, the gates should be open during inflation but not too far from its end to make these objects inside the present day horizon. This is the only tuning of the model. We also slightly modify the original χ potential (7.56) and (7.57) including the well known Coleman-Weinberg correction (Coleman and Weinberg 1973), the last term in Eq. (7.61) below, which arises as a result of summation of one-loop diagrams in scalar field theory with quartic interaction

$$U_\chi(\chi) = [(m_\chi^2 \chi^2 + h.c.) + \lambda_\chi(\chi^4 + |\chi|^4) + h.c.)] + \lambda_2 |\chi|^4 \ln \frac{|\chi|^2}{\sigma^2}, \quad (7.61)$$

7.4 Cosmological Antimatter

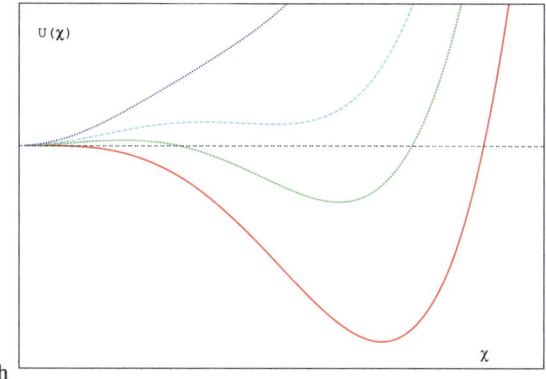

Fig. 7.1 Behavior of $U_\chi(\chi)$ for different values of $m^2_{eff}(t)$

The coupling (7.60) in addition to this potential acts as a positive time-dependent mass and thus it almost always keeps the gate to the valleys closed, except for a short period when Φ is near Φ_1.

There is a small chance for χ to reach a high value and create a large baryon asymmetry. The behavior of the potential $U_\chi(\chi) + U_{int}(\chi, \Phi)$ for different values of the effective mass $m_{eff}(t) = \lambda[\Phi(t) - \Phi_1]^2$ is shown in Fig. 7.1. The potential evolves down from the upper to the lower curve, reaching the latter when $\Phi = \Phi_1$, and then the potential returns back practically to the higher curve, when Φ drops below Φ_1. Correspondingly, the field χ rolls down toward the deeper minimum, oscillates there following the evolution of the minimum, rolls back to the origin, and starts rotating around it, as shown in Fig. 7.2.

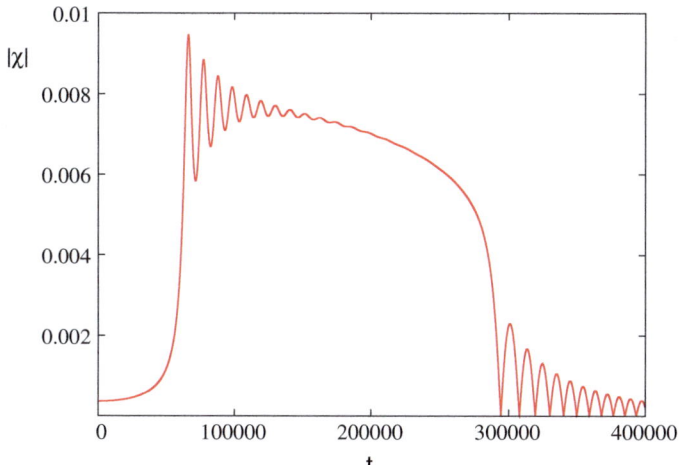

Fig. 7.2 Evolution of $|\chi|$ with time

Since the inflaton opens the gate to the deeper minimum only for a short time, the probability for χ to reach a high value is low, so in most of the space baryogenesis creates a normal tiny baryon asymmetry, but in some bubbles occupying a small fraction of the whole volume, baryon asymmetry may be huge.

After the QCD phase transition, the contrast in baryonic charge density transformed into perturbations of the energy/mass density and the bubbles with high B could form primordial black holes or compact stellar-like objects. The mass distribution of these high-B bubbles has practically the model independent shape

$$\frac{dN}{dM} = C_M \exp\left[-\gamma \ln^2 \frac{(M - M_1)^2}{M_0^2}\right], \qquad (7.62)$$

with the model dependent parameters C_M, γ, M_1, and M_0. The values of the parameters can be adjusted in such a way that superheavy black holes formed at the tail of this distribution are abundant enough to be present in every large galaxy and in some small ones. Such heavy primordial black holes could be the seeds for galaxy formation. There is no satisfactory mechanism for the creation of superheavy black holes in the framework of standard physics, but the mechanism considered here can successfully fulfill this task.

The evolved chemistry in early quasars can be explained, at least to some extend, by a more efficient production of metals during the BBN, due to a much higher value of $\eta = n_B/n_\gamma$. The standard BBN essentially stops at the synthesis of ^{4}He because of the very low value of η. However, in the model considered here η can be much larger than the canonical value, even close or larger than 1. In such conditions, much heavier primordial elements can be produced (Nakamura et al. 2015; Matsuura et al. 2004, 2005; Matsuura 2007). It is possible that some stars initiated with more metals than the usual ones and today they look older than they are, because their age is evaluated by the standard nuclear chronology. They might even look older than the Universe. This model readily explains the creation of several types of astronomical objects in the very early Universe that are difficult or even impossible to create otherwise.

Recently, several stars have been discovered in the Galaxy with an unexpectedly high age. Employing thorium and uranium abundances in comparison with each other and with several stable elements, the age of the metal-poor halo star BD+17° 3248 has been estimated 13.8 ± 4 Gyr in Cowan (2002). For comparison, the estimated age of the inner halo of the Galaxy is 11.4 ± 0.7 Gyr (Kalirai 2012). The age of the star HE 1523-0901 in the Galactic Halo has been estimated to be about 13.2 Gyr (Frebel et al. 2007). In that work, many different chronometers, such as the U/Th, U/Ir, Th/Eu and Th/Os ratios have been employed for the first time to measure the star age. Most puzzlingly, it is probably the determination of the age of the metal deficient high velocity subgiant in the solar neighborhood HD 140283, which seems to be 14.46 ± 0.31 Gyr old (Bond et al. 2013). The central value of the age exceeds the Universe age by two standard deviations, if $H_0 = 67.3$ km/s/Mpc, and if H_0 is larger, $H_0 = 74$ km/s/Mpc, as is shown by the direct astronomical determination, the star would be older than the Universe by more than six standard deviations.

7.4 Cosmological Antimatter

Galaxies at high redshifts, $z \sim 10$, cannot be observed with normal optical telescopes, which are not sensitive enough for such distant objects. Fortunately, natural gravitational lens telescopes allow to see them if the "telescope" happens to be on the light ray from the galaxy to the terrestrial observers. In such a way, a galaxy at $z \approx 9.6$ was discovered in Zheng (2012). The galaxy was formed when the Universe was about 500 Myr old. Even more striking, a galaxy at $z \approx 11$ has been observed (Coe 2013), which was formed before the Universe age was 0.41 Gyr (or even shorter with larger H_0).

Quoting Melia (2014): "Observations with WFC3/IR on the Hubble Space Telescope and the use of gravitational lensing techniques have facilitated the discovery of galaxies as far back as $z \sim 10 - 12$, a truly remarkable achievement. However, this rapid emergence of high z galaxies, barely 200 Myr after the transition from Population III star formation to Population II, appears to be in conflict with the standard view of how the early Universe evolved. This problem is very reminiscent of the better known (and probably related) premature appearance of supermassive black holes at $z \sim 6$. It is difficult to understand how $10^9 \, M_\odot$ black holes appeared so quickly after the big bang without invoking non-standard accretion physics and the formation of massive seeds, both of which are not seen in the local Universe". A quasar with $z = 7.085$ has been discovered (Mortlock 2011), i.e. it was formed at $t < 0.75$ Gyr. Its luminosity is $6.3 \cdot 10^{13} \, L_\odot$ and its mass is $2 \cdot 10^9 \, M_\odot$. Quasars are supposed to be supermassive black holes and their formation in such a short time looks problematic with conventional mechanisms.

There are strong indications that every large galaxy, as well as some relatively small ones, contain a central supermassive black hole. The mass of the black hole may be larger than 10 billions $M_\odot$ in giant elliptical and compact lenticular galaxies and about a few millions $M_\odot$ in spiral galaxies like the Milky Way. The mass of one of these object is typically 0.1 % of the mass of the stellar bulge of the host galaxy, but some galaxies seem to have huge black holes: for instance, NGC 1277 has a supermassive black hole of $1.7 \cdot 10^{10} \, M_\odot$, which corresponds to 60 % of the bulge mass of the host galaxy (Bosch et al. 2012). Another interesting example is the possible existence of a supermassive black hole in the ultra-compact dwarf galaxy M60-UCD1 (Seth 2014) with a mass of about 20 millions $M_\odot$, namely 15 % the total mass of the object. According to the conclusion of the authors, the high black hole mass and mass fraction suggest that M60-UCD1 is the stripped nucleus of a galaxy. On the other hand, the authors observed that "M60-UCD1's stellar mass is consistent with its luminosity, implying that many other ultra-compact dwarf galaxies may also host supermassive black holes. This suggests a substantial population of previously unnoticed supermassive black holes". These facts create serious problems for the standard scenario of formation of central supermassive black holes by accretion of matter in the central part of a galaxy. An inverted picture looks more plausible, when first a supermassive black hole formed and then attracted matter serving as seeds for the subsequent galaxy formation. The recent discovery (Strader 2013) of an ultra-compact dwarf galaxy older than 10 Gyr, enriched with metals, and probably with a massive black hole at its center seems to be at odds with the standard model as well. The dynamical mass of this galaxy is $2 \cdot 10^8 \, M_\odot$ and its radius is $R \sim 24$ pc, so the galaxy density is extremely high. There is a variable central X-ray source with luminosity $L_X \sim 10^{38}$ erg/s, which may be an active galactic nucleus associated with a massive black hole or a low-mass X-ray binary.

Observations of high redshift gamma ray bursts (GBRs) also indicate a high abundance of supernova at large redshifts, if GBRs are very early supernovae. The highest redshift of the observed GBR is 9.4 (Cucchiara 2011) and there are a few more GBRs with smaller but still high redshifts. The necessary star formation rate to explain these early GBRs is at odds with the canonical star formation theory.

A natural byproduct of this scenario is an abundant antimatter creation because the χ rotation can be in both directions, as explained in Sect. 7.3.6. Since the antiobjects formed in this way would be usually compact, the upper bounds on their abundance discussed in Sect. 7.1 are very much weakened, including the bounds from the BBN and the CMB, as well. The phenomenology of similar antimatter objects is discussed in Bambi and Dolgov (2007), Blinnikov et al. (2015).

Problems

7.1 How can the charge asymmetry generated in heavy particle decays vanish in equilibrium? It is stated in the literature that the inverse decay does the job. However, one can see that it is not so because, using CPT, one finds:

$$\begin{aligned}
\Gamma_{\bar{q}\bar{q} \to \bar{X}} &= (1 + \Delta_q)\Gamma_q , & \Gamma_{ql \to \bar{X}} &= (1 - \Delta_l)\Gamma_l , \\
\Gamma_{qq \to X} &= (1 - \Delta_q)\Gamma_q , & \Gamma_{\bar{q}\bar{l} \to X} &= (1 + \Delta_l)\Gamma_l .
\end{aligned} \quad (7.63)$$

Thus direct and inverse decays produce the same sign of baryon asymmetry!

References

F.C. Adams, G.L. Kane, M. Mbonye, M.J. Perry, Int. J. Mod. Phys. A **16**, 2399 (2001) [hep-ph/0009154]
P.A.R. Ade et al., Planck collaboration. Astron. Astrophys. **571**, A16 (2014) arXiv:1303.5076 [astro-ph.CO]
S.L. Adler, Phys. Rev. **177**, 2426 (1969)
I. Affleck, M. Dine, Nucl. Phys. B **249**, 361 (1985)
J. Alcaraz et al., AMS collaboration. Phys. Lett. B **461**, 387 (1999) [hep-ex/0002048]
J. Ambjorn, T. Askgaard, H. Porter, M.E. Shaposhnikov, Nucl. Phys. B **353**, 346 (1991)
N. Arkani-Hamed, S. Dimopoulos, G.R. Dvali, Phys. Lett. B **429**, 263 (1998) [hep-ph/9803315]
P.B. Arnold, L.D. McLerran, Phys. Rev. D **36**, 581 (1987)
K.R.S. Balaji, T. Biswas, R.H. Brandenberger, D. London, Phys. Lett. B **595**, 22 (2004) [hep-ph/0403014]
K.R.S. Balaji, T. Biswas, R.H. Brandenberger, D. London, Phys. Rev. D **72**, 056005 (2005) [hep-ph/0506013]
P. von Ballmoos, Hyperfine Interact. **228**, 91 (2014). arXiv:1401.7258 [astro-ph.HE]

References

C. Bambi, A.D. Dolgov, Nucl. Phys. B **784**, 132 (2007) [astro-ph/0702350]
C. Bambi, A.D. Dolgov, K. Freese, Nucl. Phys. B **763**, 91 (2007) [hep-ph/0606321]
J.S. Bell, R. Jackiw, Nuovo Cim. A **60**, 47 (1969)
S.I. Blinnikov, A.D. Dolgov, K.A. Postnov. arXiv:1409.5736 [astro-ph.HE]
M. Boezio et al., PAMELA collaboration. J. Phys. Conf. Ser. **110**, 062002 (2008)
H.E. Bond, E.P. Nelan, D.A. VandenBerg, G.H. Schaefer, D. Harmer, Astrophys. J. **765**, L12 (2013). arXiv:1302.3180 [astro-ph.SR]
R.C.E. van den Bosch, K. Gebhardt, K. Gultekin, G. van de Ven, A. van der Wel, J.L. Walsh, Nature **491**, 729 (2012). arXiv:1211.6429 [astro-ph.CO]
W. Buchmuller, P. Di Bari, M. Plumacher, New J. Phys. **6**, 105 (2004) [hep-ph/0406014]
W. Buchmuller, P. Di Bari, M. Plumacher, Annals Phys. **315**, 305 (2005a) [hep-ph/0401240]
W. Buchmuller, R.D. Peccei, T. Yanagida, Ann. Rev. Nucl. Part. Sci. **55**, 311 (2005b) [hep-ph/0502169]
T.S. Bunch, P.C.W. Davies, Proc. Roy. Soc. Lond. A **360**, 117 (1978)
B.J. Carr, S.W. Hawking, Mon. Not. Roy. Astron. Soc. **168**, 399 (1974)
M.C. Chen, (2007) hep-ph/0703087
J.H. Christenson, J.W. Cronin, V.L. Fitch, R. Turlay, Phys. Rev. Lett. **13**, 138 (1964)
S.R. Coleman, E.J. Weinberg, Phys. Rev. D **7**, 1888 (1973)
D. Coe et al., Astrophys. J. **762**, 32 (2013). arXiv:1211.3663 [astro-ph.CO]
A.G. Cohen, D.B. Kaplan, Phys. Lett. B **199**, 251 (1987)
A.G. Cohen, D.B. Kaplan, Nucl. Phys. B **308**, 913 (1988)
A.G. Cohen, D.B. Kaplan, A.E. Nelson, Ann. Rev. Nucl. Part. Sci. **43**, 27 (1993) [hep-ph/9302210]
A.G. Cohen, A. De Rujula, S.L. Glashow, Astrophys. J. **495**, 539 (1998) [astro-ph/9707087]
R. Cooke, M. Pettini, R.A. Jorgenson, M.T. Murphy, C.C. Steidel, Astrophys. J. **781**, 31 (2014). arXiv:1308.3240 [astro-ph.CO]
J.J. Cowan et al., Astrophys. J. **572**, 861 (2002) [astro-ph/0202429]
A. Cucchiara et al., Astrophys. J. **736**, 7 (2011). arXiv:1105.4915 [astro-ph.CO]
M. Dine, A. Kusenko, Rev. Mod. Phys. **76**, 1 (2003) [hep-ph/0303065]
P.A.M. Dirac, Proc. Roy. Soc. Lond. A **117**, 610 (1928)
A.D. Dolgov, Sov. Phys. JETP **52**, 169 (1980) [Zh. Eksp. Teor. Fiz. **79**, 337 (1980)]
A.D. Dolgov, Phys. Rev. D **24**, 1042 (1981)
A.D. Dolgov, Phys. Rept. **222**, 309 (1992)
A. Dolgov, J. Silk, Phys. Rev. D **47**, 4244 (1993)
A.D. Dolgov, (1997) hep-ph/9707419
A.D. Dolgov, (2005). hep-ph/0511213
A.D. Dolgov, Phys. Atom. Nucl. 73, 588 (2010). arXiv:0903.4318 [hep-ph]
A. Dolgov, K. Freese, Phys. Rev. D **51**, 2693 (1995) [hep-ph/9410346]
A.D. Dolgov, S.H. Hansen, Nucl. Phys. B **548**, 408 (1999) [hep-ph/9810428]
A.D. Dolgov, P.D. Naselsky, I.D. Novikov (2000). astro-ph/0009407
A. Dolgov, D.N. Pelliccia, Nucl. Phys. B **734**, 208 (2006) [hep-th/0502197]
A. Dolgov, K. Freese, R. Rangarajan, M. Srednicki, Phys. Rev. D **56**, 6155 (1997) [hep-ph/9610405]
R. Duperray, B. Baret, D. Maurin, G. Boudoul, A. Barrau, L. Derome, K. Protasov, M. Buenerd, Phys. Rev. D **71**, 083013 (2005) [astro-ph/0503544]
G.R. Dvali, G. Gabadadze, Phys. Lett. B **460**, 47 (1999) [hep-ph/9904221]
J.R. Ellis, J.E. Kim, D.V. Nanopoulos, Phys. Lett. B **145**, 181 (1984)
A. Frebel, N. Christlieb, J.E. Norris, C. Thom, T.C. Beers, J. Rhee, Astrophys. J. **660**, L117 (2007) [astro-ph/0703414]
V.P. Frolov, I.D. Novikov, *Black Hole Physics: Basic Concepts and New Developments* (Kluwer Academic, Dordrecht, 1998)
M. Fukugita, S. Yanagita, Phys. Lett. B **174**, 45 (1986)
M. Gell-Mann, P. Ramond, R. Slansky, in *Supergravity*, eds. by D. Freedman, P. Van Niuwenhuizen (North Holland, Amsterdam, 1979)

S.L. Glashow, in *1979 Cargése Lectures in Physics—Quarks and Leptons*, eds. by M. Lévy et al. (Plenum, New York, 1980)
G.W. Gibbons, S.W. Hawking, Phys. Rev. D **15**, 2738 (1977)
S.W. Hawking, Nature **248**, 30 (1974)
S.W. Hawking, Commun. Math. Phys. **43**, 199 (1975) [Erratum-ibid. **46**, 206 (1976)]
G. 't Hooft, Phys. Rev. Lett. **37**, 8 (1976a)
G. 't Hooft, Phys. Rev. D **14**, 3432 (1976b) [Erratum-ibid. D **18**, 2199 (1978)]
Y.I. Izotov, G. Stasinska, N.G. Guseva, Astron. Astrophys. **558**, A57 (2013). arXiv:1308.2100 [astro-ph.CO]
J. Kalirai, Nature **486**, 90 (2012). arXiv:1205.6802 [astro-ph.GA]
A. Kalweit, *Light hyper- and anti-nuclei production at the LHC measured with ALICE* (2014). https://indico.cern.ch/event/328442
M.Y. Khlopov, A.D. Linde, Phys. Lett. B **138**, 265 (1984)
D.A. Kirzhnits, JETP Lett. **15**, 529 (1972) [Pisma Zh. Eksp. Teor. Fiz. **15**, 745 (1972)]
D.A. Kirzhnits, A.D. Linde, Phys. Lett. B **42**, 471 (1972)
F.R. Klinkhamer, N.S. Manton, Phys. Rev. D **30**, 2212 (1984)
M. Kobayashi, T. Maskawa, Prog. Theor. Phys. **49**, 652 (1973)
V.A. Kuzmin, V.A. Rubakov, M.E. Shaposhnikov, Phys. Lett. B **155**, 36 (1985)
L.D. Landau, Nucl. Phys. **3**, 127 (1957)
T.D. Lee, Phys. Rept. **9**, 143 (1974)
T.D. Lee, C.N. Yang, Phys. Rev. **104**, 254 (1956)
A.D. Linde, Phys. Lett. B **116**, 335 (1982)
G. Luders, Kong. Dan. Vid. Sel. Mat. Fys. Med. **28N5**, 1 (1954)
G. Luders, Annals Phys. 2, 1 (1957) [Annals Phys. 281, 1004 (2000)]
N.S. Manton, Phys. Rev. D **28**, 2019 (1983)
N. Martin, ALICE collaboration. J. Phys. Conf. Ser. **455**, 012007 (2013)
S. Matsuura, A.D. Dolgov, S. Nagataki, K. Sato, Prog. Theor. Phys. **112**, 971 (2004) [astro-ph/0405459]
S. Matsuura, S.I. Fujimoto, M.A. Hashimoto, K. Sato, Phys. Rev. D **75**, 068302 (2007). 0704.0635 [astro-ph]
S. Matsuura, S.I. Fujimoto, S. Nishimura, M.A. Hashimoto, K. Sato, Phys. Rev. D **72**, 123505 (2005) [astro-ph/0507439]
F. Melia, Astron. J. **147**, 120 (2014). arXiv:1403.0908 [astro-ph.CO]
P. Minkowski, Phys. Lett. B **67**, 421 (1977)
D.J. Mortlock et al., Nature **474**, 616 (2011). arXiv:1106.6088 [astro-ph.CO]
R. Nakamura, M.a. Hashimoto, S.i. Fujimoto, N. Nishimura, K. Sato. arXiv:1007.0466 [astro-ph.CO]
K.A. Olive et al., Particle data group collaboration. Chin. Phys. C **38**, 090001 (2014)
R. Omnes, Phys. Rev. Lett. **23**, 38 (1969)
R. Omnes, Phys. Rev. D **1**, 723 (1970)
D.N. Page, Phys. Rev. D **13**, 198 (1976)
E.A. Paschos, Pramana **62**, 359 (2004) [hep-ph/0308261]
W. Pauli, in *Niels Bohr and the Development of Physics* (McGraw-Hill, New York, 1955)
P. Picozza, A. Morselli, J. Phys. Conf. Ser. **120**, 042004 (2008)
A. Riotto, M. Trodden, Ann. Rev. Nucl. Part. Sci. **49**, 35 (1999) [hep-ph/9901362]
V.A. Rubakov, M.E. Shaposhnikov, Usp. Fiz. Nauk **166**, 493 (1996) [Phys. Usp. **39**, 461 (1996)] [hep-ph/9603208]
A.D. Sakharov, Pisma Zh. Eksp. Teor. Fiz. **5**, 32 (1967) [JETP Lett. **5**, 24 (1967)]
M. Sasaki et al., Adv. Space Res. **42**, 450 (2008)
A. Schuster, Nature **58**, 367 (1898)
J.S. Schwinger, Phys. Rev. **82**, 914 (1951)
A. Seth et al., Nature **513**, 398 (2014). arXiv:1409.4769 [astro-ph.GA]
G. Steigman, Ann. Rev. Astron. Astrophys. **14**, 339 (1976)

References

G. Steigman, JCAP 0810, 001 (2008). 0808.1122 [astro-ph]

J. Strader et al., Astrophys. J. **775**, L6 (2013). arXiv:1307.7707 [astro-ph.CO]

A. Vilenkin, L.H. Ford, Phys. Rev. D **26**, 1231 (1982)

C.S. Wu, E. Ambler, R.W. Hayward, D.D. Hoppes, R.P. Hudson, Phys. Rev. **105**, 1413 (1957)

T. Yanagida, in *Proceedings of the Workshop on Unified Theories and Baryon Number in the Universe*, eds. by O. Sawada, A. Sugamoto (KEK, Tsukuba, Japan, 1979)

Y.B. Zeldovich, Pisma. Zh. Eksp. Teor. Fiz. **24**, 29 (1976a)

Y.B. Zeldovich, Phys. Lett. A **59**, 254 (1976b)

Y.B. Zeldovich, Zh Eksp, Teor. Fiz. **72**, 18 (1977)

Y.B. Zeldovich, I.Y. Kobzarev, L.B. Okun, Zh. Eksp. Teor. Fiz. **67**, 3 (1974) [Sov. Phys. JETP **40**, 1 (1974)]

W. Zheng et al., Nature **489**, 406 (2012). arXiv:1204.2305 [astro-ph.CO]

Chapter 8
Big Bang Nuclesynthesis

The big bang nucleosynthesis (BBN) is the production of light elements (deuterium, helium-3, helium-4, and lithium-7) in the early Universe, a few minutes after the big bang, when the temperature of the primordial plasma was between 1 MeV and 10 keV. Heavier elements were not produced for lack of time, because the temperature dropped down quickly, and those in the Universe today are mainly produced in stellar processes and supernovae explosions.

The calculation of the primordial abundances of light elements is governed by the Friedmann equations, the matter content of the Standard Model of particle physics, the properties of elementary particles, especially those of neutrinos, and nuclear reaction rates that can be measured in laboratory. Two very important features of the BBN are the freeze-out of the weak interactions, which occurred when the Universe temperature was around 1 MeV, and the deuterium bottleneck, which took place at the temperature $T \approx 70$ keV and determined the onset of the synthesis of light elements. In this framework, the abundances of deuterium (X_D), helium-3 (X_3), helium-4 (Y_4), and lithium-7 (X_7) are computed as functions of the baryon to photon number ratio, η, at the end of the nucleosynthesis (after the $e^- e^+$ annihilation). The latter is the only free parameter in the standard BBN. From the comparison of theoretical predictions and observational data, it is possible to infer η. While the primordial abundances of light elements could potentially vary by many orders of magnitude for different η, the fact that all they converge to the values which correspond to the same η in the range $10^{-10} - 10^{-9}$ supports the hypothesis of the same mechanism of their production. This is seen as a great success of the theoretical framework, and as a milestone of the Standard Model of cosmology. Today, we can obtain a more accurate and precise measurement of η from the study of the anisotropies of the CMB and its value turns out to be consistent with that inferred from the BBN. The BBN provided also the first indication for the existence of non-baryonic dark matter. From the estimate of η, we can determine the contemporary value of the ratio of the energy density of baryons to the critical energy density, Ω_B. We find $\Omega_B \approx 0.05$, which is substantially lower than the contribution from the gravitating matter, $\Omega_m \approx 0.30$, inferred from the rotational curves of galaxies, cosmological large scale structure, baryon acoustic oscillations, and more. The current interpretation is that most of the matter in the contemporary Universe may be made of weakly interacting massive particles beyond the Minimal Standard Model (MSM) of particle physics.

8.1 Light Elements in the Universe

As shown in Sect. 1.2.2, nuclear reactions inside stars have produced only a small fraction of the helium-4 present today in the Universe. Most of the helium-4 was produced in the early Universe during the BBN. However, if we want to test the predictions of the Standard Model of cosmology, it is necessary to be able to measure the primordial abundance of helium-4, as well as those of the other light elements. This is not easy, in general, because the primordially produced elements are contaminated by their production or destruction in the course of the subsequent cosmological evolution till the present time. The strategy is to find specific astrophysical sites in which the light element abundances may be close to the primordial ones. The present situation can be summarized as follows:

1. Deuterium can be easily burnt in many stellar processes, so the measurement of its abundance can be likely considered as a lower limit of the primordial one. The current best approach to estimate X_D seems to be the observation of quasar absorption line systems, namely high redshift clouds seen along the line of sight of a quasar. The light emitted by the quasar is partially absorbed by the cloud and the measurement of the deuterium absorption line provides an estimate of the deuterium abundance in the high redshift cloud. The latter is thought to be close to the deuterium abundance after the BBN. Current astronomical observations suggest a deuterium to hydrogen number density ratio at the level of $2-4 \cdot 10^{-5}$ (Kirkman et al. 2003). However, the dispersion among different measurements is usually not consistent with the errors in the single measurements. This may be prescribed to peculiar velocities in the cloud.
2. The abundance of helium-4 is traditionally expressed in terms of mass abundance and indicated by the symbol Y_4, while for the other light elements it is common to use the nucleus number abundance. Helium-4 is a very stable nucleus and it is produced inside stars, so the measurement of its abundance can be assumed to be an upper bound of the primordial one. The abundance of helium-4 is usually measured in the so called HII extragalactic regions, which are the extragalactic regions where most of the hydrogen is neutral. The abundance of helium-4 in these regions is a monotonically increasing function of the metallicity, namely the abundance of elements heavier than helium-4. The metallicity reflects the activity of stellar processes: lower is the metallicity, less nuclear reactions have taken place. The primordial abundance of helium-4 is inferred by extrapolating the measurements of low metallicity HII extragalactic regions to zero metallicity. Current observations suggest $Y_4 \approx 0.25$ (Aver et al. 2013; Izotov et al. 2007; Olive and Skillman 2004).

3. Measurements of the primordial abundances of helium-3 and lithium-7 are more uncertain. Helium-3 measurements only come from high metallicity regions, and for this reason they cannot be compared with the BBN predictions. In the case of lithium-7, the measurements are based on observations of metal-poor stars in the spheroid of our Galaxy. Stars with a metallicity lower than about 0.03 of the Solar metallicity have similar values of X_7, which is usually interpreted as an indication in favor of the hypothesis that their lithium-7 abundance is close to the primordial one. Current estimates suggest $X_7 \approx 2 \cdot 10^{-10}$ (Melendez et al. 2010). However, these measurements may be affected by systematic effects related to the model adopted to describe the stellar atmosphere.

8.2 Freeze-Out of Weak Interactions

In the Standard Model of cosmology, the Universe expands and the temperature of the primordial plasma drops down. According to the Standard Model of particle physics, when the temperature of the plasma is around 10 MeV, the relativistic particles in thermal equilibrium are photons, electrons, positrons, and all the neutrinos and antineutrinos. The non-relativistic particles in thermal equilibrium are protons and neutrons, and there are probably out-of-equilibrium dark matter particles. One could expect that the energy density of non-relativistic particles is exponentially suppressed, as it is argued in Sect. 5.2.1. However, this is not so because, due to the cosmological baryon asymmetry, the proton and neutron densities are much higher than the equilibrium ones with vanishing chemical potentials. The same is true for the dark matter particles, which were frozen at much higher temperatures, see Sect. 5.3.2. Nevertheless, the energy density of non-relativistic species is negligible at the BBN, because today the ratio between the energy density of non-relativistic and relativistic matter is approximately 10^4. Hence at the redshift $z \sim 10^9$ this ratio is about 10^{-5}. The energy density of the Universe is essentially given by that of relativistic particles, namely $\rho \approx \rho_{rel}$. Hence the first Friedmann equation can be written as

$$H = \sqrt{\frac{8\pi}{3M_{Pl}^2}\rho_{rel}} = \sqrt{\frac{8\pi}{3M_{Pl}^2}\frac{\pi^2}{30}g_*T^4}, \tag{8.1}$$

where g_* is the effective number of degrees of freedom and T is the photon temperature. For $T = 10$ MeV, the Standard Model of particle physics predicts

$$g_*^{SM}(T = 10 \text{ MeV}) = g_\gamma + \frac{7}{8}\left[g_{e^-} + g_{e^+} + N_F(g_\nu + g_{\bar{\nu}})\right]$$

$$= 2 + \frac{7}{8}[2 + 2 + 3(1+1)] = 10.75, \tag{8.2}$$

where N_F indicates the lepton generation number, assuming that all the neutrino species are light. To constrain the possible existence of additional degrees of freedom coming from new physics, say the existence of light neutrinos/antineutrinos from a hypothetical forth generation or the presence of some light particle beyond the Standard Model, we can write

$$g_* = g_*^{\text{SM}} + \frac{7}{4}\Delta N_\nu , \qquad (8.3)$$

where ΔN_ν would indicate the effective number of extra neutrino species, but it is commonly used to denote any kind of particles keeping this notation. ΔN_ν can be treated as a free parameter capable of altering the primordial abundances of light elements and to be determined from the comparison of theoretical predictions and observational data.

Though the energy densities of protons and neutrons are negligible in the Friedmann equations, their presence in the primordial plasma is very important for the light element formation. Neutrons and protons remain in thermal (kinetic) equilibrium with the primordial plasma through the following processes

$$\begin{aligned} n + e^+ &\leftrightarrow p + \bar{\nu}_e , \\ n + \nu_e &\leftrightarrow p + e^- , \\ n &\to p + e^- + \bar{\nu}_e . \end{aligned} \qquad (8.4)$$

It is known from the charge neutrality of the cosmological plasma that the chemical potential of electrons and positrons at the BBN temperatures is negligible. As for neutrinos, observations allow for a rather high value, $\mu_\nu/T \lesssim 0.1$, but this demands rather exotic scenarios of leptogenesis. So in what follows we neglect the chemical potential of relativistic particles. Thus the equilibrium with respect to the above processes implies $\mu_n \approx \mu_p$. From Eq. (5.10), we find that the ratio of the neutron to proton number densities is

$$\frac{n_n}{n_p} \approx e^{\Delta m/T} , \qquad (8.5)$$

where $\Delta m = m_n - m_p = 1.29$ MeV is the mass difference between neutrons and protons. Neutrons and protons are maintained in thermal equilibrium by the weak interactions (8.4). The reaction rate of the first two processes is $\Gamma \sim \sigma n v$, where $\sigma \sim G_F^2 T^2$ is the cross section (see Sect. 3.5), $n \sim T^3$ is the number density of the relativistic particles involved, and $v \sim 1$ is the particle's velocity. The temperature of freeze-out of the weak interactions, T_f, is found by equalling the reaction rate $\Gamma \sim G_F^2 T^5$ with the expansion rate of the Universe H. We find

$$\left.\frac{\Gamma}{H}\right|_{T=T_f} = 1 \approx \sqrt{\frac{10.75}{g_*}} \left(\frac{T_f}{0.8 \text{ MeV}}\right) . \qquad (8.6)$$

8.2 Freeze-Out of Weak Interactions

The assumption of the instant freeze-out at a well-defined temperature is clearly an approximation, but it works fairly well. More precise calculations demand the use of the kinetic equation specified in Sect. 5.2.2 with the collision integrals for all the three processes (8.4). In the limit of zero electron mass and Boltzmann statistics, the collision integrals can be taken analytically and the resulting first order differential equation for the n/p ratio allows for an approximate analytical solution and, of course, it can be easily solved numerically. The precise calculations of the n/p ratio can be, and are done, numerically without any approximation.

The neutron to baryon number density before the freeze-out of the weak interactions (8.4) is given by

$$X_n(t < t_f) = \frac{n_n}{n_p + n_n} = \frac{1}{n_p/n_n + 1} = \frac{1}{e^{\Delta m/T} + 1}, \tag{8.7}$$

which is a function of the temperature only. After the freeze-out of the weak interactions, neutrons and protons are not in thermal equilibrium any more. The processes in (8.4) are thus frozen, except for the free neutron decay. For this reason, at later times we have

$$X_n(t > t_f) = \frac{e^{-t/\tau_n}}{e^{\Delta m/T_f} + 1}. \tag{8.8}$$

We note that a variation of the freeze-out temperature, for instance due to new physics, can quite significantly change X_n, because $\Delta m/T_f \sim 1$.

8.3 Electron-Positron Annihilation

When the temperature of the Universe drops below $m_e = 0.5$ MeV, electrons and positrons become non-relativistic and their abundances start decreasing exponentially. As described in Sect. 5.2.3, the $e^{\pm}$ energy density is transferred to the cosmological plasma. However, at these temperatures neutrinos are not in thermal equilibrium any more, so the electron-positron annihilation only heats the photon gas. So after the e^+e^- annihilation the temperature of photons becomes higher than the temperature of neutrinos. The entropy conservation allows to find the relation between the photon and the neutrino temperatures (5.37)

$$T_\gamma = \left(\frac{11}{4}\right)^{1/3} T_\nu, \tag{8.9}$$

as already discussed in Sect. 5.3.1. The temperature of neutrinos is lower simply because neutrinos cannot benefit of the electron-positron annihilation. The expansion rate of the Universe now requires the following g_* factor

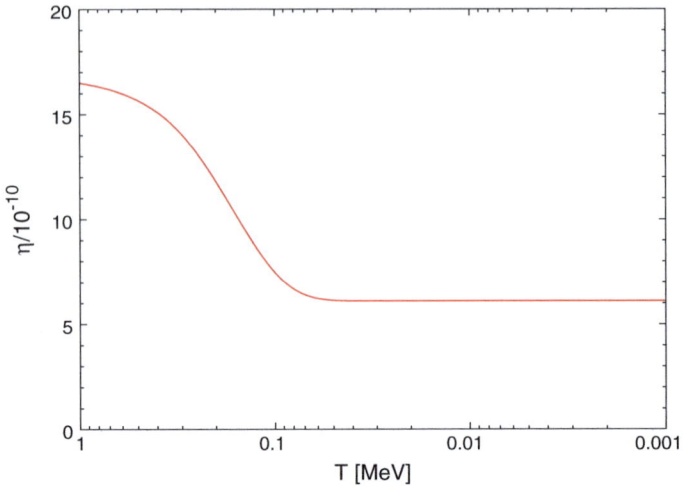

Fig. 8.1 Baryon to photon number ratio as a function of the temperature of the primordial plasma for the case $\eta = 6.1 \cdot 10^{-10}$

$$g_*^{SM}(T = 100 \text{ keV}) = g_\gamma + \frac{7}{8}\left(\frac{4}{11}\right)^{4/3} N_F (g_\nu + g_{\bar{\nu}}) = 3.36. \quad (8.10)$$

We also note that this alters the baryon to photon number ratio η. Figure 8.1 shows the result of the numerical calculations of the evolution of baryon to photon number ratio as a function of the temperature of the Universe.

8.4 Deuterium Bottleneck

The first step to produce light elements is the synthesis of deuterium. When the reaction $n + p \leftrightarrow D + \gamma$ is in thermal equilibrium, the relation between the chemical potential of protons, neutrons, and deuterium is

$$\mu_n + \mu_p = \mu_D + \mu_\gamma = \mu_D, \quad (8.11)$$

and the deuterium number density is

$$n_D = g_D \left(\frac{m_D T}{2\pi}\right)^{3/2} e^{-m_D/T} e^{(\mu_n + \mu_p)/T}. \quad (8.12)$$

8.4 Deuterium Bottleneck

Equation (8.12) can be rewritten as

$$n_D = n_n n_p \frac{n_D}{n_n n_p} = n_n n_p \frac{g_D}{g_n g_p} \left(\frac{m_D}{m_n m_p}\right)^{3/2} \left(\frac{2\pi}{T}\right)^{3/2} e^{(m_n + m_n - m_D)/T}$$

$$= n_n n_p \frac{g_D}{g_n g_p} 2^{3/2} \left(\frac{2\pi}{m_N T}\right)^{3/2} e^{W/T}, \tag{8.13}$$

where $m_N \approx m_n \approx m_p \approx m_D/2$ is the nucleon mass and $W = m_n + m_n - m_D = 2.225$ MeV is the deuterium binding energy. The deuterium to baryon number density ratio is thus

$$X_D = \frac{n_D}{n_B} = \frac{n_n}{n_B} \frac{n_p}{n_B} n_B \frac{g_D}{g_n g_p} 2^{3/2} \left(\frac{2\pi}{m_N T}\right)^{3/2} e^{W/T}, \tag{8.14}$$

where n_B is the baryon number density. If we use the neutron to baryon number density ratio, $X_n = n_n/n_B$, and the proton to baryon number density ratio, $X_p = n_p/n_B$, we can write X_D as

$$X_D = X_n X_p \eta n_\gamma \frac{g_D}{g_n g_p} 2^{3/2} \left(\frac{2\pi}{m_N T}\right)^{3/2} e^{W/T}. \tag{8.15}$$

Here $\eta = n_B/n_\gamma$ is the baryon to photon number density ratio, which is the only free parameter in the standard BBN. Since $g_D = 3$, $g_n = g_p = 2$, and $n_\gamma \approx 0.24 T^3$, Eq. (8.15) becomes

$$X_D \approx 8 X_n X_p \eta \left(\frac{T}{m_N}\right)^{3/2} e^{W/T}. \tag{8.16}$$

The synthesis of deuterium effectively starts when its abundance is not negligible any more. At first approximation, we can demand that in Eq. (8.16) $X_D \sim X_n \sim X_p \sim 1$. This determines the temperature of the production of deuterium

$$T_D \approx \frac{W}{-\ln \eta - \frac{3}{2} \ln \frac{T_D}{m_N}}. \tag{8.17}$$

For $\eta \approx 6 \cdot 10^{-10}$, the temperature is $T_D \approx 70$ keV. This is roughly the correct value, as can be checked by a comparison with numerical calculations. We note that $T_D \ll W$; that is, the production of deuterium becomes possible only when the temperature of the plasma is significantly lower than the deuterium binding energy. The reason is that $\eta \ll 1$, namely there are many photons with respect to baryons, and therefore even when the temperature of the plasma is lower than the deuterium binding energy, there are still many photons with energy higher than W and they can thus destroy deuterium nuclei. It is necessary to wait that the photon temperature becomes much lower than W, so the photons with such an energy would be sufficiently rare. At that point, the production of deuterium turns to be efficient.

8.5 Primordial Nucleosynthesis

The production of light elements can only start at the temperature T_D, when the abundance of deuterium is not negligible any more. The synthesis of deuterium is indeed the necessary first step for that of heavier elements and therefore the BBN is inhibited by the difficulty to form deuterium. At lower temperatures, energetic photons in the primordial plasma are not abundant any more, and the deuterium synthesis becomes efficient. However, deuterium is immediately burnt to form heavier elements. At first approximation, we can assume that all the neutrons survived till the deuterium synthesis eventually form helium-4, while only a very small fraction of them goes to form other nuclei. Within this approximation, the primordial abundance of helium-4 is

$$Y_4 = \frac{2n_n}{n_n + n_p}\bigg|_{T=T_D} \approx 0.25. \tag{8.18}$$

Helium-4 is very stable, being a doubly magic nucleus with two protons and two neutrons. Moreover, the production of heavier elements during the BBN is difficult because of lack of time (the temperature of the Universe is dropping down quickly) and the absence of stable nuclei with atomic number between 5 and 8. For instance, two nuclei of helium-4 may produce a nucleus of beryllium-8, but the latter is unstable

$$^4\text{He} +\,^4\text{He} \rightarrow\,^8\text{Be} \rightarrow\,^4\text{He} +\,^4\text{He}. \tag{8.19}$$

Helium-5 produced by the reaction

$$^4\text{He} + n \rightarrow\,^5\text{He} \tag{8.20}$$

is also very unstable, and its lifetime is $\tau \sim 10^{-23}$ s. Helium-3 is not very abundant, so the reaction

$$^4\text{He} +\,^3\text{He} \rightarrow\,^7\text{Be} + \gamma \tag{8.21}$$

$$^7\text{Be} \rightarrow\,^7\text{Li} + e^+ + \nu_e \tag{8.22}$$

producing lithium-7 cannot be efficient.

Except for helium-4, the abundances of the other light elements (deuterium, helium-3, lithium-7) has to be computed numerically. At the end, one obtains the primordial abundances as functions of the baryon to photon ratio η, shown in Fig. 8.2. The primordial abundance of helium-4 rather weakly depends upon η, only logarithmically. On the contrary, deuterium is exponentially sensitive to η and before the accurate CMB data became available, measurements of the deuterium abundance were the best method to determine η, that is why deuterium got the name "baryometer".

The fact that all the observations are consistent with the baryon to photon ratio at the level of $\eta \sim 5 - 7 \cdot 10^{-10}$ is a great success of the standard BBN and a milestone

8.5 Primordial Nucleosynthesis

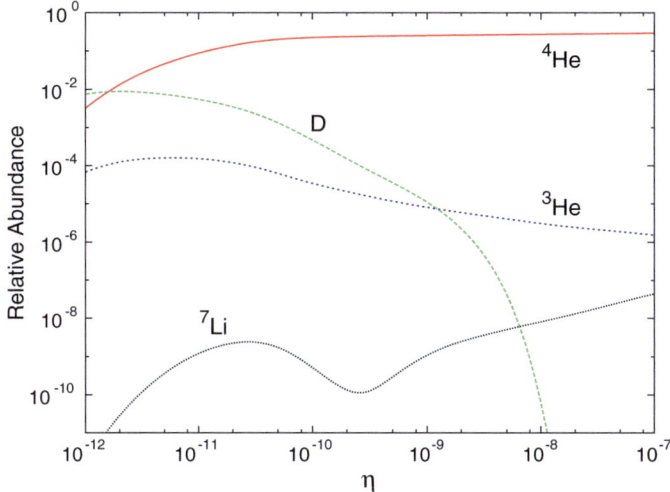

Fig. 8.2 Relative abundances of helium-4, deuterium, helium-3, and lithium-7 as functions of the baryon to photon number ratio η. The comparison of theoretical predictions and observations of the abundances of primordial elements requires $\eta \sim 5 - 7 \cdot 10^{-10}$. This value is consistent with more accurate measurements by the CMB, $\eta = 6.1 \cdot 10^{-10}$

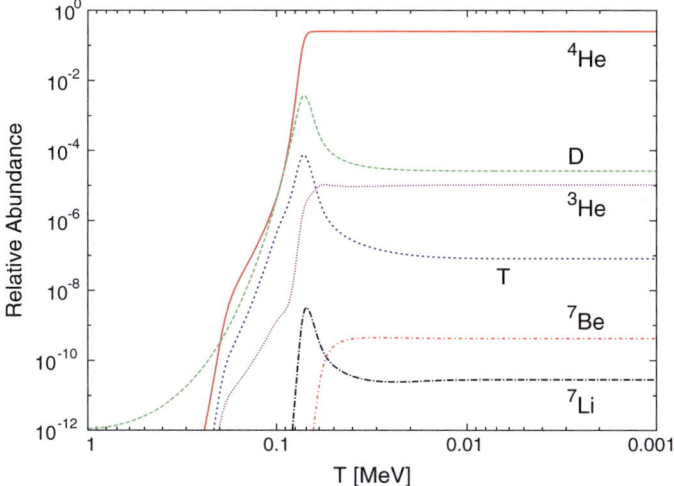

Fig. 8.3 Relative abundances of helium-4, deuterium, helium-3, tritium, beryllium-7, and lithium-7 as functions of the temperature of the primordial plasma for $\eta = 6.1 \cdot 10^{-10}$. Tritium and beryllium-7 are unstable and eventually decay, respectively, into helium-3 and lithium-7

of the Standard Model of cosmology. Today, CMB data provide a more accurate measurement of η: the accepted value is $\eta = 6.1 \cdot 10^{-10}$ (Olive et al. 2014) and this is consistent with the estimate of η from the BBN.

Figure 8.3 shows the abundances of deuterium, tritium, helium-3, helium-4, lithium-7, and beryllium-7 as functions of the temperature of the Universe. As we can see, the first element to be produced is the deuterium, and only when the latter is relatively abundant the production of the other elements is possible. Tritium and beryllium-7 are unstable and decay, respectively, into helium-3 and lithium-7. We note that Fig. 8.3 shows the evolution of the primordial abundances for $\eta = 6.1 \cdot 10^{-10}$. For different values of η, there may be some qualitative differences. For instance, with $\eta = 6.1 \cdot 10^{-10}$ most of the primordial lithium-7 comes from the decay of beryllium-7. With a slightly lower value of η, for instance $\eta = 2 \cdot 10^{-10}$, the contribution from the decay of beryllium-7 to the final abundance of lithium-7 is negligible.

8.6 Baryon Abundance

From the BBN measurement of η, it is possible to determine the present day ratio of the baryon energy density to critical energy density, $\Omega_B^0 = \rho_B^0/\rho_c^0$. The contemporary baryon energy density ρ_B^0 is given by

$$\rho_B^0 = m_N n_B^0 = m_N \eta n_\gamma^0 , \qquad (8.23)$$

where $n_\gamma^0 \approx 400$ photons/cm^3 is the present day number density of CMB photons. We assume that there have been no influx of energy into the CMB from the BBN epoch till today and thus $\eta = n_B^{BBN}/n_\gamma^{BBN} = n_B^0/n_\gamma^0$.

The absence of extra heating is expected within the Standard Model of cosmology, but it may not be true in the presence of new physics. An analysis of observational data shows no signs of that, so one can put rather restrictive bounds on possible new physics.

As follows from the first Friedmann equation, the value of the critical density today can be written as $\rho_c^0 = 3 M_{Pl}^2 H_0^2/(8\pi)$ and therefore

$$\Omega_B^0 h_0^2 = \frac{8\pi}{M_{Pl}^2} \frac{m_N \eta n_\gamma^0}{30,000 \left(\frac{\text{km}}{\text{s·Mpc}}\right)^2} , \qquad (8.24)$$

where we have used Eq. (1.8). Adopting the value $\eta = 6 \cdot 10^{-10}$, we find $\Omega_B^0 h_0^2 \approx 0.02$. The value of h_0 inferred from different observations, namely the CMB Planck data, the study of type Ia supernovae, and the traditional astronomical measurements, is around 0.7. There is some tension between different measurements at the level of less than 10%.

8.6 Baryon Abundance

Using these results, we conclude that the amount of baryonic matter in the contemporary Universe is $\Omega_B^0 \approx 0.04$. This is significantly lower than the amount of all the clustered gravitating matter, $\Omega_m^0 \approx 0.30$, which is found from multiple independent astronomical data: study of the rotational curves of galaxies, gravitational lensing, abundances of galactic clusters, angular fluctuations of the CMB temperature, baryon acoustic oscillations and other data on the large scale structure of the Universe.

8.7 Constraints on New Physics

The BBN is a powerful tool to constrain new physics. Sometimes it is even called a vacuum cleaner eliminating extension of the MSM. Theoretical predictions of the primordial abundances of light elements depend on the magnitude of the cosmological parameters and, in particular, on the expansion rate of the Universe and on the reaction rates in particle and nuclear physics. New physics can easily alter the final result.

The BBN has provided the first indications in support for the existence of only three generations of fermions and in this sense made an essential contribution in favor of the MSM. If there were a fourth generation and the neutrino associated to the new charged lepton were light, in the sense that it would have been relativistic at the beginning of the BBN, the effective number of degrees of freedom would be ($\Delta N_\nu = 1$)

$$g_*(T = 10 \text{ MeV}) = 10.75 + \frac{7}{4}\Delta N_\nu = 12.5 . \tag{8.25}$$

This would alter the expansion rate of the Universe and thus change the freeze-out temperature of the weak reactions of $p \rightarrow n$ transformations (8.4), see Eq. (8.6). A fourth generation with a light neutrino would increase T_f, so the neutron abundance X_n would be higher, and eventually the BBN would produce a larger amount of helium-4. Figure 8.4 shows the theoretical prediction of the abundance of primordial helium-4 Y_4 as a function of the number of neutrinos N_ν. From the measurement of the primordial abundance of helium-4, we can argue that there are only three generations of fermions, with the caveat that such a conclusion assumes that the neutrino of the possible new generation is light and with properties similar to those of the three generations of the Standard Model (Cyburt et al. 2005). Still there are indications both from the BBN and the CMB that the effective number of the neutrino species is larger than three, though it is compatible with three at one or two sigma level.

We can use the same ideas to constrain the abundances of other light particles predicted in theories beyond the MSM of particle physics. If these particles interact very weakly with ordinary matter, their presence only alters the expansion rate of the Universe through a contribution to ΔN_ν. Since the baryon to photon ratio can be determined today with a very good precision from the angular fluctuations of the CMB, the comparison between theoretical predictions and observations can be used to constrain ΔN_ν.

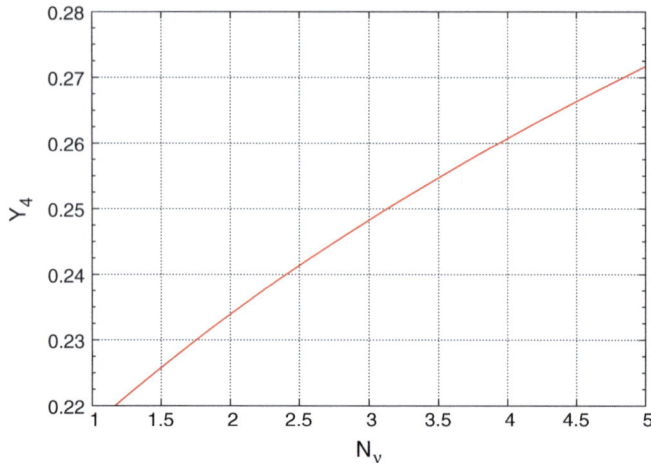

Fig. 8.4 Theoretical prediction of the abundance of primordial helium-4 Y_4 as a function of the number of neutrinos N_ν for the case $\eta = 6.1 \cdot 10^{-10}$

The expansion rate of the Universe during the BBN epoch may also deviate from the one expected from General Relativity in the case of alternative theories of gravity or in models with extra dimensions. For instance, in the case of one extra dimension compactified on a circle, the effective 4-dimensional first Friedmann equation takes the form (Cline et al. 1999)

$$H^2 = \frac{8\pi}{3M_{\text{Pl}}^2}\rho\left(1 + \frac{\rho}{2\sigma}\right), \tag{8.26}$$

where ρ is the energy density of ordinary matter on the brane, while σ is the brane tension. Here the study of the primordial abundances of the light elements provide a lower bound to σ.

The list of models that can be constrained, or even ruled out, by the BBN is long and any new scenario beyond the Standard Model of particle physics or beyond General Relativity must pass the BBN test to be seriously taken in consideration as a viable framework. The BBN has also been used to constrain possible time variations of fundamental constants (expected in many extensions of General Relativity), lepton asymmetries (introducing non-vanishing lepton chemical potentials), new particles that may decay and destroy the nuclei produced during the BBN, inhomogeneities during the BBN, etc. For a review, see e.g. Dolgov (2002); Malaney and Mathews (1993).

Problems

8.1 What is the effect on the abundance of primordial helium-4 in the following cases?

(a) The lifetime of neutrons is longer/shorter.
(b) The tauon neutrino has a mass $m_{\nu_\tau} \approx 10$ MeV.
(c) The deuterium binding energy is higher/lower.

References

E. Aver, K.A. Olive, R.L. Porter, E.D. Skillman, JCAP **1311**, 017 (2013). arXiv:1309.0047 [astro-ph.CO]
J.M. Cline, C. Grojean, G. Servant, Phys. Rev. Lett. **83**, 4245 (1999) [hep-ph/9906523]
R.H. Cyburt, B.D. Fields, K.A. Olive, E. Skillman, Astropart. Phys. **23**, 313 (2005) [astro-ph/0408033]
A.D. Dolgov, Phys. Rept. **370**, 333 (2002) [hep-ph/0202122]
Y.I. Izotov, T.X. Thuan, G. Stasinska, Astrophys. J. **662**, 15 (2007) [astro-ph/0702072]
D. Kirkman, D. Tytler, N. Suzuki, J.M. O'Meara, D. Lubin, Astrophys. J. Suppl. **149**, 1 (2003) [astro-ph/0302006]
R.A. Malaney, G.J. Mathews, Phys. Rept. **229**, 145 (1993)
J. Melendez, L. Casagrande, I. Ramirez, M. Asplund, W. Schuster, Astron. Astrophys. **515**, L3 (2010). arXiv:1005.2944 [astro-ph.SR]
K.A. Olive, E.D. Skillman, Astrophys. J. **617**, 29 (2004) [astro-ph/0405588]
K.A. Olive et al., Particle data group collaboration. Chin. Phys. C **38**, 090001 (2014)

Chapter 9
Dark Matter

The fact that a significant fraction of the matter in the contemporary Universe cannot be inside stars was already noted by Fritz Zwicky in 1933. Using the virial theorem, Zwicky estimated the total mass of the Coma cluster from the motion of some of its galaxies. He measured a mass much higher than that obtained from the brightness of the galaxies. Subsequent studies of galaxies and galaxy clusters confirmed that the main contribution to their masses was done by some invisible matter. This was the first indication that most of the matter in the Universe was not in the form of ordinary baryonic matter. However, at that time the cosmological fraction of the energy density of matter, Ω_m, was very poorly known.

Now we know that the BBN data fix the present day fraction of the baryon energy density at the level of $\Omega_B \approx 0.05$, see Chap. 8, while the total energy density of non-relativistic matter is approximately five times larger, $\Omega_m \approx 0.3$, as it is found from the analysis of the large scale structure of the Universe and the angular fluctuations of the CMB. These data provide very strong support to the idea that most of the matter in the Universe is not the usual baryonic one.

At the beginning of the 1990s, it was believed that $\Omega_{tot} < 1$ and that the Universe was open, in contradiction with the inflationary prediction of a 3D flat universe. There were even attempts to modify the inflationary scenario in such a way that it could naturally lead to $\Omega_{tot} < 1$, but they were not particularly successful. On the other hand, there was an accumulation of data indicating the inconsistencies of the open universe model with low Ω_{tot}. In particular, the calculated Universe age in such models was smaller than the estimates obtained by the nuclear chronology and by the ages of old stellar clusters. All such problems were eliminated after the discovery of the accelerating expansion of the Universe. Now it is established that the Universe is practically flat with $\Omega_{tot} = 1$, where the necessary 0.7 comes from the contribution of a quite mysterious form of energy, currently called dark energy.

More details on this topic can be found in Bergstrom (2000), which provides an extended review on the observational evidence of dark matter and possible detection methods.

9.1 Observational Evidence

The amount of matter in a galaxy can be inferred from the study of the rotational velocity curve of the gas clouds around the galaxy. The velocity v as a function of the radial distance from the galactic center r can be measured from the Doppler shift of spectral lines. In Newtonian mechanics, we have

$$\frac{v(r)^2}{r} = \frac{G_N M(r)}{r^2} \Rightarrow v(r) = \sqrt{\frac{G_N M(r)}{r}}, \qquad (9.1)$$

where $M(r)$ is the total mass within the radius r. If the stars provided the main contribution to the mass of a galaxy, at large radii, beyond the visible galaxy, one should expect $v \sim r^{-1/2}$. However, this is not what we observe: at large radii $v \sim const$, which implies $M(r) \sim r$, namely a galaxy extends to larger radii with respect to that is seen by optical observations.

Systematic and accurate measurements of galactic rotation curves started in the 1970s (Freeman 1970) with spiral galaxies. Spiral galaxies are a class of galaxies consisting of a central bulge and a thin disk. In the case of spiral galaxies, we find that v increases linearly at small radii until it reaches a typical value of about 200 km/s, and then it remains constant. On the contrary, the surface luminosity of the disk falls off exponentially. Today we know the rotational curves of thousands of galaxies. The measurements suggest the existence of a dark matter halo surrounding every galaxy and with the mass about ten times larger than the mass of the visible stars in the disk. It is worth noting that other types of galaxies seem also to be dark matter dominated with even larger fraction of dark matter. For instance, this is the case of dwarf spiral galaxies. Their rotational curve continues rising well beyond the radius of the luminous disk. Figure 9.1 shows the rotational curve of the galaxy M33, which belongs to the Local Group. v does not reach a constant value but continues rising. The dark matter contribution to the total galaxy mass is higher than in the case of normal spiral galaxies. Strictly speaking, these methods only measure local density inhomogeneities. Moreover, the observational identification of the dark matter halo is difficult. Eventually, an estimate of the cosmological fraction of the matter density suggests $\Omega_m \approx 0.2 - 0.4$. The study of virial velocities in galaxy clusters demonstrates the same features and provides similar results.

The measurement of Ω_m inferred from the study of galactic rotation curves is supported by a combination of independent results. The study of the supernovae Ia at high redshift provides the strongest evidence for the present accelerating expansion of the Universe. If we assume that the Universe is only made of non-relativistic matter and with a non-vanishing cosmological constant, these data can be used to constrain the allowed area on the $(\Omega_\Lambda, \Omega_m)$ plane. The analysis of the CMB anisotropies leads to the conclusion that the Universe is almost flat, namely $\Omega_\Lambda + \Omega_m \approx 1$. The combination of supernovae and CMB data leads to $\Omega_m \approx 0.30$.

All the above methods can provide an estimate of the total gravitating matter in the Universe. They do not really tell us anything about its nature. The stellar

9.1 Observational Evidence

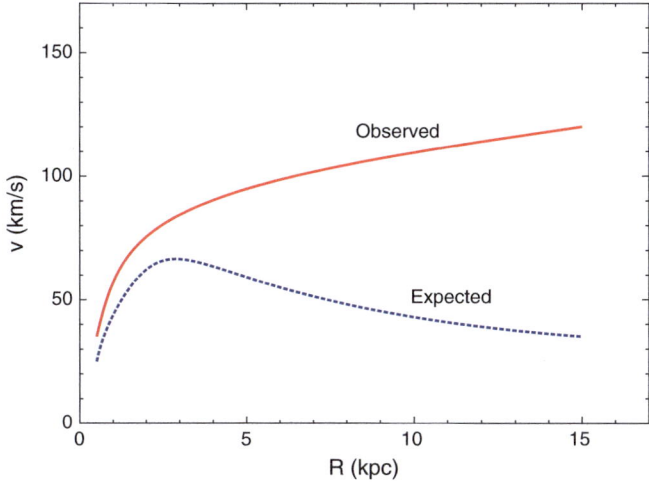

Fig. 9.1 Sketch of the observed HI rotation *curve* of the dwarf galaxy M33 (*red solid line*) and of that expected from the stellar distribution (*blue dashed line*)

contribution is $\Omega_{stars} \sim 0.003 - 0.01$, which is thus a small fraction of Ω_m. A much larger contribution to the mass can be made by the gas in the interstellar medium or non-luminous bodies like planets. As shown in Chap. 8, the study of the primordial abundances of light elements requires that the fraction of baryons is $\Omega_B \approx 0.05$ while any other form of gravitating matter does not make any contribution into it.

While there are scenarios in which only a fraction of the total amount of baryons were available for the nucleosynthesis in the early Universe, they definitively require exotic mechanisms. The most natural interpretation is that most of the dark matter is not made of baryons. Non-baryonic dark matter is also required to explain the formation of large scale structures (see Chap. 12). The combination of measurements of the CMB temperature fluctuations, which probe very large scales, and measurements of the galaxy power spectrum, which probe smaller scales, proves that non-baryonic matter is necessary to explain observational data, because baryons were locked in with photons until recombination, which prevented a quick growth of fluctuations.

The most convincing evidence for the existence of non-baryonic dark matter probably comes from the Bullet Cluster (Clowe et al. 2004). This is a system consisting of a subcluster that passed through a cluster about 150 Myr ago. The key-point is that the collision between the two clusters seems to have caused a separation of the dark matter component from the baryonic one. Observations show that stars, baryonic matter in the form of gas, and dark matter have different collision properties and seem to rule out the possibility of explaining dark matter as a modification of gravity at kpc scale. Using optical observations, we can study the distributions of the stars, which are not strongly affected by the cluster collision. X-ray measurements track the distribution of the hot gas, which represents the main component of the baryonic matter. Because of electromagnetic interactions among the particles of the gas, the cluster collision made the baryonic matter concentrate at the center of the system. Lastly, gravitational lensing studies map the distribution of gravitating matter. Observations show that most of the mass in the cluster was not affected by the collision. The interpretation of these observations is that most of the mass consists of weakly interacting dark matter and, unlike the galaxy curves, it is independent of possible modifications of Newton's law at kpc scales. We note that in the past there was a controversial issue concerning the initial infall velocity of the clusters, which seemed to be beyond that expected within the Standard Cosmological Model. If so, a modification of gravity might have been necessary. However, this tension seems to be solved by more recent studies (Lage and Farrar 2015).

9.2 Dark Matter Candidates

After it was realized that most matter in the Universe is not luminous, astronomical observations focused on the search for objects such as black holes, neutron stars, faint old white dwarfs, planets, and similar bodies, collectively called massive astrophysical compact halo objects (MACHOs). In the 1970s, the BBN studies pointed out the discrepancy between baryonic matter, $\Omega_B \approx 0.05$, and gravitating matter inferred by dynamical methods, $\Omega_m \approx 0.2 - 0.4$. While there could be scenarios in which only a small fraction of the total baryons in the Universe were available at the BBN, they definitively require quite exotic mechanisms. Astronomical surveys for gravitational microlensing attempting to find MACHOs (Afonso 2003; Alcock 2000; Tisserand 2007) have succeeded in the discovery of such objects with masses of the order of the Solar mass, but their amount is very small, and definitively too small to make all the necessary invisible matter. The current data on the abundance of MACHOs in the Galactic Halo are shown in Fig. 9.2.

Dark matter candidates can be grouped into three classes, namely cold dark matter (CDM), warm dark matter (WDM), and hot dark matter (HDM). The key-point to belong to one or another group is the distance that the particle travelled during the Universe history. By definition, CDM particles have the free-streaming length much shorter than the typical size of a protogalaxy. WDM candidates have the free-streaming length of the order the typical size of a protogalaxy, while in the HDM case the free-streaming length is much larger than the size of a protogalaxy. The particles' free-streaming length is a crucial parameter in structure formation theory, because primordial density fluctuations with wave length shorter than the free-streaming length are washed out by the particle motion from the overdense regions to the underdense ones.

The free-streaming length of dark matter particles that were in thermal equilibrium in the early Universe is determined by the ratio of their decoupling temperature to their mass, T_f/m. The process of decoupling (or freezing) is described in Chap. 5. For example, neutrinos decoupled at $T_f \sim 1$ MeV, so the length of their travel in the FRW background till they became non-relativistic is equal to

9.2 Dark Matter Candidates

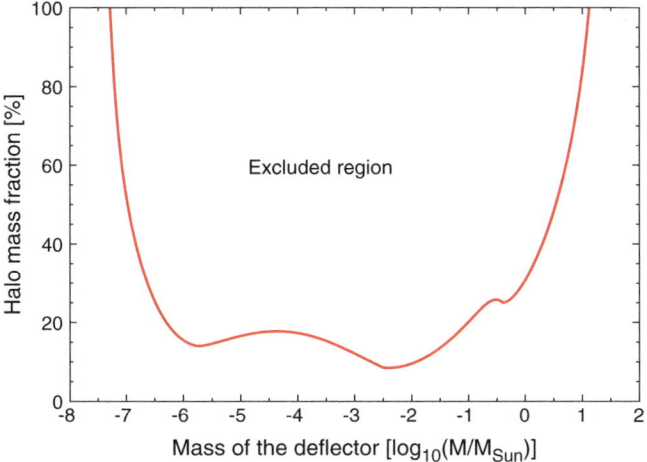

Fig. 9.2 Sketch of the constraints on the halo mass fraction of MACHOs as a function of the MACHO mass

$$\ell_{fs} = a(t) \int_{t_f}^{t_{fs}} \frac{dt'}{a(t')} + \text{(nonrel)} \approx 2 t_{fs}, \qquad (9.2)$$

where the second term is the length of the neutrino propagation after it became non-relativistic, which is small and can be neglected. It is assumed that the cosmological expansion regime is relativistic with $a(t) \sim \sqrt{t}$. The upper limit of the integration is taken at the moment when neutrinos became non-relativistic, i.e., their temperature dropped down e.g., to $m_\nu/3$.

The mass inside the free streaming radius can be estimated as

$$M_{fs} = \frac{32\pi}{3} \rho t_{fs}^3 = M_{\text{Pl}}^2 t_{fs}, \qquad (9.3)$$

where we took for ρ the radiation dominated stage expression $\rho = 3 M_{\text{Pl}}^2/(32\pi t^2)$. The free streaming time can be roughly estimated as $t_{fs} \sim 0.1 M_{\text{Pl}}/T^2 \approx M_{\text{Pl}}/m_\nu^2$. see Eq. (5.14). Thus finally we obtain for particles (not necessarily neutrinos) that were relativistic at decoupling

$$M_{fs} \sim \frac{M_{\text{Pl}}^3}{m_\nu^2} \approx 10^{18} M_\odot \left(\frac{\text{eV}}{m}\right)^2. \qquad (9.4)$$

Evidently, for neutrinos with $m_\nu < 1$ eV, the free streaming mass is much larger than the galactic mass. So neutrinos for sure make HDM. Particles with $m \sim 1$ keV would make WDM, while heavier ones would form CDM.

On the other hand, models with dark matter particles produced in non-thermal processes are also possible and an example of that is the axion, which is briefly presented in Sect. 9.2.2. Despite a very small mass, axions form CDM because they were created at rest.

CDM, WDM, and HDM predict different scenarios for the large scale structure formation in the Universe, namely for the formation of galaxies, galaxy clusters, and superclusters. In the case of CDM, smaller structures are formed first and then they congregate to form larger structures. In the case of HDM, density fluctuations at small scales are washed out and therefore the first structures are large. The latter must then fragment into galaxies. Observations favor CDM candidates, because looking at high redshift we see that galaxies formed first, while clusters and superclusters formed later. There is recent accumulated evidence in favor of some fraction of WDM as well.

An interesting class of dark matter candidates are the so-called weakly interacting massive particles (WIMPs). All these candidates interact through the weak nuclear force or through a force with a similar strength. They weakly interact with ordinary matter, but not too weakly to make a direct detection impossible. Moreover, they may have a mass in the GeV or TeV range, which makes also their number density around us not too low for a direct detection. Such dark matter candidates may be produced in particle physics colliders, because their mass is not too high. Lastly, they are an appealing class of candidates even because of the so-called "WIMP miracle". If we consider particles with a mass of order 100 GeV−1 TeV subject to the weak nuclear force in the primordial plasma, we see that they should have decoupled at a temperature of order 10 GeV. Interestingly, their abundance today would be consistent with $\Omega_m \sim 0.3$, which is the value requested by observations.

9.2.1 Lightest Supersymmetric Particle

As discussed in Sect. 3.4.1, Supersymmetry is mainly motivated by the hierarchy problem, namely the necessity to protect the Higgs mass from quantum corrections that would make it huge. Supersymmetric models have the appealing feature of having potentially good dark matter candidates.

In the minimal supersymmetric extension of the Standard Model of particle physics, the Lagrangian of the theory admits dangerous terms that would predict the non-conservation of the baryon and lepton numbers. For instance, these terms would make proton unstable, in disagreement with the experimental constraints. The problem can be fixed by imposing a new symmetry called R-parity. With such a symmetry, the lightest supersymmetric particle, or LSP, is stable. If the LSP is electrically neutral, it would be a good dark matter candidate. In many supersymmetric models, the LSP is the lightest neutralino, which is a superposition of the bino, i.e., the fermionic super-partner of the Standard Model gauge boson associated with the $U_Y(1)$ field, the neutral wino, i.e., the fermionic super-partner of the electrically-neutral Standard Model gauge boson associated with the $SU_L(2)$ field, and the neutral higgsinos, i.e., the fermionic super-partners of the supersymmetric Higgs scalars. The lightest neutralino is often considered one of the best dark matter candidates, even though supersymmetric particles at LHC have not been so far discovered.

In other models, the LSP dark matter candidate may be the axino, fermionic super-partner of the axion, or the gravitino, spin-3/2 super-partner of the graviton in Supergravity models where gravity is also taken into account and supersymmetrized. The lightest sneutrino (the scalar super-partner of the Standard Model neutrinos) is not a good dark matter candidate in the minimal supersymmetric extension of the Standard Model, because of its large cross section with nucleons, which is today experimentally excluded. However, sneutrinos can still be good dark matter candidates in more sophisticated models.

9.2.2 Axion

As briefly mentioned in Sect. 3.4.4, the strong CP problem in QCD may be solved with the introduction of a new global $U(1)$ symmetry, which is spontaneously broken at low energies, about 100 MeV. The theory predicts the existence of a spin-0 particle called axion, which gets a non-vanishing mass after the formation of the vacuum condensate of gluon fields at the QCD phase transition. Above this phase transition, axions would be massless Goldstone bosons, see Sect. 7.3.5 for some explanation, but below it an explicit symmetry breaking is induced by the condensate and axions acquire a small mass due to non-perturbative QCD effects:

$$m_a \approx 0.62 \left(\frac{10^7 \text{ GeV}}{f_a} \right) \text{ eV}, \qquad (9.5)$$

where f_a is the $U(1)$ symmetry breaking scale. Constraints on the axion mass come from direct laboratory search and astrophysical observations (stellar cooling and supernova dynamics) (Raffelt 1997). Axions with a mass at the level of a few μeV might still be viable dark matter candidates. Despite such a low mass, they would be CDM particles, because they would have been produced at rest and never been in thermal equilibrium.

9.2.3 Super-Heavy Particles

In the Standard Model of particle physics, fermions, quarks, and some gauge bosons get a mass after the electroweak symmetry breaking. The masses generated in this way should be of the order of the electroweak symmetry breaking scale, which is about a few hundred GeV, multiplied by the coupling constants of their interaction with the Higgs boson. In particular, that is how the masses of the intermediate bosons $\sim$100 GeV are generated. In the same way, the GUT scenarios naturally predict super-heavy gauge or Higgs-like bosons with masses of the order of the GUT scale $M_{\rm GUT} \sim 10^{14} - 10^{16}$ GeV. In principle there could be other super-heavy particles, which happen to be stable or very much long-lived due to some (quasi) conserved quantum number. If these particles do not have long range electromagnetic and strong interactions, they could be good dark matter candidates. Their direct and indirect detections may be extremely difficult, if not impossible, since their large mass implies a very low cosmological number density. Moreover, it is not clear if so heavy particles can be stable against gravitational decay. While we do not have any reliable quantum gravity theory to describe particle processes at the Planck scale, from heuristic arguments we may expect the possibility of a decay via a virtual black hole (Bambi et al. 2007) with the lifetime

$$\tau \sim \frac{M_{\rm Pl}^4}{M_{\rm GUT}^5} \sim 10^{-13} \text{ s}, \tag{9.6}$$

see the discussion at the end of Sect. 7.3.4. The decay may be forbidden by some unknown symmetry, but broken symmetries or global symmetries cannot do it, which makes difficult to have these particles stable.

9.2.4 Primordial Black Holes

Primordial black holes have been considered for a long time as viable dark matter candidates. They may have been produced in the early Universe, well before the advent of the first stars, from the collapse of overdense regions, gravitational collapse of cosmic strings or domain walls, during first or second order phase transitions, etc. For a review, see e.g., Carr (2003). In most scenarios, relative energy perturbations of order unity stopped expanding and recollapsed as soon as they crossed the cosmological horizon. In this case, the maximum mass of primordial black holes is set by the total mass within the cosmological horizon, namely $M_{\rm hor} = M_{\rm Pl}^3/E^2$ where E is the energy scale at which primordial black holes formed, and it turns out to be

$$M_{BH} \approx M_{\rm Pl}^2 \, t_{\rm f} \approx 5 \cdot 10^{26} \frac{1}{\sqrt{g_*}} \left(\frac{1 \text{ TeV}}{T_{\rm f}}\right)^2 \text{ g}, \tag{9.7}$$

where g_* is the effective number of relativistic degrees of freedom at the time $t_{\rm f}$ of the formation of primordial black holes, when the temperature of the Universe was $T_{\rm f}$. In this way, M_{BH} may range from the Planck mass $M_{\rm Pl}$, for black holes formed at the Planck epoch, to $M_\odot$, for black holes formed at the QCD phase transition. Primordial black holes formed after the QCD phase transition may have much larger masses, as it is argued in Sect. 7.4, Eq. (7.62).

Low mass black holes are extremely compact objects. For example, a black hole with the mass $M_{BH} = 10^{15}$ g has the radius $r_g \approx 10^{-13}$ cm. Because of that,

9.2 Dark Matter Candidates

they behave as super-heavy particles possessing only gravitational interactions. This makes their possible detection very difficult.

However, this is true only in the limit of classical physics. At semiclassical level, black holes are not really black and stable, but emit thermal radiation with the equilibrium black body spectrum[1] and temperature $T_{BH} = M_{Pl}^2/8\pi M_{BH}$, see Eq. (7.24). This expression for the temperature is true for a non-rotating and electrically neutral black hole. This process is called the Hawking radiation. The evaporation timescale is $\tau_{evap} \sim M_{BH}^3/M_{Pl}^4$; more accurate expression is given by Eq. (7.26). Primordial black holes with the initial mass $M_{BH} \sim 5 \cdot 10^{14}$ g would have the lifetime of the order of the Universe age. Primordial black holes with larger masses could survive to our time and may be registered by their Hawking radiation. However, the black hole temperature quickly decreases with the black hole mass and the Hawking emission for macroscopic black holes become completely negligible.

While primordial black holes may still represent a fraction of dark matter, their cosmological abundance is strongly constrained. Primordial black holes with an initial mass $M_{BH} \lesssim 5 \cdot 10^{14}$ g would have already evaporated (τ_{evap} is shorter than the age of the Universe). However, it is possible that quantum gravity effects make Planck mass black holes stable (Adler et al. 2001), and in this case they may form the whole dark matter in the Universe. For $M_{BH} \sim 10^{15} - 10^{16}$ g, there is a strong bound on their possible abundance, at the level of $\Omega_{BH} \lesssim 10^{-8}$ (Page and Hawking 1976), derived from the observed intensity of the diffuse γ-ray background. So they may contribute only a tiny fraction of the non-relativistic matter in the Universe. The abundance of primordial black holes in the mass range $10^{17} - 10^{26}$ g might be constrained from the observations of old neutron stars in regions in which the density of dark matter is supposed to be high (Pani and Loeb 2014). For higher mass, $M_{BH} \gtrsim 10^{26}$ g, the most stringent constraints come from the search for MACHOs (Afonso 2003; Alcock 2000; Tisserand 2007).

9.3 Direct Search for Dark Matter Particles

Direct detection experiments look for signals from the passage of dark matter particles through specially designed very sensitive detectors. Most of these experiments are aimed at the detection of WIMPs scattering off nuclei of the detector. They typically operate in deep underground laboratories to reduce the cosmic ray background. A partial list of past, present, and future direct detection experiments is presented in Table 9.1. For a recent overview on the status of direct searches for dark matter, see e.g., Schumann (2015).

The interaction rate of WIMPs with a detector mainly depends on their masses and cross section. For this reason, the results of experiments are commonly expressed

---

[1] In fact the spectrum is not really black but is distorted by the effects of the particle propagation in the gravitational field of the black hole after the emission from the horizon. For more detail see Sect. 7.3.4.

Table 9.1 Partial list of direct detection experiments

Experiment	Target	Location
ADMX	Axion	University of Washington (Washington)
CDMS	WIMPs	Soudan Underground Laboratory (Minnesota)
CoGeNT	WIMPs	Soudan Underground Laboratory (Minnesota)
COUPP	WIMPs	Fermilab (Illinois)
CRESST	WIMPs	Gran Sasso National Laboratory (Italy)
DAMA	WIMPs	Gran Sasso National Laboratory (Italy)
DarkSide	WIMPs	Gran Sasso National Laboratory (Italy)
DEAP	WIMPs	SNOLAB (Canada)
DRIFT	WIMPs	Boulby Underground Laboratory (UK)
EDELWEISS	WIMPs	Modane Underground Laboratory (France)
EURECA	WIMPs	Modane Underground Laboratory (France)
LUX	WIMPs	Sanford Underground Laboratory (South Dakota)
PICASSO	WIMPs	SNOLAB (Canada)
PVLAS	Axion	Legnaro National Laboratory (Italy)
SIMPLE	WIMPs	Laboratoire Souterrain à Bas Bruit (France)
WARP	WIMPs	Gran Sasso National Laboratory (Italy)
XENON	WIMPs	Gran Sasso National Laboratory (Italy)
ZEPLIN	WIMPs	Boulby Underground Laboratory (UK)

as constraints in the WIMP mass-cross section plane. The WIMP interaction rate is $\Gamma = nv\sigma$, where n is the WIMP number density, v is the WIMP velocity, and σ is the cross section of WIMP scattering off nucleus. The local dark matter energy density is estimated to be $\rho \approx 0.4$ GeV/cm^3 and therefore $n = \rho/m$ depends on the unknown WIMP mass m. The WIMP velocity distribution is usually assumed to be of Maxwellian form with the mean velocity close to the velocity of the stars in the Galaxy, namely around 200 km/s in the Solar System. Cross sections can be grouped into two classes, spin-independent and spin-dependent cross sections. Theoretically motivated scenarios usually have WIMPs with spin-independent cross sections, but generally speaking models with spin-dependent cross sections cannot be excluded.

A possible observational signature of dark matter is an annual modulation of the signal due to the variation of the relative velocity of Earth and WIMPs. The Solar System moves with a velocity of about 220 km/s with respect to the Galactic rest-frame and the motion of the Earth around the Sun is along the same direction in June and in the opposite direction in December. As a result, we should expect the variation of the WIMP scattering rate by about 3 %, with the maximum rate in June and the minimum rate in December. The daily rotation of Earth may cause a daily forward/backward asymmetry of the nuclear recoil direction, which can also be used as an experimental signature. Another interesting signature would be the measurement of the propagation direction of the colliding particles. In this case,

9.3 Direct Search for Dark Matter Particles

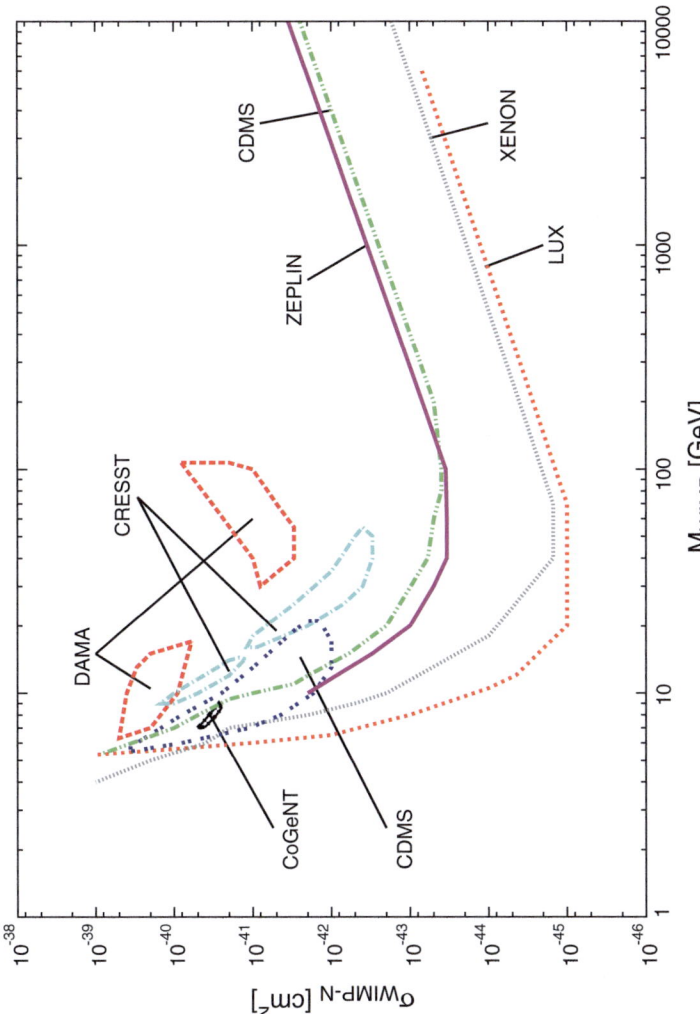

Fig. 9.3 Schematic summary of direct search for dark matter particles reporting the current limits from the experiments ZEPLIN (*magenta-solid line*), CDMS (*green-dashed double dotted line*), XENON (*gray-dotted line*), and LUX (*orange-dotted line*) and the possible dark matter signals from the experiments DAMA (*red-dashed closed curve*), CRESST (*light blue-dashed dotted closed curve*), CDMS (*blue-dotted closed curve*), and CoGeNT (*black-solid closed curve*). Dark matter candidates from the Minimal Supersymmetric Standard Model of particle physics are expected in the region $M_{WIMP} \sim 100$ GeV-1 TeV and $\sigma_{WIMP-N} \sim 10^{-49} - 10^{-45}$ cm^2, which is not yet explored. Possible dark matter signals have been found in the so-called low-mass WIMP region, but there is no agreement among different experiments

we exploit the relative motion of the Sun with respect to the Galaxy. The signal should be stronger in the direction of the motion of the Solar System, which could be distinguished from the background noise, since the latter is produced on Earth and should be isotropic.

Figure 9.3 shows the current status of the WIMP search. Most experiments provide limits on the WIMP detection in the mass-cross section plane. The shape of the constraints can be easily explained. At low masses, the sensitivity of the detector is limited by the detector energy threshold. For a WIMP mass in the range 10 GeV to 10 TeV, the expected nuclear recoil energies are usually in the range $1 - 100$ keV. At high masses, the sensitivity decreases because of the decreasing WIMP number flux, since ρ is fixed and $n = \rho/m$.

There are not only upper bounds but also statements of WIMP detection, which are however difficult to reconcile with the negative results of other experiments. The strongest claim comes from the DAMA/LIBRA collaboration: they have observed for several years an annual modulation in the event rate which would be consistent with the expected signal from WIMPs (see Fig. 9.4). More recently, CDMS, CRESST,

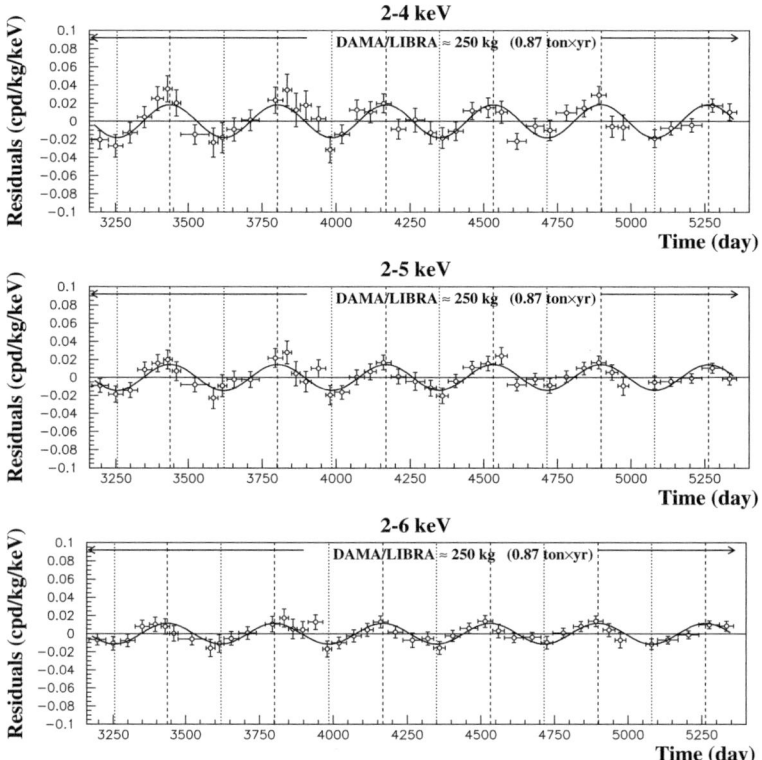

Fig. 9.4 Residual signal measured by DAMA/LIBRA $2-4$, $2-5$, and $2-6$ keV energy intervals as a function of time. From Bernabei (2010), under the terms of the creative commons attribution noncommercial license

and CoGeNT reported evidence of possible detections of WIMPs in their detectors, but there is no common consensus on the interpretation of these results, which seem also to be in conflict with the limits from other collaborations.

Direct search for axion dark matter proceeds in a completely different way. The idea is to observe the axion-photon transformation, $a \to \gamma$, in a strong magnetic field. The only running experiment is ADMX at the University of Washington Seattle. It is providing upper bounds on the $a\gamma\gamma$ coupling constant $g_{a\gamma\gamma}$ for μeV mass axions.

9.4 Indirect Search for Dark Matter Particles

Indirect quest for dark matter particles are performed by astronomical observations of possible products of their annihilation or decay. For instance, if there are equal densities of dark matter particles and antiparticles or if particles and antiparticles are the same, as expected in some scenarios, they may annihilate and produce γ-rays or e^+e^- and $\bar{p}p$ pairs. However, in the case of the so-called asymmetric dark matter, there is a dominant excess of particles over antiparticles (or vice versa) and these effects are absent.

Anyhow, an observation of an excess of γ-rays, antiprotons, positrons, or high energy neutrinos-antineutrinos in the cosmic ray background or from specific sources (e.g., the Sun or the Center of the Galaxy, where the dark matter density is expected to be higher) may be an indication of dark matter. Such a detection clearly requires a very good knowledge of the contribution from astrophysical processes and of the propagation of cosmic rays in the Galaxy, which is not usually the case. Indirect search for dark matter particles can be seen as complementary to direct detection experiments, since they may test different regions of the parameter space, where dark matter particles have different masses and coupling constants.

Some astronomical observations might have already registered dark matter signals, but systematics effects, in particular the contributions from astrophysical processes, are not really under control and there is no consensus on the interpretation of these data. In 2009, the PAMELA collaboration reported the observation of an excess of positrons in cosmic rays in the range 10 – 100 GeV (see Fig. 9.5) (Adriani 2009). Their measurement was confirmed by other experiments. The observations by ATIC, FERMI/LAT, and H.E.S.S. also reported an excess of electrons and positrons in the range 100 – 1000 GeV. However, the origin of these positrons is not yet clear. The required cross section to explain this excess is not consistent with that expected for thermal WIMPs. Some specific WIMP scenarios have been proposed in the literature to do it, but they seem now to be ruled out by the FERMI/LAT measurements of the flux of high energy photons. On the contrary, some astrophysical explanations, like positron production from pulsars of the Galaxy (Profumo 2011), appear more convincing.

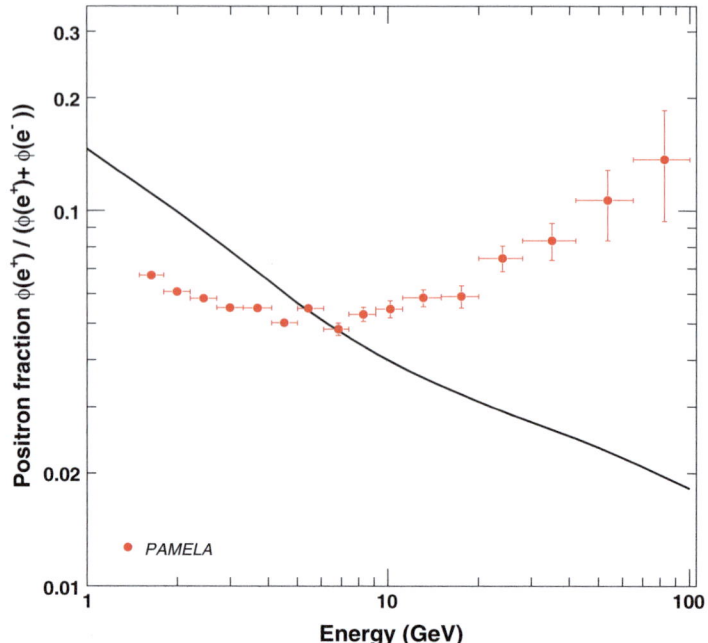

Fig. 9.5 The PAMELA positron fraction compared with a theoretical model (*black solid line*). Reprinted by permission from Macmillan Publishers Ltd: O. Adriani et al., Nature **458**, 607–609, copyright 2009. http://www.nature.com/

Problems

9.1 The dark matter energy density in the Solar System is estimated to be around 0.4 GeV/cm^3. Let us assume that dark matter consists of particles with a mass of 100 GeV, that they only interact via the weak nuclear force (exchange of W- and Z-bosons), and that their typical velocity is $v/c \sim 10^{-3}$.

(a) Estimate the dark matter particle flux (number of particles/cm^2/s) on Earth.
(b) Estimate the interaction rate of dark matter particles with a human being.
(c) How do the previous estimates change in the case of super-heavy dark matter particles with a mass of 10^{15} GeV?

References

R.J. Adler, P. Chen, D.I. Santiago, Gen. Rel. Grav. **33**, 2101 (2001) [gr-qc/0106080]
O. Adriani et al., PAMELA collaboration. Nature **458**, 607 (2009). arXiv:0810.4995 [astro-ph]
C. Afonso et al., EROS collaboration. Astron. Astrophys. **400**, 951 (2003) [astro-ph/0212176]
C. Alcock et al., MACHO collaboration. Astrophys. J. **542**, 281 (2000) [astro-ph/0001272]

References

C. Bambi, A.D. Dolgov, K. Freese, Nucl. Phys. B **763**, 91 (2007) [hep-ph/0606321]

L. Bergstrom, Rept. Prog. Phys. **63**, 793 (2000) [hep-ph/0002126]

R. Bernabei *et al.* [DAMA and LIBRA Collaborations], Eur. Phys. J. C **67**, 39 (2010). arXiv:1002.1028 [astro-ph.GA]

B.J. Carr, Lect. Notes Phys. **631**, 301 (2003) [astro-ph/0310838]

D. Clowe, A. Gonzalez, M. Markevitch, Astrophys. J. **604**, 596 (2004) [astro-ph/0312273]

K.C. Freeman, Astrophys. J. **160**, 811 (1970)

C. Lage, G.R. Farrar, arXiv:1406.6703 [astro-ph.GA]

D.N. Page, S.W. Hawking, Astrophys. J. **206**, 1 (1976)

P. Pani, A. Loeb, JCAP **1406**, 026 (2014). arXiv:1401.3025[astro-ph.CO]

S. Profumo, Central. Eur. J. Phys. **10**, 1 (2011). arXiv:0812.4457 [astro-ph]

G.G. Raffelt, in *Beyond the Desert 1997: Accelerator and Non-Accelerator Approaches* (Institute of Physics, London, 1998), pp. 808-815 [astro-ph/9707268]

M. Schumann. arXiv:1501.0120 [astro-ph.CO]

P. Tisserand et al., EROS-2 collaboration. Astron. Astrophys. **469**, 387 (2007) [astro-ph/0607207]

Chapter 10
Cosmic Microwave Background

As we know, the temperature and density of the primeval plasma dropped down in the course of the cosmological expansion. About 370,000 years after the big bang, protons and electrons bound together forming neutral hydrogen. So charged particles disappeared from the plasma and photons decoupled from matter and started freely propagating through the whole Universe. These photons are observed today as the so-called cosmic microwave background (CMB) radiation. This radiation is an inevitable prediction of the big bang theory. The discovery of the CMB was a milestone in the establishment of the Standard Model of cosmology. At the decoupling, the temperature of the Universe was about 3000 K, but the Universe expansion has made these photons redshift and the CMB temperature measured today is around 2.7 K. The CMB was predicted by Gamow and collaborators in the 1940s. The calculations were repeated later by other groups, obtaining controversial estimates for the present day CMB temperature because of the poor knowledge of the cosmological parameters at that time.

The CMB was discovered accidentally by Penzias and Wilson in 1964. They were performing some satellite communication experiments at the Bell Laboratories, in New Jersey. They observed an unexpected isotropic background signal corresponding to a black body temperature of about 3.5 K. For this discovery, Penzias and Wilson received the Nobel Prize in 1978.

The next important step in CMB physics has been done by the COBE satellite, which was launched in 1989 and measured the CMB temperature with high precision. Temperature anisotropies at the level of one part in 10^5 were detected for the first time and this was the beginning of modern cosmology as a precise science. Better and better measurements have been possible with some balloon-borne experiments, the WMAP satellite, and more recently the Planck satellite. The last 20 years have really been a golden age for CMB physics, thanks to which most of the cosmological parameters are now measured with a precision at the percent level. In particular, there is robust evidence in support of the so-called ΛCDM model, in which about 5 % of the energy density of the Universe is made of ordinary matter (protons, neutrons, electrons), about 25 % is made of non-baryonic dark matter, and 70 % is made of dark energy. The latter may possibly be a tiny but non-zero positive vacuum energy, or, what is the same, the cosmological constant. Note, however, that the ΛCDM model encounters nowadays some problems that might demand a modification of the minimal standard scenario.

The subject of CMB physics is quite complicated technically and for this reason we provide here only a very simple overview without many details. The interested reader can find more material on the subject in advanced textbooks, like Dodelson (2003); Weinberg (2008), or specialized review articles, like Hu et al. (2002); White et al. (1994).

10.1 Recombination and Decoupling

After the BBN, the primordial plasma predominantly consisted of photons, protons, helium-4 nuclei, and electrons. The abundance of other light nuclei was quite low. Neutrons had either been bound in nuclei or decayed. Neutrinos had decoupled from the primordial plasma at the freeze-out of the weak interactions, when the Universe temperature was about 1 MeV. Dark matter particles were presumably present, but they also did not interact with the primordial plasma. The Universe continued its expansion and the temperature decreased. Electrons and protons could eventually form neutral hydrogen: this event is called *recombination*. The ionization fraction of electrons X_e as a function of the plasma temperature T can be described by the Saha equation (Saha et al. 1921) (compare with analogous Eq. (8.12) at the BBN)

$$\frac{1 - X_e}{X_e} = \frac{4\sqrt{2}\,\zeta(3)}{\sqrt{\pi}} \eta \left(\frac{T}{m_e}\right)^{3/2} \exp\left(\frac{E_{\text{ion}}}{T}\right), \quad (10.1)$$

where $\zeta(3) \approx 1.20206$ is the Riemann zeta function, η is the baryon to photon ratio, and $E_{\text{ion}} \approx 13.6$ eV is the hydrogen ionization energy. The Saha equation is valid in the case of thermal equilibrium and takes into account only the reaction

$$p + e^- \leftrightarrow H + \gamma. \quad (10.2)$$

It works around the epoch of recombination, while at later times it is necessary to consider even the 2-photon reaction $p + e^- \rightarrow H^* + \gamma$ and $H^* \rightarrow H + 2\gamma$, where H^* is an excited state of the hydrogen atom.

In the Standard Model of cosmology, the baryon to photon ratio appearing in Eq. (10.1) is the same as the η at the end of the BBN, but in extensions of the standard theory this may not be true. In the case of new physics, the primordial plasma may have been reheated by the decay/annihilation of new particles, like the electron-positron annihilation after the freeze-out of the weak interactions reheated photons but not neutrinos.

The moment of recombination may be defined as the time at which $X_e = 0.5$, but the exact definition is not very important, because the transition from $X_e \sim 1$

10.1 Recombination and Decoupling

to $X_e \ll 1$ was very fast. From Eq. (10.1), we find the recombination temperature $T_{\text{rec}} \approx 0.26$ eV. We note that T_{rec} is significantly smaller than E_{ion} because $\eta = 6.1 \cdot 10^{-10} \ll 1$. At higher temperatures, the population of high energy photons with $E \gtrsim E_{\text{ion}}$ was still quite abundant and these energetic photons were able to destroy neutral hydrogen. A similar situation was found for the onset of the synthesis of light elements in Sect. 8.4. Since the present day CMB temperature is 2.7 K, the recombination redshift is $1 + z_{\text{rec}} = T_{\text{rec}}/T_0 \approx 1100$.

Before recombination, photons and matter were in thermal equilibrium through elastic Thomson scattering of photons off free electrons, because of the large Thomson cross section $\sigma_{\text{Th}} = 8\pi\alpha^2/3m_e^2 \approx 7 \cdot 10^{-25}$ cm^2. The photon interaction rate is $\Gamma = \sigma_{\text{Th}} n_e$, where $n_e = X_e \eta n_\gamma$ is the number density of free electrons and $n_\gamma \approx 0.24 T^3$ is the number density of photons. The decoupling temperature can be estimated from the condition $\Gamma = H$ or, equivalently, from the optical depth for Thomson scattering, namely the Thomson scattering probability of a photon from the time t till today

$$\tau = \int_t^{t_0} n_e \sigma_{\text{Th}}\, dt = \int_z^0 n_e \sigma_{\text{Th}} \left(\frac{dt}{dz}\right) dz \approx 0.37 \left(\frac{z}{1000}\right)^{14.25}. \quad (10.3)$$

After recombination, the optical depth drastically rose, because the cross section of the photon elastic scattering on a neutral atom is by far smaller than the Thomson cross section. The photon decoupling marks the transition from an opaque ($\tau \gg 1$) to a transparent ($\tau \ll 1$) Universe. The events of recombination and photon decoupling are strongly related and occurred more or less at the same time, so the decoupling temperature and redshift are $T_{\text{dec}} \approx T_{\text{rec}}$ and $z_{\text{dec}} \approx z_{\text{rec}}$, with a minor dependence on the cosmological model, which is encoded in dt/dz in Eq. (10.3). Of course, both events are not instantaneous, but they last for a finite time $\Delta z_{\text{rec}} \approx \Delta z_{\text{dec}} \approx 100$. However, the transition is rapid. After that, the photons of the CMB started freely propagating in the Universe. The CMB photons reaching us today decoupled at the so-called last scattering surface, which is an ideal spherical surface around us, situated at $z \approx 1100$.

10.2 Formalism for the Description of Fluctuations

The CMB is almost ideally isotropic over the sky. It has almost precise black body spectrum with the temperature 2.725 K. However, accurate measurements show small fluctuations of the temperature and of the polarization. Since these fluctuations are seen on a 2-dimensional spherical surface, our sky, they are conveniently described by spherical harmonics. We define the temperature fluctuation in the direction $\hat{\mathbf{n}} = (\theta, \phi)$ as

$$\Theta(\hat{\mathbf{n}}) = \frac{T(\hat{\mathbf{n}}) - T_0}{T_0}, \quad (10.4)$$

where $T_0 = 2.725$ K is the average temperature. Θ can be expanded in spherical harmonics as

$$\Theta(\hat{\mathbf{n}}) = \sum_{l=0}^{+\infty} \sum_{m=-l}^{l} a_{lm} Y_{lm}(\theta, \phi), \tag{10.5}$$

$$Y_{lm}(\theta, \phi) = \sqrt{\frac{(2l+1)}{4\pi} \frac{(l-m)!}{(l+m)!}} P_l^m(\cos\theta) e^{im\phi}, \tag{10.6}$$

where a_{lm}s are the amplitudes of the corresponding harmonics, Y_{lm}s, and P_l^ms are the Legendre polynomials. a_{lm}s can be determined by exploiting the fact the spherical harmonics Y_{lm}s form a complete orthonormal set on the unit sphere and thus

$$a_{lm} = \int_0^{2\pi} d\phi \int_{-\pi}^{\pi} d\theta \, \cos\theta \, \Theta(\hat{\mathbf{n}}) Y_{lm}^{\star}(\theta, \phi). \tag{10.7}$$

Since a_{lm} is real, $a_{lm}^* = a_{l\,-m}$. We also note that any multipole l represents an angular scale on the sky of about π/l.

It should be clear that the fluctuation map we observe in the sky is a particular realization of a stochastic function that depends on our position. We cannot predict the precise form of such a particular realization, but only its statistical properties in terms of a specific cosmological model. We assume the validity of the *ergodic hypothesis*, which means that an average over all spatial positions within a given realization is equivalent to an average over the ensemble. These averages are commonly denoted by $\langle \ldots \rangle$. For instance, $\langle a_{lm} \rangle = 0$. Cosmological data should thus be compared with average values of the proper quantities found from the explored cosmological models. Assuming an isotropic sky (no preferred axis) and a Gaussian statistics (no correlation among modes), the power spectrum completely characterizes the anisotropies. The power spectrum of the temperature fluctuations, or *TT power spectrum*, is given by

$$C_l = \frac{1}{2l+1} \sum_{m=-l}^{l} \langle |a_{lm}|^2 \rangle, \tag{10.8}$$

and the sum over m is done because there is no preferred direction. Since we are considering the TT power spectrum, C_l is sometimes indicated by C_l^{TT} or $C_l^{\Theta\Theta}$. The typical shape and features of the TT power spectrum are shown in Fig. 10.1, while current measurements are reported in Fig. 10.2. Since we have only our Universe, there is an intrinsic statistical error in the estimate of these coefficients. This is given by

$$\frac{\Delta C_l}{C_l} = \sqrt{\frac{2}{2l+1}}, \tag{10.9}$$

10.2 Formalism for the Description of Fluctuations 195

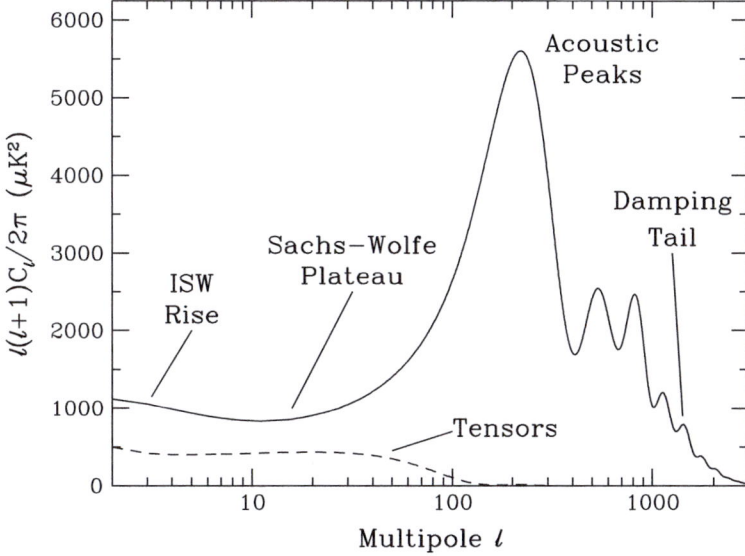

Fig. 10.1 Example of TT power spectrum for the standard ΛCDM model and its main features. From Olive et al. (2014)

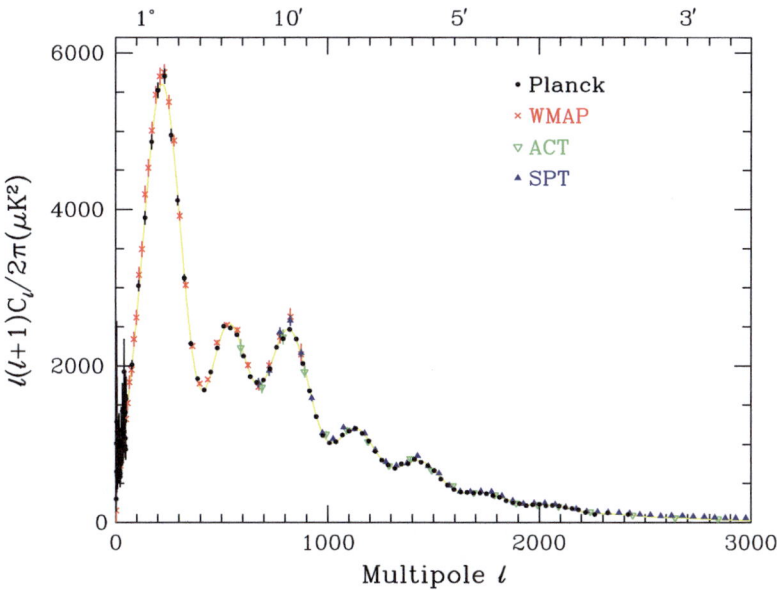

Fig. 10.2 TT power spectrum data from the Planck, WMAP, ACT (Atacama Cosmology Telescope), and SPT (South Pole Telescope) experiments. From Olive et al. (2014)

and it is refereed to as *cosmic variance*. Cosmological models predict their C_l and it is possible to measure C_l from the average of a_{lm} over m.

In most theoretical models, a_{lm}s are nearly Gaussian random fields. A small level of non-Gaussianity should be expected, both of primordial origin and from some post-recombination effects, like non-linear growth of fluctuations on small scales. However, there are inflationary models that predict a significant level of non-Gaussianity, and, since current observations are consistent with no primordial non-Gaussianity, these models can be either ruled out or constrained. The *temperature 2-point function* is the expectation value of the correlation of the temperature fluctuations between two points in the sky, which is related to the power spectrum by

$$C(\vartheta) = \langle \Theta(\hat{\mathbf{n}})\, \Theta(\hat{\mathbf{n}}') \rangle = \frac{1}{4\pi} \sum_{l=0}^{+\infty} (2l+1) C_l P_l(\cos \vartheta), \qquad (10.10)$$

where $\cos \vartheta = \hat{\mathbf{n}} \cdot \hat{\mathbf{n}}'$, namely ϑ is the angle between the two directions. We note that $C(\vartheta)$ only depends on the the angular separation ϑ, and not on the orientation of $\hat{\mathbf{n}}$ and $\hat{\mathbf{n}}'$, because it is assumed that there is no preferred direction. If the temperature fluctuations are Gaussian, all the higher-point correlation functions vanish. In the last years, there has been a lot of interest in the possibility of detecting non-Gaussianity, but current measurements only provide an upper bound on it. A popular way to constrain a non-Gaussian signal is through the temperature 3-point function. The interest in the search for primordial non-Gaussianity is motivated by the quite appealing possibility of getting information on the physics during or even before inflation.

As a consequence of the Thomson scattering of an anisotropic radiation field, the CMB is polarized at the level of $\sim 5\,\%$ of the temperature anisotropies, corresponding to a few μK. The CMB polarization field can be divided into two types, which are usually called E-modes and B-modes (Kamionkowski et al. 1997; Zaldarriaga et al. 1997). They are defined in terms of the second derivatives of the polarization amplitude, as it is explained below.

The 2×2 polarization density matrix (in the direction orthogonal to the photon propagation) can be expanded in the full set of 2×2 matrices

$$\rho_{ij} = J\left(I/2 + \xi_k \sigma_k\right), \qquad (10.11)$$

where I is the unit matrix, σ_k ($k = 1, 2, 3$) are the Pauli matrices, and the coefficients ξ_i are the so-called Stokes parameters. Such a matrix has two well known algebraic invariants: the trace, which is equal (or proportional) to the radiation intensity

$$J = \delta_{ij} \rho_{ij} = |E_x|^2 + |E_y|^2, \qquad (10.12)$$

and the helicity

$$V = \varepsilon_{ij} \rho_{ij}. \qquad (10.13)$$

10.2 Formalism for the Description of Fluctuations

The parity conservation in electromagnetic interactions demands that circular polarization (or helicity) vanishes, which implies $\xi_2 = 0$.

There are two more (now differential) invariants of the polarization matrix: the scalar $S = \partial_i \partial_j \rho_{ij}$ (E-mode) and the pseudoscalar $P = \varepsilon_{ik} \partial_i \partial_j \rho_{jk}$ (B-mode). For purely scalar perturbations, the only way to write the polarization matrix is

$$\rho_{ij} = \left(2\partial_i \partial_j - \delta_{ij}\partial^2\right)\Psi, \tag{10.14}$$

where Ψ is a scalar function. Correspondingly, $P = 0$. $P \neq 0$ is an indication for something extra beyond scalar perturbations.

Apart from scalars, there could be:

1. Vector perturbations created by photon scattering in magnetized interstellar/intergalactic medium created, e.g., by large scale magnetic fields,

$$\rho_{ij} = \partial_i V_j - \partial_j V_i, \quad P = \varepsilon_{ij}\partial^2 \partial_i V_j. \tag{10.15}$$

2. Tensor perturbations, e.g. gravitational waves,

$$\rho_{ij} \sim \partial^{-2}(\partial_i h_{3j} - \partial_j h_{3i}), \quad P \sim \varepsilon_{ik}\partial_i h_{3k}. \tag{10.16}$$

3. Second order scalar perturbations, e.g. for $\Psi_2 = \partial_t \Psi_1$,

$$\rho_{ij} \sim \partial_i \Psi_1 \partial_j \Psi_2 - \partial_i \Psi_2 \partial_j \Psi_1, \quad P = \varepsilon_{ik}\partial_i(\Delta\Psi_1 \partial_k\Psi_2 - \Delta\Psi_2 \partial_k \Psi_1). \tag{10.17}$$

All such types of perturbations result in $P \neq 0$ and thus they can create B-mode of polarization.

In analogy with the power spectrum of the temperature fluctuations, we can introduce cross power spectra of temperature and polarization fluctuations. Since E-modes have $(-1)^l$ parity and B-modes have $(-1)^{l+1}$ parity, some cross power spectra identically vanish. In addition to the TT-spectrum, the non-vanishing spectra are the TE-spectrum, the EE-spectrum, and the BB-spectrum. Their 2-point functions are

$$\langle \Theta\, E \rangle = \frac{1}{4\pi} \sum_{l=0}^{+\infty} (2l+1) C_l^{TE} P_l(\cos\vartheta),$$

$$\langle E\, E \rangle = \frac{1}{4\pi} \sum_{l=0}^{+\infty} (2l+1) C_l^{EE} P_l(\cos\vartheta),$$

$$\langle B\, B \rangle = \frac{1}{4\pi} \sum_{l=0}^{+\infty} (2l+1) C_l^{BB} P_l(\cos\vartheta), \tag{10.18}$$

where here E and B indicate, respectively, the polarization fluctuations in the E- and B-modes.

10.3 Anisotropies of the CMB

The temperature fluctuations in Eq. (10.5) are formally expanded in terms of spherical harmonics, from $l = 0$ to $l = +\infty$. However, the monopole term ($l = 0$) just provides the average temperature over the whole sky and it is affected by the cosmic variance: we can measure the average value at our position, not the average value in the Universe. We find $T_0 = 2.7255 \pm 0.0006$ K (Olive et al. 2014), which implies

$$\begin{aligned} n^0_{CMB} &= 411 \text{ photons/cm}^3, \\ \rho^0_{CMB} &= 4.64 \cdot 10^{-34} \text{ g/cm}^3 = 2.60 \cdot 10^{-10} \text{ GeV/cm}^3, \\ \Omega^0_{CMB} h_0^2 &= 2.47 \cdot 10^{-5}. \end{aligned} \quad (10.19)$$

The dipole term ($l = 1$) represents temperature fluctuations with an angular scale in the sky of order π. Here, the dominant contribution comes from our proper motion with respect to the CMB reference frame: photons are blueshifted on the one side and redshifted on the other side. The amplitude of the dipole term is 3.355 ± 0.008 mK and it corresponds to the Solar System velocity $v \approx 370$ km/s. Eventually, information on the cosmological parameters can be extracted from the power spectrum from $l = 2$ to some $l = l_{\max}$, where $l_{\max}$ is determined by the resolution of the observations.

Temperature fluctuations are usually grouped into primary and secondary. Primary anisotropies are produced at redshifts $z \geq z_{\text{dec}}$, namely at or before the last scattering surface, and they clearly carry information on the pre-recombination Universe. Secondary anisotropies are produced later, at redshift $z < z_{\text{dec}}$, and they carry information on the physics of the post-recombination Universe. We note that the mechanism responsible for the primordial perturbations may generate scalar, vector, and tensor modes. However, vector modes decay due to the expansion of the Universe. Tensor modes decay as soon as they enter the cosmological horizon, so their contribution should be strongly suppressed for angular scales smaller than the one associated to the last scattering surface, which is about $1°$ (see Fig. 10.1). Tensor modes may be produced by primordial gravitational waves, but their detection in the TT power spectrum seems unlikely, because of both their small contribution and the cosmic variance affecting the spectrum at low ls. Tensor modes may instead be detected in the BB power spectrum.

10.3.1 Primary Anisotropies

The main feature in the TT power spectrum is the presence of *acoustic peaks* for $l \gtrsim 100$, as clearly shown in Fig. 10.1. They were created by acoustic oscillations of the photon-baryon fluid. Before recombination, photons were tightly coupled to the proton-electron plasma. Because of perturbations in the gravitational field caused

10.3 Anisotropies of the CMB

by the dark matter component, the baryon component tended to collapse, forming rising inhomogeneities, while the photon component provided the pressure to oppose to it. The result was an oscillation of this photon-baryon fluid. The amplitude of these perturbations was small, at the level of $\delta\rho/\rho \sim 10^{-5}$, and therefore they evolved linearly and every mode was independent from the others. Oscillation started as soon as the inhomogeneity of a certain wave length entered the cosmological horizon. Since the Universe is homogeneous and isotropic at first approximation, inhomogeneities of the same wave length entered the cosmological horizon at the same time and thus they were in phase.

The first acoustic peak (the highest peak in Fig. 10.1) was generated by perturbations that entered the cosmological horizon at the photon decoupling. The second and higher-orders peaks were produced by perturbations that entered the cosmological horizon at earlier times. For the flat spectrum of the primordial perturbations, see Sect. 12.2.6, the amplitudes of the density perturbations were the same when they entered the cosmological horizon. However, the amplitude of the observed higher peaks in the temperature fluctuations typically drops down with larger l. The reason for that is the redshift of the oscillations inside the horizon because of the expansion.

Of course, no peaks could be produced after recombination, because photons decoupled from matter and therefore oscillations of the photon-baryon fluid were not possible any more. The location and the height of the peaks depend on the cosmological parameters and therefore their measurement can be used to determine the latter. The first acoustic peak is particularly important. The height of the first peak can be used to determine Ω_B, while its position to infer the geometry of the Universe (open, flat, closed), since

$$l_{\text{peak}} \approx \frac{220}{\sqrt{\Omega_{tot}^0}}. \tag{10.20}$$

Current CMB data require an almost flat Universe, namely $\Omega_{tot}^0 \approx 1$. The sensitivity of the first peak position to the geometry of the Universe can be understood by the following simple arguments. The physical size of the wave length corresponding to the first peak is known: it is equal to the cosmological horizon (more accurately to the sound horizon) at recombination. The angle under which the peak is observed at the present day depends upon the space geometry: in an open geometry, a fixed length is observed at smaller angle than that in a closed geometry. The angle at which the first highest peak is observed corresponds to the flat geometry with a percent precision.

The even peaks correspond to underdense regions and in general have smaller height than the odd peaks, corresponding to overdense regions (with a proper account of the redshift). The ratio of the heights of neighboring peaks allows the determination of the cosmological baryon to photon ratio η. In the acoustic "oscillators", the baryons play the role of mass, while the photon pressure is a kind of a spring. Thus a larger baryon density leads to a larger enhancement of the odd peaks over the even peaks.

As shown in Fig. 10.1, at $l \gtrsim 1000$ the acoustic peaks are exponentially suppressed. This is due to the so-called *Silk damping*. The latter is a consequence of the photon diffusion, which caused damping of small scale anisotropies after the corresponding fluctuation entered the cosmological horizon. The photon mean free path is determined by the Thomson scattering and is equal to

$$l_\gamma = \frac{1}{\sigma_{Th} n_e}. \tag{10.21}$$

The effect of diffusion damping is strongly amplified by the finite duration of the decoupling (see discussion after Eq. 10.3) when the number density of electrons drastically dropped down.

10.3.2 Secondary Anisotropies

Secondary anisotropies were produced after the photon decoupling and they thus carry information about the Universe at lower redshift. Three secondary signals of particular importance are associated to the Sachs-Wolfe effect, reionization, and the Sunyaev-Zeldovich effect.

The *Sachs-Wolfe effect* is responsible for temperature fluctuations at large angular scales, say $l \lesssim 100$. The *intrinsic Sachs-Wolfe effect* is induced by the photon redshift/blueshift due to the gravitational potential at the last scattering surface. The *integrated Sachs-Wolfe effect* is due to the passage of photons through time varying gravitational potentials.

The first stars formed at redshift $z \approx 10$ and provided new high energy photons capable of separating protons and electrons bound in neutral hydrogen atoms. For this reason, this epoch is called *reionization*. The new free electrons reopened the possibility of Thomson scattering of CMB photons. The interaction of CMB photons with free electrons inevitably affects pre-recombination small scale anisotropies and produces polarization anisotropies at large angular scales.

The *Sunyaev-Zeldovich effect* is a distortion in the CMB spectrum created by the inverse Compton scattering $e^- + \gamma \rightarrow e^- + \gamma$ of the CMB photons off hot electrons. The magnitude of the effect is independent of the redshift of the source. The effect allows the study of the properties of galaxy clusters and the measurement of the cosmological parameters, in particular the Hubble parameter.

10.3.3 Polarization Anisotropies

E-modes are mainly generated by Thomson scattering in an inhomogeneous plasma. Its magnitude is proportional to the quadrupole asymmetry of the CMB. E-mode must inevitably exist in the standard CMB theory.

10.3 Anisotropies of the CMB

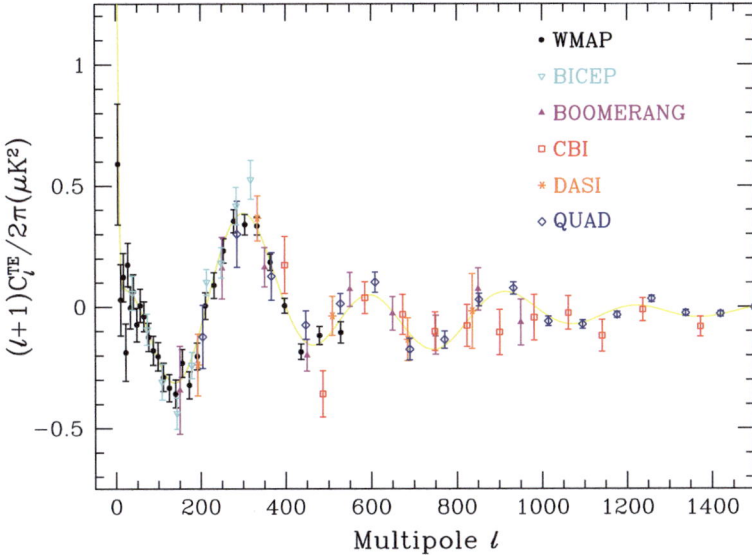

Fig. 10.3 TE power spectrum data from the WMAP, BICEP, BOOMERANG, CBI, DASI, and QUAD experiments. From Olive et al. (2014)

The existence of B-mode is questionable and this makes it especially interesting. B-modes may be generated by weak gravitational lensing of E-modes or by gravitational waves.

Since only a small fraction of the CMB radiation is polarized, an accurate measurement of the polarization power spectra is quite challenging. The TE and EE power spectra were detected for the first time in 2002 by the DASI experiment and they are now measured with relatively good accuracy. Both spectra exhibit a series of acoustic peaks produced by the oscillation of the primordial photon-baryon fluid before decoupling, see Figs. 10.3 and 10.4. A direct measurement of the BB power spectrum was obtained in 2014 by POLARBEAR and BICEP2. The results are shown in Fig. 10.5. The recent interest in the B-modes was created by the announcement of the BICEP2 collaboration in March 2014 of the discovery of primordial gravitational waves in the BB power spectrum (Ade et al. 2014a). They found a ratio of tensor to scalar perturbation amplitude $r \approx 0.2$. This result has been strongly criticized and it seems that the observed signal comes neither from primordial gravitational waves nor from lensed E-modes, but it is induced by the foreground created by dust in the interstellar medium (Flauger et al. 2014).

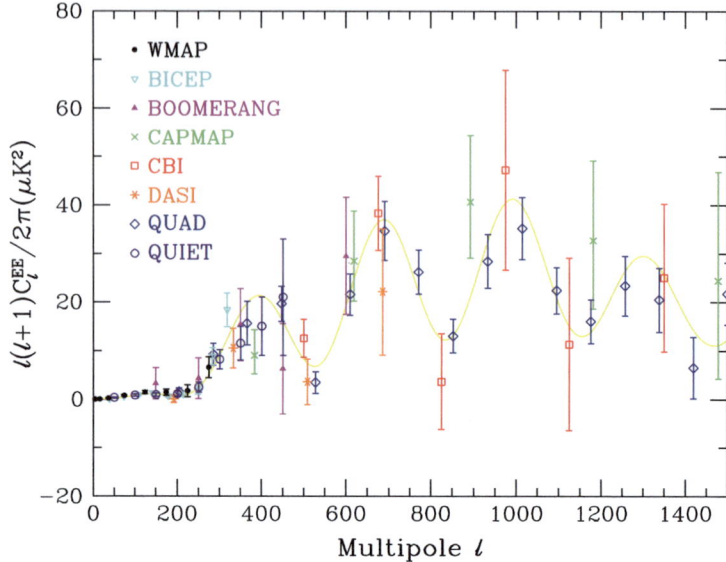

Fig. 10.4 EE power spectrum data from the WMAP, BICEP, BOOMERANG, CAPMAP, CBI, DASI, QUAD, and QUIET experiments. From Olive et al. (2014)

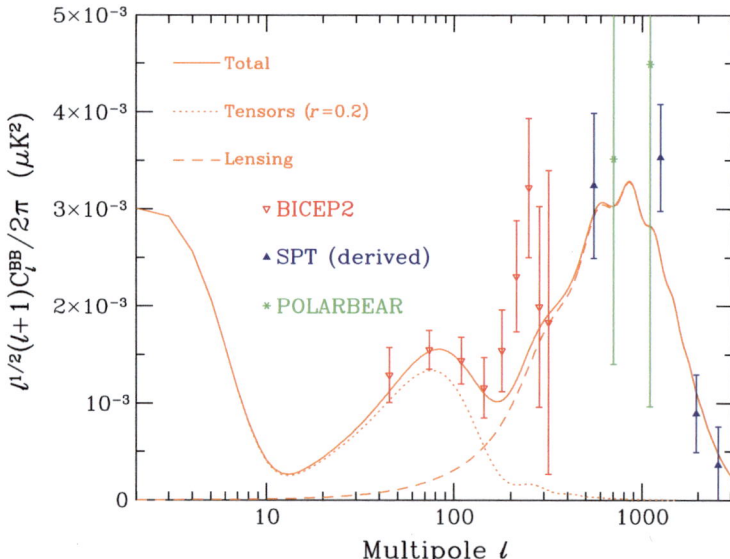

Fig. 10.5 BB power spectrum data from the BICEP, SPT, and POLARBEAR experiments. In the case of SPT, the measurements are derived from a lensing correlation analysis. Previous measurements were only able to report upper limits. From Olive et al. (2014).

10.4 Primordial Perturbations

Primordial perturbations can be decomposed in scalar, vector, and tensor modes. Scalar modes can be of different types, but the most important ones are adiabatic and isocurvature perturbations. *Adiabatic density perturbations* describe fluctuations that do not alter the comoving entropy per unit mass, which can be taken as $s \sim T^3/\rho_m \sim \rho_\gamma^{3/4}/\rho_m$ because the entropy is carried almost entirely by photons and the mass is carried by non-relativistic matter. In this case

$$\frac{\delta s}{s} = \frac{3}{4}\frac{\delta\rho_\gamma}{\rho_\gamma} - \frac{\delta\rho_m}{\rho_m} = 0. \tag{10.22}$$

If $\delta_\gamma = \delta\rho_\gamma/\rho_\gamma$ is the fluctuation of the radiation component and $\delta_m = \delta\rho_m/\rho_m$ is that of the matter component, an adiabatic perturbation requires

$$\delta_m = \frac{3}{4}\delta_\gamma. \tag{10.23}$$

Most of the inflationary models generate adiabatic perturbations. *Isocurvature perturbations* correspond to fluctuations that do not alter the local spatial curvature.[1] This is possible, for instance, in the case the density perturbation of a component, say $\delta\rho_1$, is compensated by that of another component, namely $\delta\rho_2 = -\delta\rho_1$, so that the total energy density is unperturbed. Thus isocurvature perturbations can be considered as perturbations in chemical content with constant total energy density. Such perturbations could be generated, e.g., in inhomogeneous scenarios of baryogenesis. If primordial perturbations were generated by topological defects like cosmic strings, they would also be of isocurvature type.

The nature of the primordial perturbations can be inferred from the position of the acoustic peaks in the TT power spectrum. Adiabatic density perturbations predict a series of acoustic peaks at the l-positions with the ratio 1:2:3, while in the case of isocurvature perturbations the ratio is 1:3:5. CMB data clearly favor adiabatic primordial perturbations, but a contribution at the level of a few percent from isocurvature perturbations is still allowed. Inflation may thus be the mechanism responsible for the creation of the primordial perturbations. Models of structure formation involving cosmic strings are not consistent with observations. Multi-field inflationary models predicting a mixture of adiabatic and isocurvature perturbations can also be ruled out, or at least strongly constrained.

In the case of perfect adiabatic perturbations, the variation of comoving curvature perturbations $\mathcal{R}$ does not change with time. The variance of these perturbations can be rewritten in Fourier space

---

[1] The spatial curvature ^3R is the scalar curvature computed from the 3-metric γ_{ij}. In terms of the line element of the FRW metric, $ds^2 = dt^2 - \gamma_{ij}dx^i dx^j$.

$$\Delta_{\mathcal{R}}^2 = \frac{1}{2\pi^2} \int \mathcal{P}_{\mathcal{R}}(k) \, k^2 \, dk. \quad (10.24)$$

The power spectrum $\mathcal{P}_{\mathcal{R}}(k)$ is commonly cast in the following form

$$\mathcal{P}_{\mathcal{R}}(k) = A_s \left(\frac{k}{k_0}\right)^{n_s - 1}, \quad (10.25)$$

where A_s and n_s are, respectively, the amplitude and the spectral index of scalar perturbations, and k_0 is a reference wavelength. The so-called Harrison-Zeldovich spectrum, or scale-invariant spectrum, has $n_s = 1$, see also Chap. 12. If the primordial perturbations were produced by inflation, the measurement of A_s and n_s can constrain the inflationary potential $V(\phi)$.

Tensor perturbations can be treated in a similar way. The associated power spectrum $\mathcal{P}_h(k)$ can be written in terms of a tensor amplitude A_t and a tensor spectral index n_t. Since it seems unlikely to have high quality data capable of measuring n_t in the near future, one usually simplifies the picture and, instead of considering the cosmological parameters n_t and A_t, data are fitted with the ratio r of the tensor to scalar power at some small value of k.

10.5 Determination of the Cosmological Parameters

The existing excellent measurements of the TT power spectrum allow the determination of the basic cosmological parameters with a precision at the level of percent. The other power spectra have been measured: there is now the effort to improve these data and it is a promising research field for the future. The cosmological parameters affect these power spectra since they enter the equations of the evolution of linear perturbations on the FRW background. CMBFAST[2] (Seljak et al. 1996) and CAMB[3] (Lewis et al. 2000) are two publicly available codes to numerically compute the CMB temperature and polarization power spectra over a wide range of cosmological parameters.

From the comparison of theoretical predictions and CMB data, it is possible to measure some cosmological parameters that determine the evolution of the anisotropies. The "basic" set of these parameters includes the amplitude of scalar perturbations A_s, the spectral index of scalar perturbations n_s, the tensor to scalar perturbation ratio r, the dimensionless Hubble constant h_0, the baryon density of the Universe $\Omega_B^0 h_0^2$, the cold dark matter density of the Universe $\Omega_{CDM}^0 h_0^2$, the total energy density of the Universe Ω_{tot}^0, and the reionization optical depth τ. However, the set of cosmological parameters may change according to the physics we want to explore. For instance, we may set $\Omega_{tot}^0 = 1$ and infer the cosmological constant

---

[2]http://lambda.gsfc.nasa.gov/toolbox/tb_cmbfast_ov.cfm.
[3]http://camb.info/.

contribution from $\Omega_\Lambda^0 = 1 - \Omega_m^0$, where $\Omega_m^0 = \Omega_{CDM}^0 + \Omega_B^0$. The contribution from radiation Ω_γ^0 is always ignored, because the radiation energy density is negligible at the present time. Still, it is possible to consider other free parameters, like the contribution from neutrinos, $\Omega_\nu^0 h_0^2$, to derive a very precise upper bound on the neutrino mass, or the dark energy equation of state, w_{DE}.

Eventually, from the study of the CMB we learn that the Standard Model of cosmology works fairly well. Recombination occurred at $z_{\text{rec}} \approx 1100$ and the Universe was reionized by the first stars at $z_{\text{rei}} \approx 10$. The Universe is almost flat, namely $\Omega_{tot}^0 \approx 1$, with a small amount of baryonic matter, a larger fraction of cold dark matter, and the dominant contribution comes from a mysterious vacuum-like energy, which is cosmologically significant but tiny for particle physics standards, see Sect. 11.2. Primordial perturbations are almost Gaussian and purely adiabatic, and this supports the inflationary paradigm, while the non-detection of isocurvature perturbations strongly constrains the presence of topological defects, which are ruled out as the main mechanism for structure formation. Current data support the standard ΛCDM model with (Ade et al. 2014b)

$$\Omega_B^0 \approx 0.05 \quad \Omega_{CDM}^0 \approx 0.27 \quad \Omega_\Lambda^0 \approx 0.68. \tag{10.26}$$

The dimensionless Hubble constant is $h_0 \approx 0.67$, thought there is a tension between the CMB (Planck) data and the traditional astronomical measurements, and there is strong evidence that $n_s < 1$, which is predicted by some inflationary models.

Problems

10.1 Derive n_{CMB}^0, ρ_{CMB}^0, and Ω_{CMB}^0 in Eq. (10.19).

10.2 Compute n_{CMB}, ρ_{CMB}, and Ω_{CMB} at the time of photon decoupling.

References

P.A.R. Ade et al. [BICEP2 Collaboration], Phys. Rev. Lett. **112**, 241101 (2014a). arXiv:1403.3985 [astro-ph.CO]
P.A.R. Ade et al. [Planck Collaboration], Astron. Astrophys. **571**, A16 (2014b). arXiv:1303.5076 [astro-ph.CO]
S. Dodelson, *Modern Cosmology*, 1st edn. (Academic Press, San Diego, 2003)
R. Flauger, J.C. Hill, D.N. Spergel, JCAP **1408**, 039 (2014). arXiv:1405.7351 [astro-ph.CO]
W. Hu, S. Dodelson, Ann. Rev. Astron. Astrophys. **40**, 171 (2002). [astro-ph/0110414]
M. Kamionkowski, A. Kosowsky, A. Stebbins, Phys. Rev. D **55**, 7368 (1997). [astro-ph/9611125]
A. Lewis, A. Challinor, A. Lasenby, Astrophys. J. **538**, 473 (2000). [astro-ph/9911177]
K.A. Olive et al., Particle data group collaboration. Chin. Phys. C **38**, 090001 (2014)
M.N. Saha, Roy. Soc. London Proc. Series A **99**, 135 (1921)
U. Seljak, M. Zaldarriaga, Astrophys. J. **469**, 437 (1996). [astro-ph/9603033]

S. Weinberg, *Cosmology*, 1st edn. (Oxford University Press, Oxford, 2008)
M.J. White, D. Scott, J. Silk, Ann. Rev. Astron. Astrophys. **32**, 319 (1994)
M. Zaldarriaga, U. Seljak, Phys. Rev. D **55**, 1830 (1997). [astro-ph/9609170]

Chapter 11
Dark Energy

11.1 Cosmological Acceleration

One of the most impressive discoveries in astronomy, made during approximately the last two decades, was an accumulated evidence proving that the cosmological expansion is not slowing down, as one would expect for the motion of matter moving in a self-gravity field. On the opposite, the expansion speed is growing and this increase started at a relatively recent cosmological epoch, at a redshift of order unity. The astonishment created by this discovery can be better understood if we use the following, though not precise, analogy. Let us consider the motion of a stone thrown vertically up from the Earth surface. As everybody knows, the stone moves up with a decreasing velocity, then at some stage stops and falls down, now with rising velocity. If the initial stone velocity were larger than a certain value, it would never come back but still would move with smaller and smaller speed. This picture very closely describes the main features of the Universe expansion. However, the recently established character of the cosmological expansion corresponds to the picture in which the stone initially moving with deceleration later starts accelerating, as if a rocket engine, attached to it, was switched on. So the stone would never return, independently on the initial velocity. In the cosmological case, a small initial velocity of the stone corresponds to a geometrically closed universe, while a high speed, which allows the stone to reach infinity, corresponds to an open universe.

It was believed until recently that the final destiny of the Universe and its 3-dimensional geometry were rigidly connected. An open universe will expand forever and a closed universe with the geometry of a 3-dimensional sphere at some future moment will stop expanding and will start contracting to a hot singularity, as it is in the example with a stone with a small initial velocity. With dark energy, the Universe behaves as a stone equipped with a rocket and will never collapse, independently of its geometry. However, let us keep in mind that a change of the equation of state of dark energy is possible and it may restore the old one-to-one correspondence between the universe geometry and its destiny.

The analogy between the stone flight and the Universe expansion can be extended further to the initial push. It is practically established that at the very beginning the Universe expanded with acceleration very probably induced by the inflaton field, which created something analogous to the contemporary dark energy but operating at a much higher energy scale and during a much shorter time. After this initial push, the subsequent motion, both of the stone and of the Universe expansion, is simply an inertial one.

On the other hand, this antigravitating expansion is not unreasonable from the point of view of General Relativity. According to the second Friedman equation (4.11)

$$\frac{\ddot{a}}{a} = -\frac{4\pi}{3M_{\text{Pl}}^2}(\rho + 3P), \tag{11.1}$$

not only energy density but also pressure gravitates and since pressure may be negative, the cosmological acceleration becomes positive if $P < -\rho/3$. Of course, if the energy density is allowed to be negative, an antigravity force inducing the accelerated expansion might also arise, but theories with $\rho < 0$ are pathological and usually are not considered.

Note that life is possible only due to such an antigravity induced by negative pressure, because otherwise the Universe would have never expanded and would have remained very small, with a curvature of the order of the Planck value. In Newtonian theory, antigravity does not exist and life would not be possible. So we need antigravity at the beginning to create the initial push (inflation), but "who ordered antigravity now"?

To avoid misunderstanding, let us note that antigravity in General Relativity is possible only for infinitely large objects. Any finite object with positive energy density can create only gravitational attraction. This is essentially the Jebsen-Birkhoff theorem (Birkhoff 1923; Jebsen 1921), well known in General Relativity. However, in infrared modified gravity, considered in Sect. 11.1.3, this theorem is invalid and finite objects may be able to create gravitational repulsion.

An accelerated expansion will be eternal for any 3-dimensional geometry if the equation of state of dark energy is not changed in the future. This is in contrast with the inflationary prediction for a non-accelerated universe at late time. Inflation says that our part of the Universe will eventually collapse back to a singularity, because the density perturbations $\delta\rho/\rho$ with increasing scale would inevitably lead to $\Omega > 1$ and thus this part of the Universe, which may be far outside of the present day horizon, would be geometrically closed. Hence, the accelerated cosmological expansion will save us from burning. Maybe this is the reason for its existence. The source that creates the accelerated expansion is unknown. Two mechanisms are mainly discussed. Firstly, it could be the so-called dark energy, which has negative pressure: with a sufficiently large absolute value of P, i.e. $|P| > \rho/3$, we have $\ddot{a} > 0$. A possible form of dark energy could be the vacuum energy (or, which is the same, a cosmological constant or Λ-term) with $P = -\rho$. Another form of dark energy could be a quasi-constant scalar field ϕ analogous to that responsible for inflation. In this case, the difference between the exponential expansions at the dawn of the Universe and today is only in the energy and time scales, but it is huge.

The density of the vacuum energy remains constant as the Universe expands. Thus if dark energy is some vacuum energy, the expansion will last forever for any 3-dimensional geometry, as it is mentioned above. However, if dark energy is the energy of a very light scalar field or of a field with very flat potential, then in a distant future, when the Hubble parameter drops down becoming comparable with the mass of the field or with the slope of its potential, the expansion will become decelerated again and the field will evolve down to zero because of the cosmological redshift or production of very light or massless particles. So ultimately the Universe destiny will be again determined by its geometry as it is in the good old FRW cosmology.

An accelerated expansion could be also induced if the gravitational interaction is modified at small curvature. Instead of the usual General Relativity action, which is linear in the scalar curvature R, an additional non-linear term may be introduced, so the Lagrangian turns into $\mathscr{L} \to R + F(R)$. In principle, more complicated scalars can also be considered, as e.g. those constructed from the square of the Ricci or Riemann tensors, $R_{\mu\nu}R^{\mu\nu}$, $R_{\mu\nu\alpha\beta}R^{\mu\nu\alpha\beta}$, or more complicated invariants. However, due to the non-linearity of the Lagrangian, the emerging equations of motion become of higher order and may have tachyonic or ghost type solutions, or the solutions may be strongly unstable or singular. The version with $F(R)$ is safer but the condition of stability and/or of absence of singularities impose some obligatory restrictions on the form of $F(R)$, see Sect. 11.1.3.

11.1.1 Astronomical Data

Phenomenologically, dark energy can be described by an antigravitating substance with the equation of state $P = w\rho$ and $w = -1.10^{+0.08}_{-0.07}$ (Olive et al. 2014). There are several independent pieces of data, based on completely different cosmological and/or astrophysical phenomena, proving that the speed of expansion indeed started growing with time at the redshift $z_{acc} \approx 0.65$. The observational data proving the existence of the cosmological acceleration include:

1. The Universe age crisis, which arose in the 1980s. With $H_0 \geq 70$ km/s/Mpc, the Universe would be too young, $t_U < 10$ Gyr, while stellar evolution and nuclear chronology demand $t_U \geq 13$ Gyr. The necessity of dark energy is seen from the expression for the Universe age which can be calculated by integration of the first Friedmann equation (4.9)

$$t_U = \frac{1}{H} \int_0^1 \frac{dx}{\sqrt{1 - \Omega_{tot}^0 + \Omega_m^0 x^{-1} + \Omega_r^0 x^{-2} + x^2 \Omega_\Lambda^0}}, \quad (11.2)$$

where we used the notations $\Omega_m^0 = \Omega_{CDM}^0 + \Omega_B^0$ for the density of non-relativistic matter, Ω_Λ^0 for the density of vacuum energy with $w = -1$, or vacuum-like energy with $w \approx -1$, and Ω_r for the density of relativistic matter. The values of all the parameters are taken at the present time, as indicated by the upper index 0. For $\Omega_{tot}^0 = 1$, $\Omega_m^0 \approx 0.3$, and $\Omega_\Lambda^0 \approx 0.7$, the calculated Universe age well agrees with the observationally deduced value. There is some contributions of semi-relativistic neutrinos if their mass is non-zero and larger than $1.6 \cdot 10^{-4}$ eV. According to neutrino oscillation data, at least two neutrino mass eigenstates may have that large mass, but the neutrino impact on t_U is minor.
2. The low magnitude of the matter density parameter, $\Omega_m^0 = 0.3$. It was measured by several independent methods: through the mass-to-light ratio of galaxies, gravitational lensing of distant objects, galactic clusters evolution (number of clusters for different redshifts z), spectrum of the angular fluctuations of the CMB, etc.
3. On the other hand, inflation predicts $\Omega_{tot}^0 = 1$ and it is indeed proven by the position of the first highest peak in the spectrum of the angular fluctuations of the CMB. The position of the first peak quite accurately determines $\Omega_{tot}^0 = 1 \pm 0.03$.
4. The data on the large scale structure formation and on the angular fluctuations of the CMB well fit the theory if $\Omega_\Lambda^0 \approx 0.7$. In an accelerated universe, the fluctuations of the matter density and of temperature at large scales are suppressed and this effect is clearly observed.
5. Last but not least, there are direct measurements of acceleration by the dimming of high z supernovae of type Ia. There exist persuasive arguments that these supernovae are so-called standard candles, namely sources with known luminosity. If this is true, then being dimmer means that these supernovae are at a larger distance than they would be if the expansion was the normal decelerated one. So a possible conclusion is that the Universe expands faster than expected. The dimming could be created by a light absorption on the way from the supernovae to the observer due to some unknown agent. However, the observed non-monotonic dependence on z excludes this explanation of the dimming. Indeed, if the dimming is induced by the accelerated expansion, then at higher z the observed dimming should decrease, because $\rho_m \sim 1/a^3$, while $\rho_\Lambda = const$ and equilibration between cosmological repulsion and attraction take place at $z \approx 0.65$, while at larger z the usual attractive gravity operated. Hence the brightness of supernovae observed at these higher z would tend to the normal value expected in the standard decelerated cosmology. Evidently if the dimming is explained by some absorptive medium, the effect would increase with rising z, but this is not observed.

For the discovery of this striking effect, in 2011 Saul Perlmutter, Brian Schmidt, and Adam Riess received the Nobel Prize in physics, as stated: "for the discovery of the accelerating expansion of the Universe through observations of distant supernovae".

11.1.2 Acceleration by a Scalar Field

The simplest form of dark energy can be represented by a scalar field with the canonical kinetic term and a very small mass or, more precisely, with a very slowly varying potential. Such a field satisfies Eq. (6.12), which, in the homogeneous limit, i.e. $\phi = \phi(t)$ independent of the space coordinates, is reduced to

$$\ddot{\phi} + 3H\dot{\phi} + U'(\phi) = 0. \tag{11.3}$$

This is equivalent to the equation of motion of a point-like body in Newtonian mechanics with the potential $U(\phi)$ and the liquid friction term $H\dot{\phi}$. If the Hubble parameter is large, as it is specified below, then the Newtonian "acceleration" $\ddot{\phi}$ can be neglected and Eq. (11.3) reduces to the first order equation

$$\dot{\phi} = -\frac{U'(\phi)}{3H}. \tag{11.4}$$

This is the so-called slow-roll approximation, see also Sect. 6.3 about this approximation for the description of inflation. If the cosmological energy density is dominated by a slowly varying ϕ, then $\rho_\phi \approx U(\phi)/2$, see Eq. (6.13), and according to the expression (4.9) we have $H^2 = 4\pi U/3M_{Pl}^2$. Note that, for a slowly varying ϕ, the vacuum-like condition $P = -\rho$ is approximately fulfilled. As we know, such an equation of state leads to an accelerated quasi-exponential expansion.

The slow roll approximation is valid if $\ddot{\phi} \ll 3H\dot{\phi}$ and $\dot{\phi}^2 \ll 2U(\phi)$. These conditions are realized if $U''/U \ll 8\pi/3M_{Pl}^2$, which, in turn, demands a very large magnitude of ϕ. For example, for the harmonic potential $U = m^2\phi^2/2$, the amplitude of ϕ should be larger than the Planck mass: $\phi^2 > (4\pi/3) M_{Pl}^2$. If we demand that the energy density of ϕ is of the order of the present day cosmological energy density, the mass of ϕ should be tiny, $m_\phi < 1/t_U \approx 10^{-42}$ eV.

It is usually assumed that the field ϕ, though slowly, is dropping down. A constant $\phi = \phi_0$ with $U(\phi_0) \neq 0$ is equivalent to a cosmological constant, which is a viable candidate as a driving force of the cosmological acceleration, but the idea of the quasi-dynamical phenomenology with a scalar field is to invent something that is not just a trivial subtraction constant of the vacuum energy. By construction, the potential $U(\phi)$ is chosen in such a form that it vanishes at the equilibrium point, where $U' = 0$. This condition eliminates a trivial vacuum energy. There are several suggestions for $U(\phi)$ smoothly tending to zero when $\phi \to \pm\infty$. For such potentials, U' automatically tends to zero in this limit. Simple examples of such potentials are $U \sim 1/\phi^q$ or $U \sim \exp(-\phi/\mu)$, where μ is a constant parameter with dimension of mass (Lucchin and Matarrese 1985; Sahni et al. 1992). These potentials were specially introduced for a phenomenological description of the accelerated expansion. As for the fundamental reasons for their existence, they are rather vague.

The motion of $\phi(t)$ in such potentials is quite different from those with minima at a finite $\phi = \phi_0$, for instance $U(\phi) = m_\phi^2(\phi - \phi_0)^2/2$ or $U(\phi) = \lambda(\phi - \phi_0)^4/4$. Such potentials are natural in quantum field theory, because they lead to renormalizable theory. When ϕ is sufficiently close to ϕ_0 that the square of the Hubble parameter H^2 becomes comparable or smaller than m_ϕ^2, or $\lambda(\phi - \phi_0)^2$, the quasi-exponential accelerated expansion turns into the good old decelerated one. At the onset of this regime, ϕ starts oscillating around its minimum, producing massless elementary particles, and the expansion turns back into a decelerated regime. On the other hand, e.g. for an exponential potential, the cosmological scale factor would evolve as

$$a \sim t^{16\pi\mu/M_{\rm Pl}}, \qquad (11.5)$$

and for $\mu > M_{\rm Pl}/16\pi$ the expansion regime would always be accelerated.

11.1.3 Modified Gravity

A competing description of the accelerated expansion might be through a gravity modification at large scales. As we have already mentioned, this can be done by adding a non-linear function of curvature $F(R)$ to the usual Einstein-Hilbert action

$$S = \frac{M_{\rm Pl}^2}{16\pi} \int d^4x \sqrt{-g}[R + F(R)] + S_m, \qquad (11.6)$$

where S_m is the matter action. The function $F(R)$ is chosen in such a way that the gravitational equations of motion, which replace the usual Einstein equations, have an accelerated de Sitter-like solution with a constant curvature R even in the absence of matter. The choice of $F(R)$ is by no way unique and several possibilities are explored in the literature.

The equations of motion in such a theory have the form

$$\left(1 + F'\right) R_{\mu\nu} - \frac{1}{2}(R + F) g_{\mu\nu} + \left(g_{\mu\nu} \nabla_\alpha \nabla^\alpha - \nabla_\mu \nabla_\nu\right) F' = \frac{8\pi}{M_{\rm Pl}^2} T_{\mu\nu}, \quad (11.7)$$

where $F' = dF/dR$ and ∇_μ is the covariant derivative. It is often sufficient to consider only the equation for the trace of Eq. (11.7)

$$3\nabla^2 F'_R - R + R F'_R - 2F = \frac{8\pi}{M_{\rm Pl}^2} T^\mu_\mu, \qquad (11.8)$$

where $\nabla^2 \equiv \nabla_\mu \nabla^\mu$ is the covariant D'Alember operator. A second order equation for the metric appears only in the classical Einstein theory, when the action is linear in R. A non-linear function $F(R)$ leads to a higher order equation of motion. Such an equation may give rise to undesirable consequences for the theory: emergence of ghosts, tachyons, singular behavior of solutions, instability, etc., so special care should be taken to avoid these problems.

11.1 Cosmological Acceleration

In the first papers in which gravity modifications were proposed for a description of the accelerated expansion (Capozziello et al. 2003; Carroll et al. 2004), the function $F(R)$ was taken as $F(R) = -\mu^4/R$. However, it was shown in Dolgov et al. (2003) that a similar $F(R)$ leads to an exponential instability in the presence of matter, so the usual gravitational interactions would be strongly distorted. To cure this "disease", further modifications have been suggested. There are several proposals (Appleby and Battye 2007; Hu and Sawicki 2007; Starobinsky 2007) of $F(R)$ with similar properties. For instance, the proposal of Starobinsky (2007) is

$$F(R) = \lambda R_0 \left[\left(1 + \frac{R^2}{R_0^2} \right)^{-n} - 1 \right] - \frac{R^2}{6m^2}. \tag{11.9}$$

The last term is added to prevent a past singularity in cosmology. It can also eliminate the future singularity in systems with rising energy/mass density found in Frolov (2008), Arbuzova and Dolgov (2011).

The suggested theories of modified gravity possess some peculiar features. It was found in Arbuzova and Dolgov (2011), Arbuzova et al. (2012, 2013) that, in systems with rising mass/energy density, high frequency and large amplitude oscillations of the curvature are induced. These oscillations lead to the production of elementary particles, which may be observable in the spectra of energetic cosmic rays. In the background of this oscillating solution, gravitational repulsion between objects of finite size is possible (Arbuzova et al. 2014). Such a repulsion might be responsible for the creation of the observed cosmic voids.

11.2 Problem of Vacuum Energy

The problem of vacuum energy is quite a unique example of disagreement between theoretical expectations and data by 50–100 orders of magnitude. The story began almost a century ago, when Einstein introduced into his equations an additional term proportional to the metric tensor (Einstein 1918)

$$R_{\mu\nu} - \frac{1}{2} g_{\mu\nu} R - \Lambda g_{\mu\nu} = \frac{8\pi}{M_{\text{Pl}}^2} T_{\mu\nu}. \tag{11.10}$$

The coefficient Λ must be constant to satisfy the constraints of general covariance and energy-momentum conservation, namely $\nabla^\mu G_{\mu\nu} = 0$, $\nabla^\mu g_{\mu\nu} = 0$, and $\nabla^\mu T_{\mu\nu} = 0$ (see Chap. 2). Λ is usually called cosmological constant. The Λ-term is evidently equivalent to the energy-momentum tensor of vacuum

$$T_{\mu\nu}^{(vac)} = \rho^{(vac)} g_{\mu\nu}. \tag{11.11}$$

There are several theoretically expected natural contributions into $\rho^{(vac)}$ and there is one among them, which is not just theory but practically an experimental fact. Though this term is not the largest but still huge. It is 45 orders of magnitude larger than the cosmological energy density. There are many reviews (Binetruy 2000; Burgess 2004; Dolgov 1989, 1998; Fujii 2000; Kim 2004; Martin 2012; Peebles and Ratra 2003; Sahni 2002; Sahni and Starobinsky 2000; Straumann 2002; Vilenkin 2001; Weinberg 1989, 2000) on the problem and suggestions for its solution, not very successful so far, so we will not dwell on the theoretical constructions but we will only describe this most striking contribution to $\rho^{(vac)}$. QCD certainly demonstrates that something mysterious happens in the vacuum. u- and d-quarks making protons, $p = uud$, and neutrons, $n = udd$, have very small masses, around 5 MeV. So the nucleon mass should be 15 MeV minus its binding energy, instead of approximately 940 MeV. The solution to this problem suggested by QCD is that the vacuum is not empty but filled with a quark (Gell-Mann et al. 1968) and a gluon (Shifman et al. 1979) condensates, $\langle \bar{q}q \rangle \neq 0$ and $\langle G_{\mu\nu}G^{\mu\nu} \rangle \neq 0$, which have altogether the negative vacuum energy

$$\rho_{vac}^{(QCD)} \approx -0.01\,\text{GeV}^4 \approx -10^{45} \rho_0, \qquad (11.12)$$

where ρ_0 is the present day energy density of the Universe. The vacuum condensates turn out to be destroyed around quarks, and the result is that the proton mass becomes

$$m_p = 2m_u + m_d - \rho_{vac}^{(QCD)} l_p^3 \sim 1\,\text{GeV}, \qquad (11.13)$$

where $l_p \sim$ a few GeV^{-1} is the proton size.

The value of the vacuum energy of the quark and gluon condensates (11.12) is practically established by experiments. To adjust the total vacuum energy down to the observed magnitude, $\sim 10^{-47}\,\text{GeV}^4$, there must exist another contribution to the vacuum energy of the opposite sign and equal to the QCD one with the precision of one part to 10^{45}. This new field cannot have any noticeable interactions with quarks and gluons, otherwise it would be observed in direct experiments, and though it does not know (almost) anything about QCD, it must have the same vacuum energy density as the condensates mentioned above. This is one of the greatest mysteries of Nature.

The problems of vacuum and dark energies are surely closely connected and there is little hope to understand the nature of dark energy without the solution of the vacuum energy problem. There is another mystery, namely why the energy densities of matter and vacuum are so close to each other at the present time despite different laws of their evolution in the course of the cosmological expansion: $\rho_m \sim 1/t^2$ and $\rho_{vac} = const$. All these problems may be solved by a dynamical adjustment mechanism, but unfortunately a satisfactory model is not yet found, see e.g. Dolgov (1982), the reviews Binetruy 2000; Burgess 2004; Dolgov 1989, 1998; Fujii 2000; Kim 2004; Martin 2012; Peebles and Ratra 2003; Sahni 2002; Sahni and Starobinsky 2000; Straumann 2002; Vilenkin 2001; Weinberg 1989, 2000, or the lectures Dolgov 2008.

References

E.V. Arbuzova, A.D. Dolgov, Phys. Lett. B **700**, 289 (2011). arXiv:1012.1963 [astro-ph.CO]
E.V. Arbuzova, A.D. Dolgov, L. Reverberi, Eur. Phys. J. C **72**, 2247 (2012). arXiv:1211.5011 [gr-qc]
E.V. Arbuzova, A.D. Dolgov, L. Reverberi, Phys. Rev. D **88**(2), 024035 (2013). arXiv:1305.5668 [gr-qc]
E.V. Arbuzova, A.D. Dolgov, L. Reverberi, Astropart. Phys. **54**, 44 (2014). arXiv:1306.5694 [gr-qc]
S.A. Appleby, R.A. Battye, Phys. Lett. B **654**, 7 (2007). arXiv:0705.3199 [astro-ph]
P. Binetruy, Int. J. Theor. Phys. **39**, 1859 (2000) [hep-ph/0005037]
G.D. Birkhoff, *Relativity and Modern Physics* (Harvard University Press, Cambridge, 1923)
C.P. Burgess, Ann. Phys. **313**, 283 (2004) [hep-th/0402200]
S. Capozziello, S. Carloni, A. Troisi, Recent Res. Dev. Astron. Astrophys. **1**, 625 (2003) [astro-ph/0303041]
S.M. Carroll, V. Duvvuri, M. Trodden, M.S. Turner, Phys. Rev. D **70**, 043528 (2004) [astro-ph/0306438]
A.D. Dolgov, in *The Very Early Universe*, ed. by G. Gibbons, S.W. Hawking, S.T. Tiklos (Cambridge University Press, Cambridge, 1982)
A.D. Dolgov, in *Proceedings of the XXIVth Rencontre de Moriond*, eds. J. Adouse, J. Tran Thanh Van (Les Arcs, France, 1989)
A.D. Dolgov, in *Fourth Paris Cosmology Colloquium*, ed. by H.J. De Vega, N. Sanchez (World Scientific, Singapore, 1998)
A.D. Dolgov, Phys. Atom. Nucl. **71**, 651 (2008) [hep-ph/0606230]
A.D. Dolgov, M. Kawasaki, Phys. Lett. B **573**, 1 (2003) [astro-ph/0307285]
A. Einstein, Sitzgsber. Preuss. Acad. Wiss. **1**, 142 (1917)
A.V. Frolov, Phys. Rev. Lett. **101**, 061103 (2008). arXiv:0803.2500 [astro-ph]
Y. Fujii, Grav. Cosmol. **6**, 107 (2000) [gr-qc/0001051]
M. Gell-Mann, R.J. Oakes, B. Renner, Phys. Rev. **175**, 2195 (1968)
W. Hu, I. Sawicki, Phys. Rev. D **76**, 064004 (2007). arXiv:0705.1158 [astro-ph]
J.T. Jebsen, Norsk Matematisk Tidsskrift (Oslo) **3**, 21 (1921)
J.E. Kim, Mod. Phys. Lett. A **19**, 1039 (2004) [hep-ph/0402043]
F. Lucchin, S. Matarrese, Phys. Rev. D **32**, 1316 (1985)
J. Martin, C. R. Phys. **13**, 566 (2012). arXiv:1205.3365 [astro-ph.CO]
K.A. Olive et al., [Particle Data Group Collaboration], Chin. Phys. C **38**, 090001 (2014)
P.J.E. Peebles, B. Ratra, Rev. Mod. Phys. **75**, 559 (2003) [astro-ph/0207347]
V. Sahni, Class. Quant. Grav. **19**, 3435 (2002) [astro-ph/0202076]
V. Sahni, H. Feldman, A. Stebbins, Astrophys. J. **385**, 1 (1992)
V. Sahni, A.A. Starobinsky, Int. J. Mod. Phys. D **9**, 373 (2000) [astro-ph/9904398]
M.A. Shifman, A.I. Vainshtein, V.I. Zakharov, Nucl. Phys. B **147**, 385 (1979)
A.A. Starobinsky, JETP Lett. **86**, 157 (2007). arXiv:0706.2041 [astro-ph]
N. Straumann, astro-ph/0203330 (2002)
A. Vilenkin, hep-th/0106083 (2001)
S. Weinberg, Rev. Mod. Phys. **61**, 1 (1989)
S. Weinberg, astro-ph/0005265 (2000)

Chapter 12
Density Perturbations

The behavior of instabilities of self-gravitating systems was first investigated in the case of non-relativistic Newtonian gravity by Jeans (1902). It was later extended to General Relativity by Lifshitz (1946). Nowadays, this theory is widely used in cosmology to study the rise of perturbations in the Universe (Zeldovich and Novikov 1983; Mukhanov 2005; Weinberg 2008; Gorbunov and Rubakov 2011). The comparison between theoretical calculations and astronomical data is a very powerful tool for testing the Standard Model of cosmology.

12.1 Density Perturbations in Newtonian Gravity

The original Jeans approach is based on the well known Poisson equation, which relates the Newtonian potential Φ to the matter density ρ

$$\Delta\Phi = \frac{4\pi}{M_{\text{Pl}}^2}\rho\,. \tag{12.1}$$

The evolution of the matter density ρ, the pressure P, and the velocity $\mathbf{v}$ in a self-created gravitational field is governed by two hydrodynamic equations, namely the Euler equation and the continuity equation, given, respectively, by

$$\partial_t(\rho\mathbf{v}) + \rho(\mathbf{v}\nabla)\mathbf{v} + \nabla P + \rho\nabla\Phi = 0\,, \tag{12.2}$$

$$\partial_t\rho + \nabla(\rho\mathbf{v}) = 0\,. \tag{12.3}$$

At this stage, there are three equations and four unknowns: ρ, P, $\mathbf{v}$, and Φ. To obtain one more necessary equation, we need a physical input, namely an information about the properties of matter, which are specified below by the equation of state (12.5).

The original version of this chapter was revised: The errors in this chapter have been corrected. The correction to this chapter can be found at https://doi.org/10.1007/978-3-662-48078-6_13.

This system of equations is usually solved perturbatively under the assumption of infinitesimally small fluctuations over the known background quantities:

$$\rho = \rho_b + \delta\rho, \quad \mathbf{v} = \mathbf{v_b} + \delta\mathbf{v}, \quad P = P_b + \delta P, \quad \Phi = \Phi_b + \delta\Phi. \quad (12.4)$$

The result is a system of three linear differential equations for the perturbations $\delta\rho$, $\delta\mathbf{v}$, δP, and $\delta\Phi$. To close the system, one usually imposes the "acoustic" equation of state

$$\delta P = c_s^2 \delta\rho, \quad (12.5)$$

where c_s is the speed of sound.

In the Jeans theory, it is assumed that the background mass density is homogeneous and time independent and that the background pressure and the background velocity vanish, namely $P_b = 0$ and $\mathbf{v}_b = 0$. One can immediately see that these assumptions are nor self-consistent. From Eq. (12.2), it follows that the background potential must be spatially constant, $\nabla\Phi_b = 0$, but this contradicts the Poisson equation (12.1) at the zeroth order, i.e. for the background quantities.

This problem is discussed in Zeldovich and Novikov (1983), who argued that in a time dependent but spatially constant background, $\rho_b(t)$, the theory can be formulated in a self-consistent way. Physically, such a case is realized in cosmology.

On the other hand, to cure this shortcoming in flat spacetime, Mukhanov (2005) suggested the addition of an artificial antigravitating substance, such as a vacuum-like energy, that can counterbalance the gravitational attraction of the background. In this way Eq. (12.1) can be satisfied at the zeroth order. An alternative possibility employed in Eingorn et al. (2014) is that the background density vanishes, so Eq. (12.1) becomes a relation between first order quantities.

We note that this problem is absent in relativistic cosmology, where the zeroth order background equations are satisfied, see e.g. Zeldovich and Novikov (1983), Gorbunov and Rubakov (2011). Contrary to the case of perturbations in flat spacetime, the background quantities in cosmology are solutions of the equations of motion at the zeroth order approximation. If we neglect this problem and assume that the background energy/mass density is homogeneous and time independent and that the background gravitational potential vanishes, or it is constant, we arrive at the classical Jeans result. These two conditions, $\rho_b = const$ and $\Phi_b = const$, are in clear contradiction with Eq. (12.1).[1] Nevertheless, we proceed further, make the Fourier transformation, $\sim\exp(-i\mu t + i\mathbf{k}\mathbf{r})$, and expand Eqs. (12.1)–(12.3) up to the first order in terms of the Fourier amplitudes of the δ quantities, such as $\delta\rho_k$, etc. (to simplify the notation, we omit the sub-k indices below):

---

[1] The assumption that $\rho_b = const$ is technically essential because it allows to reduce the differential equations governing the evolution of perturbations to algebraic ones by Fourier transformation. For time dependent $\rho(t)$, as it takes place in cosmology, one has either to find analytical solutions of an ordinary differential equation for $\delta\rho_k(t)$ or to solve it numerically.

12.1 Density Perturbations in Newtonian Gravity

$$-k^2 \delta \Phi = \frac{4\pi}{M_{\text{Pl}}^2} \delta\rho, \tag{12.6}$$

$$-i\mu\rho_b \delta \mathbf{v} + i\mathbf{k} c_s^2 \delta\rho + i\mathbf{k}\delta\Phi = 0, \tag{12.7}$$

$$-i\mu \delta\rho + \rho_b (\mathbf{k}\delta\mathbf{v}) = 0. \tag{12.8}$$

This system is reduced to a single equation for $\delta\rho$

$$\delta\rho \left(-\mu^2 + k^2 c_s^2 - \frac{4\pi}{M_{\text{Pl}}^2} \rho_b \right) = 0, \tag{12.9}$$

which has a non-trivial solution if

$$\mu = \pm \sqrt{k^2 c_s^2 - \frac{4\pi}{M_{\text{Pl}}^2} \rho_b}. \tag{12.10}$$

If $k^2 c_s^2 > 4\pi\rho_b / M_{\text{Pl}}^2$, the pressure force dominates over the gravity force and the density perturbations oscillate propagating as sound waves. In the opposite case, gravity is stronger and the density perturbations exponentially rise

$$\delta\rho/\rho_b \sim \exp\left[t \sqrt{\frac{4\pi}{M_{\text{Pl}}^2} \rho_b - k^2 c_s^2} \right]. \tag{12.11}$$

The Jeans wave vector

$$k_J = \frac{\sqrt{4\pi\rho_b}}{M_{\text{Pl}} c_s} \tag{12.12}$$

is the boundary value of the wave number separating acoustic oscillations and rising perturbations.

The corresponding wavelength $\lambda_J = 2\pi/k_J$ is called the Jeans wavelength. The mass inside the Jeans radius λ_J is

$$M_J = \frac{4\pi \rho_b \lambda_J^3}{3} = \frac{4\pi^{5/2} c_s^3 M_{\text{Pl}}^3}{3\rho_b^{1/2}} \tag{12.13}$$

and it is called the *Jeans mass*. Objects with $M > M_J$ continue collapsing until, and if, the equation of state becomes more rigid. If this never happens, they turn into a black hole.

We may suggest a more accurate approach to the problem in such a way that the zeroth order equations for the background are satisfied (Arbuzova et al. 2014). As an example, we consider a spherically symmetric cloud of particles with initially vanishing pressure and velocities, and we study the classical non-relativistic Jeans problem in Newtonian gravity. We will not confine ourselves to the case of a time independent background. We want instead to consider the time dependent scenario taking as initial condition that of a homogeneous distribution $\rho_b(t=0) \equiv \rho_0 = const$ inside a sphere of radius r_m, while outside this sphere $\rho = 0$. The initial values for the particle velocities and pressure are taken to be zero and the initial potential Φ at $t = 0$ is supposed to be solution of the Poisson equation (12.1)

$$\Phi_b(t=0, r > r_m) = -\frac{M}{r M_{\text{Pl}}^2},$$

$$\Phi_0 \equiv \Phi_b(t=0, r < r_m) = \frac{2\pi}{3 M_{\text{Pl}}^2} \rho_0 r^2 + C_0, \qquad (12.14)$$

where the total mass of the gravitating sphere is

$$M = \frac{4\pi}{3} \rho_0 r_m^3 \qquad (12.15)$$

and $C_0 = -2\pi \rho_0 r_m^2 / M_{\text{Pl}}^2$ to make the potential continuous at $r = r_m$ (the value of C_0 is not important for us).

In what follows we are interested in the internal solution $r < r_m$. Now we can find how the background quantities ρ, $\mathbf{v}$, and P evolve with time for small t. From Eq. (12.2), it follows that

$$\mathbf{v}_b(r, t) = -\nabla \Phi_0 \, t = -\frac{4\pi}{3 M_{\text{Pl}}^2} \rho_0 \, t \, \mathbf{r}. \qquad (12.16)$$

From the continuity equation, we find

$$\rho_b(t, r) = \rho_0 + \rho_1 = \rho_0 \left(1 + \frac{2\pi}{3 M_{\text{Pl}}^2} \rho_0 t^2\right). \qquad (12.17)$$

It is interesting that ρ rises with time in such a way that it remains constant in space. Because of the homogeneity of ρ, the pressure remains zero, $P_1 = 0$. The time variation of the background potential is found using Eq. (12.1)

$$\Phi_b(r, t) = \Phi_0 + \Phi_1 = \frac{2\pi}{3 M_{\text{Pl}}^2} r^2 \rho_0 \left(1 + \frac{2\pi}{3 M_{\text{Pl}}^2} \rho_0 t^2\right). \qquad (12.18)$$

Now we can study the evolution of perturbations over this time dependent background. We proceed as usually, writing $\rho = \rho_b(r, t) + \delta\rho$, $\Phi = \Phi_b(r, t) + \delta\Phi$, $\mathbf{v} = \mathbf{v}_1(r, t) + \delta\mathbf{v}$, and $\delta P = c_s^2 \delta\rho$, where c_s is the speed of sound. Here all the δ quantities are infinitesimal and are neglected beyond the first order. At the first step, we also neglect the products of small sub-one quantities with the infinitesimal delta's, i.e. such terms as, for instance, the product of $\rho_1 = 2\pi \rho_0^2 t^2 / (3 M_{\text{Pl}}^2)$ by $\delta\rho$, assuming that the time elapsed is sufficiently short. Thus we obtain

12.1 Density Perturbations in Newtonian Gravity

$$\Delta(\delta\Phi) = \frac{4\pi}{M_{Pl}^2}\delta\rho, \tag{12.19}$$

$$\partial_t \delta\mathbf{v} + \nabla\delta\Phi + \frac{\delta\rho}{\rho_0}\nabla\Phi_b + \frac{\nabla\delta P}{\rho_0} = 0, \tag{12.20}$$

$$\partial_t \delta\rho + \rho_0 \nabla(\delta\mathbf{v}) = 0. \tag{12.21}$$

Within this approximations, the time dependence of the coefficients disappears. Later we will include time dependent corrections to the background.

The problem of the perturbation evolution is usually studied for the Fourier transformed quantities, which allows to reduce a system of partial differential equations containing time and space derivatives to a system of ordinary differential equations for functions of time only. Moreover, if the coefficients of the latter are constant, making the Fourier transformation with respect to time we come to a system of algebraic equations, as it is done in the standard Jeans analysis performed above.

Equation (12.20) contains the term $(\delta\rho/\rho_0)\nabla\Phi_b$, which explicitly depends on the coordinate $\mathbf{r}$ through the background potential $\nabla\Phi_b = (4\pi/3)r\rho_0/M_{Pl}^2$. So, strictly speaking, the spatial Fourier transformation would not lead to a system of ordinary differential equations with respect to time for the Fourier amplitudes such as, e.g., $\delta\rho_\mathbf{k}(t)$ with an algebraic dependence of its coefficients on the wave number, $\mathbf{k}$. Instead the term containing r_j Fourier-transforms as

$$\int d^3r\, r_j e^{i\mathbf{kr}} \delta X(\mathbf{r}) = -i \frac{\partial \delta X(\mathbf{k})}{\partial k_j}, \tag{12.22}$$

where δX is one or other infinitesimal perturbation. As a result, we obtain a differential equation in terms of derivatives over k, which is by no means simpler than the original equations in the coordinate space. However, in the two opposite limits of very small r and large kr, the Fourier transformation makes sense and the equations can be practically reduced to the case of the coordinate independent situation.

To see if such terms are essential, we need to compare the Fourier transform of the last term in Eq. (12.20), namely

$$\int \frac{d^3k}{(2\pi)^3} \frac{\nabla\delta P}{\rho_0} e^{-i\lambda t + i\mathbf{kr}} \sim kc_s^2 \frac{\delta\rho(\lambda, \mathbf{k})}{\rho_0}, \tag{12.23}$$

with the third term in the same equation. In other words, we have to compare kc_s^2 with $\nabla\Phi_b = (4\pi/3)r\,\rho_0/M_{Pl}^2$, see Eq. (12.18). For a homogeneous matter sphere with radius r_m, this term is equal to

$$\frac{4\pi r\rho_0}{3M_{Pl}^2} = \frac{r_g r}{2r_m^3}, \tag{12.24}$$

where $r_g = 2M/M_{\text{Pl}}^2$ is the gravitational (Schwarzschild) radius and $M = (4\pi/3)\rho_0 r_m^3$ is the total mass of the spherical cloud under scrutiny. This term would be subdominant for $r < 2c_s^2 r_m^3 k/r_g$.

We should keep in mind, however, that the wave number k cannot be arbitrary small. For an object with size r_m the treatment is valid for the wave lengths smaller than r_m. It means that $k > 2\pi/r_m$. Still there is quite a large parameter space if, as it is usually true, $r_g \ll r_m$.

We have also to check if the Jeans wave length satisfies the condition $\lambda_J < r_m$. It would be fulfilled if

$$\frac{3c_s^2}{8\pi^2} \frac{r_g}{r_m} < 1. \tag{12.25}$$

In this parameter range, we can neglect the r dependent term, $(\delta\rho/\rho_0)\nabla\Phi_b$, in comparison to $\nabla\delta P/\rho_0$. Within this approximation, the system (12.19)–(12.21) coincides with the classical system (12.1)–(12.3). Thus we obtain the usual Jeans result, which justifies the original approach. The case of non-negligible dependence on **r**, which occurs for large r, is considered in Sect. 12.3, where the evolution of perturbations in modified gravity is studied.

Let us now estimate the effect of time variations of the background potential, velocity, and energy density. The characteristic rising time of small perturbations is of the order $(4\pi\rho_0/M_{\text{Pl}}^2)^{-1/2}$. It is the same as the time of the classical rise of ρ_1 (12.17). Hence an account of time variation of the background quantities would noticeably change the evolution of perturbations. We can estimate the impact of a rising background energy density on the rise of perturbations in adiabatic approximation replacing the exponent in Eq. (12.11) with the integral

$$\frac{\delta\rho_{J1}}{\rho_0} \sim \exp\left\{\int_0^t dt \left[\frac{4\pi}{M_{\text{Pl}}^2}\rho_b(t,r) - k^2 c_s^2\right]^{1/2}\right\}. \tag{12.26}$$

where $\rho_b(t,r)$ is given by Eq. (12.17). Estimating the above integral for small k, we find that the enhancement factor $\delta\rho_{J1}/\delta\rho_J$ is equal to 1.027 after a time $t = t_{grav}$, where $t_{grav} = M_{\text{Pl}}/\sqrt{4\pi\rho_0}$, while for $t = 2t_{grav}$ it is 1.23, for $t = 3t_{grav}$ it is 1.89, and for $t = 5t_{grav}$ it is 11.9. Here $\delta\rho_J$ is the classical Jeans perturbation in the time independent background (12.11). Note that to derive (12.11) and (12.26) we assume that $t \leq t_{grav}$, so we should not treat these factors as numerically accurate. Still we can interpret them as an indication that the rise of fluctuations is indeed faster then in the usual Jeans scenario. For more precise evaluation, one can solve the Fourier transformed ordinary differential equations numerically in time dependent background.

12.2 Density Perturbations in General Relativity

In this section, we consider the evolution of scalar perturbations in General Relativity. We study the perturbation rise in a spherically symmetric and asymptotically flat spacetime and in the cosmological FRW background. First, we present the necessary expressions for the metric, the curvature tensors, and the energy-momentum tensor of matter. The latter is assumed to have the form of that of a perfect fluid. Second, we discuss the choice of gauge for perturbations in the coordinate dependent background. We then study the rise of perturbations in a spherically symmetric and asymptotically flat background, which may depend on both time and space coordinates. This is a generalization of the classical Jeans problem with an account of General Relativity effects. We follow here our paper (Arbuzova et al. 2014). Lastly, we describe the density evolutions in the cosmological FRW background. More detailed considerations of perturbation evolutions in cosmology can be found in Zeldovich and Novikov (1983), Gorbunov and Rubakov (2011).

12.2.1 Metric and Curvature

As in the previous section, we consider a spherically symmetric cloud of matter with an initially constant energy density inside the boundary radius $r = r_m$. We choose isotropic coordinates in which the line element takes the form

$$ds^2 = A dt^2 - B \delta_{ij} dx^i dx^j, \qquad (12.27)$$

where the functions A and B may depend upon r and t. The corresponding Christoffel symbols are

$$\Gamma^t_{tt} = \frac{\dot{A}}{2A}, \quad \Gamma^t_{jt} = \frac{\partial_j A}{2A}, \quad \Gamma^j_{tt} = \frac{\delta^{jk}\partial_k A}{2B}, \quad \Gamma^t_{jk} = \frac{\delta_{jk}\dot{B}}{2A},$$

$$\Gamma^k_{jt} = \frac{\delta^k_j \dot{B}}{2B}, \quad \Gamma^k_{lj} = \frac{1}{2B}(\delta^k_l \partial_j B + \delta^k_j \partial_l B - \delta_{lj}\delta^{kn}\partial_n B). \qquad (12.28)$$

For the Ricci tensor, including terms quadratic in Γs, we obtain

$$R_{tt} = \frac{\Delta A}{2B} - \frac{3\ddot{B}}{2B} + \frac{3\dot{B}^2}{4B^2} + \frac{3\dot{A}\dot{B}}{4AB} + \frac{\partial^j A \partial_j B}{4B^2} - \frac{\partial^j A \partial_j A}{4AB}, \qquad (12.29)$$

$$R_{tj} = -\frac{\partial_j \dot{B}}{B} + \frac{\dot{B}\partial_j B}{B^2} + \frac{\dot{B}\partial_j A}{2AB}, \qquad (12.30)$$

$$R_{ij} = \delta_{ij}\left(\frac{\ddot{B}}{2A} - \frac{\Delta B}{2B} + \frac{\dot{B}^2}{4AB} - \frac{\dot{A}\dot{B}}{4A^2} - \frac{\partial^k A \partial_k B}{4AB} + \frac{\partial^k B \partial_k B}{4B^2}\right)$$
$$- \frac{\partial_i \partial_j A}{2A} - \frac{\partial_i \partial_j B}{2B} + \frac{\partial_i A \partial_j A}{4A^2} + \frac{3\partial_i B \partial_j B}{4B^2} + \frac{\partial_i A \partial_j B + \partial_j A \partial_i B}{4AB}. \quad (12.31)$$

Here and in what follows, the upper space indices are raised with the Kronecker delta, namely $\partial^j A = \delta^{jk}\partial_k A$. The curvature scalar is

$$R = \frac{\Delta A}{AB} - \frac{3\ddot{B}}{AB} + \frac{2\Delta B}{B^2} + \frac{3\dot{A}\dot{B}}{2A^2 B} - \frac{\partial^j A \partial_j A}{2A^2 B} - \frac{3\partial^j B \partial_j B}{2B^3} + \frac{\partial^j A \partial_j B}{2AB^2}. \quad (12.32)$$

The Einstein tensor $G_{\mu\nu} = R_{\mu\nu} - 1/2\, g_{\mu\nu} R$ is

$$G_{tt} = -\frac{A\Delta B}{B^2} + \frac{3\dot{B}^2}{4B^2} + \frac{3A\partial^j B \partial_j B}{4B^3}, \quad (12.33)$$

$$G_{tj} = R_{tj}, \quad (12.34)$$

$$G_{ij} = \delta_{ij}\left(\frac{\Delta A}{2A} + \frac{\Delta B}{2B} - \frac{\ddot{B}}{A} + \frac{\dot{B}^2}{4AB} + \frac{\dot{A}\dot{B}}{2A^2} - \frac{\partial^k A \partial_k A}{4A^2} - \frac{\partial^k B \partial_k B}{2B^2}\right)$$
$$- \frac{\partial_i \partial_j A}{2A} - \frac{\partial_i \partial_j B}{2B} + \frac{\partial_i A \partial_j A}{4A^2} + \frac{3\partial_i B \partial_j B}{4B^2} + \frac{\partial_i A \partial_j B + \partial_j A \partial_i B}{4AB}. \quad (12.35)$$

12.2.2 Energy-Momentum Tensor

The energy-momentum tensor is taken with the form of that of a perfect fluid without dissipative corrections

$$T_{\mu\nu} = (\rho + P)U_\mu U_\nu - P g_{\mu\nu}, \quad (12.36)$$

where ρ and P are, respectively, the energy and pressure densities of the fluid, and the 4-velocity is

$$U^\mu = \frac{dx^\mu}{ds} \quad \text{and} \quad U_\mu = g_{\mu\alpha}U^\alpha. \quad (12.37)$$

The infinitesimal physical (or proper) distance is $dl^2 = B dr^2$. Correspondingly, we define the vector of the physical velocity as $v^j = \sqrt{B}\, d\mathbf{r}/dt$. We assume that the 3-velocity is small and we thus neglect terms quadratic in $\mathbf{v}$. The result is

$$U_j = -\frac{Bv_j}{\sqrt{A}\sqrt{1 - \mathbf{v}^2/A}} \approx -\frac{Bv_j}{\sqrt{A}}, \quad (12.38)$$

12.2 Density Perturbations in General Relativity

where $v_j = v^j$. From the condition

$$1 = g^{\mu\nu}U_\mu U_\nu = \frac{1}{A}U_t^2 - \frac{1}{B}\delta^{kj}U_k U_j \approx \frac{1}{A}U_t^2, \qquad (12.39)$$

we find $U_t \approx \sqrt{A}$. Now we can write

$$T_{tt} = (\rho + P)U_t^2 - PA \approx \rho A, \qquad (12.40)$$
$$T_{jt} = (\rho + P)U_t U_j \approx -(\rho + P)v_j B/\sqrt{A}, \qquad (12.41)$$
$$T_{ij} = (\rho + P)U_i U_j - P g_{ij} \approx P B \delta_{ij}. \qquad (12.42)$$

12.2.3 Choice of Gauge

In a cosmological situation, the spatially flat FRW metric only depends on time, but not on space coordinates

$$ds^2_{cosmo} = dt^2 - a^2(t)d\mathbf{r}^2. \qquad (12.43)$$

As shown in several textbooks, see e.g. Mukhanov (2005), Weinberg (2008), Gorbunov and Rubakov (2011), this allows to impose the *Newtonian gauge condition* on the perturbed metric. For scalar perturbations, the line element takes the form

$$ds^2_{pert} = (1 + 2\Phi)dt^2 - a^2(t)(1 - 2\Psi)\delta_{ij}dx^i dx^j, \qquad (12.44)$$

where Φ and Ψ are the metric perturbations or, in other words, stochastic deviations from the cosmological background metric.

We consider a spherically symmetric and asymptotically flat background metric in isotropic coordinates, with the line element equal to (see e.g. Chap. 16 in Lightman et al. (1975) or Sect. 12.3)

$$ds^2_{sph} = A\, dt^2 - B\, \delta_{ij}\, dx^i dx^j, \qquad (12.45)$$

where A and B are functions of space and time in the form

$$A(t,r) = 1 + A_1(t)r^2, \quad B(t,r) = 1 + B_1(t)r^2. \qquad (12.46)$$

Calculations are greatly simplified if we assume that deviations from the Minkowski metric are sufficiently weak and so $A \approx 1$ and $B \approx 1$. The dependence of the background on the space coordinates creates serious problems when one tries to impose the Newtonian gauge condition, as we illustrate in what follows.

For scalar fluctuations, the general form of the perturbed metric is

$$ds^2_{scalar} = (A + 2\Phi)dt^2 + (\partial_j C)dt\, dx^j - \left[(B - 2\Psi)\delta_{ij} - \partial_i\partial_j E\right] dx^i dx^j.\tag{12.47}$$

The Newtonian gauge condition implies $C = E = 0$, which can be easily realized in cosmology by a proper change of coordinates. Under the coordinate transformation $\tilde{x}^\alpha = x^\alpha + \xi^\alpha$, the metric tensor transforms as

$$\tilde{g}_{\alpha\beta}(\tilde{x}) = g^b_{\alpha\beta}(\tilde{x}) + \delta g_{\alpha\beta} - g^b_{\alpha\mu}\partial_\beta\xi^\mu - g^b_{\beta\mu}\partial_\alpha\xi^\mu,\tag{12.48}$$

where $g^b_{\alpha\beta}$s are the "old" background metric coefficients at the point $\tilde{x}$ and the $\delta g_{\alpha\beta}$s are the fluctuations around this metric. Fluctuations around the new metric are defined as $\delta\tilde{g}_{\alpha\beta} = \tilde{g}_{\alpha\beta}(\tilde{x}) - g^b_{\alpha\beta}(\tilde{x})$. Taking into account that $g^b_{\alpha\beta}(\tilde{x}) = g^b_{\alpha\beta}(x) + (\partial_\mu g^b_{\alpha\beta})\xi^\mu$, we finally find

$$\delta\tilde{g}_{\alpha\beta} = \delta g_{\alpha\beta} - (\partial_\mu g^b_{\alpha\beta})\xi^\mu - g^b_{\alpha\mu}\partial_\beta\xi^\mu - g^b_{\beta\mu}\partial_\alpha\xi^\mu.\tag{12.49}$$

This gives

$$\delta\tilde{g}_{00} = \delta g_{00} - (\xi^t\partial_t A + \xi^k\partial_k A) - 2A\partial_t\xi^t,\tag{12.50}$$

$$\delta\tilde{g}_{0j} = \delta g_{0j} - A\partial_j\xi^t + B\delta_{jk}\partial_t\xi^k,\tag{12.51}$$

$$\delta\tilde{g}_{ij} = \delta g_{ij} + \delta_{ij}(\xi^t\partial_t B + \xi^k\partial_k B) + B(\delta_{kj}\partial_i\xi^k + \delta_{ki}\partial_j\xi^k).\tag{12.52}$$

For scalar perturbations, we restrict ourselves to "longitudinal" coordinate changes, that is

$$\xi^i = \partial^i\zeta = -\left(\frac{\partial_j\zeta}{B}\right).\tag{12.53}$$

To eliminate $\delta\tilde{g}_{0j}$, we have to impose the following condition

$$\partial_j C - A\,\partial_j\xi^t - B\,\partial_t\left(\frac{\partial_j\zeta}{B}\right)$$
$$\equiv \partial_j\left[C - B\,\partial_t\left(\frac{\zeta}{B}\right) - A\,\xi^t\right] + \partial_j B\,\partial_t\left(\frac{\zeta}{B}\right) + \xi^t\partial_j A = 0.\tag{12.54}$$

The sum of the last two terms in this equation vanishes if we choose

$$\xi^t = \frac{B'}{A'}\partial_t\left(\frac{\zeta}{B}\right),\tag{12.55}$$

where the prime denotes the derivative with respect to r. Evidently, the term in square brackets can be cancelled out with a proper choice of ζ.

12.2 Density Perturbations in General Relativity

Now we need to get rid of the gradient terms in Eq. (12.47), i.e. to impose the condition

$$\partial_i \partial_j E - 2\partial_i \partial_j \zeta + \frac{\partial_i B}{B} \partial_j \zeta + \frac{\partial_j B}{B} \partial_i \zeta = 0. \tag{12.56}$$

Unfortunately, there is no way to satisfy this equation. Firstly, we have already used all the freedom to eliminate g_{tj} and, secondly, there are terms of two different kinds. The first two terms are purely longitudinal ones, while the last two contain both transverse and longitudinal contributions and it is impossible to eliminate both with a single function ζ.

Let us note that with the "scalar" coordinate change there appear vector and tensor metric perturbations due to the dependence of the background metric functions on the spatial coordinates. This is an artifact of the coordinate choice. Probably these vector and tensor modes could be eliminated if one allows for a "transverse" coordinate transformation $\xi^j = \xi_\perp^j + \partial^j \zeta$. We will not pursue this issue further and in what follows we will assume, as we have mentioned above, that deviations from the flat metric are small and thus $A \approx B \approx 1$. In this approximation the problems with the gauge do not appear.

12.2.4 Evolution of Perturbations in Asymptotically Flat Spacetime

Usually the relativistic equations are taken in the weak field limit, so the terms proportional to Γ^2 in the expressions for the Ricci tensor are omitted. Differentiating the equation for G_{tt} over time and that for G_{jt} over x^j, we derive the continuity equation, while taking the time derivative of the equation for G_{jt} and the derivative over x^i of the equation for G_{ij} we obtain the Euler equation. However, if we restrict ourselves to the first order in Γ in the Ricci tensor, we do not obtain self-consistent equations. So, the second order terms in $R_{\mu\nu}$ are necessary and we derived the continuity and Euler equations with this procedure. On the other hand, one can take a simpler path, deriving the Euler and continuity equations from the conditions $\nabla_\mu T_j^\mu = 0$ and $\nabla_\mu T_t^\mu = 0$. Since we have four unknown functions, we need two more equations. We can take the equation for G_{tt} and the $\partial_i \partial_j$-component of the equation for G_{ij} at linear order in Γs. Correspondingly, we only keep terms linear in the derivatives of A and B and take $A = B = 1$ otherwise.

The Einstein equations written in terms of the Einstein tensor $G_{\mu\nu} = R_{\mu\nu} - g_{\mu\nu} R/2$ have the form

$$G_{\mu\nu} = \frac{8\pi}{M_{\text{Pl}}^2} T_{\mu\nu} \equiv \tilde{T}_{\mu\nu}, \tag{12.57}$$

where we introduced $\tilde{T}_{\mu\nu}$, which will be convenient in the future. The equations for G_{tt} and for the $\partial_i \partial_j$-component of the equation for G_{ij} are

$$-\Delta B = \tilde{\rho}, \tag{12.58}$$

$$\partial_i \partial_j (A + B) = 0. \tag{12.59}$$

The continuity and Euler equations are, respectively,

$$\dot{\rho} + \partial_j [(\rho + P) v^j] + \frac{3}{2} \rho \dot{B} = 0, \tag{12.60}$$

$$\rho \dot{v}_j + \partial_j P + \frac{1}{2} \rho \partial_j A = 0. \tag{12.61}$$

We assume that the background metric changes slowly as a function of space and time and we study small fluctuations around the background quantities: $\rho = \rho_b + \delta\rho$, $\delta P = c_s^2 \delta\rho$, $\mathbf{v} = \delta\mathbf{v}$, $A = A_b + \delta A$, $B = B_b + \delta B$. The corresponding linear equations for the infinitesimal perturbations are

$$-\Delta \delta B = \delta\tilde{\rho}, \tag{12.62}$$

$$\partial_i \partial_j (\delta A + \delta B) = 0, \tag{12.63}$$

$$\delta\dot{\rho} + \rho \, \partial_j \delta v^j + \frac{3}{2} \rho \, \delta\dot{B} = 0, \tag{12.64}$$

$$\rho \, \delta\dot{v}_j + \partial_j \delta P + \frac{1}{2} \rho \, \partial_j \delta A = 0. \tag{12.65}$$

Equations (12.62)–(12.65) coincide with the corresponding equations in Weinberg (2008), Mukhanov (2005), Gorbunov and Rubakov (2011) for a static universe, i.e. for $a(t) = 1$ and $H = 0$. We note that with our definitions $\delta A \equiv 2\Phi$ and $\delta B \equiv -2\Psi$.

We look for a solution in the form $\sim \exp[-i\lambda t + i\mathbf{k}\cdot\mathbf{x}]$ and we obtain the following expressions for the frequency eigenvalues

$$\lambda^2 = \frac{c_s^2 k^2 - \tilde{\rho}/2}{1 + 3\tilde{\rho}/(2k^2)}. \tag{12.66}$$

This result almost coincides with the Newtonian one (12.11). An extra term in the denominator is induced by the relativistic volume variation, and it is small for $k \sim k_J$.

12.2.5 Evolution of Perturbations in Cosmology

The description of perturbations in cosmology is simpler than in the previous case, because the background metric and the energy density do not depend on the spatial coordinates x^is (in the 3-dimensional flat metric) and because the background quantities satisfy the zeroth order equations. Below we essentially repeat the considerations of the previous section with the concrete form of the metric and ρ_b.

12.2 Density Perturbations in General Relativity

The perturbed metric is written as

$$g_{\mu\nu} = g^{(b)}_{\mu\nu} + h_{\mu\nu}, \tag{12.67}$$

where $g^{(b)}_{\mu\nu}$ is the flat FRW background

$$g^{(b)}_{tt} = 1, \quad g^{(b)}_{tj} = 0, \quad g^{(b)}_{ij} = -a^2(t)\delta_{ij}, \tag{12.68}$$

while $h_{\mu\nu}$ describes small perturbations. It turns out that any perturbation can be decomposed into three parts that evolve independently and they are called, respectively, scalar, vector, and tensor perturbations (Lifshitz 1946). In what follows, we will consider the simplest and cosmologically most interesting case of scalar perturbations.

The time-time component of the metric perturbation is a scalar in the 3-dimensional space and we write it as

$$g_{tt} = 1 + 2\Phi, \quad \text{or} \quad h_{tt} = 2\Phi, \tag{12.69}$$

In what follows, we see that Φ becomes the Newtonian potential in the non-relativistic limit. The space-time components of the metric make a 3-vector and so, as the components of any vector, they can be written as the sum of a gradient and of a transversal vector

$$h_{tj} = a(t)\left(\partial_j f + W_j\right), \tag{12.70}$$

where the factor $a(t)$ is separated for convenience, f is a scalar function, and $\partial_j W_j = 0$ by definition of transversal vector. The index contraction here and below is done with δ_{ij}. Since we are only interested in scalar perturbations, we will not consider W_j in what follows. The space-space components of the metric can be written as

$$g_{ij} = -a^2(t)\left[\delta_{ij}(1-2\Psi) + \partial_i\partial_j S + \partial_i Q_j + \partial_j Q_i + Y_{ij}\right], \tag{12.71}$$

where, as we see below, Ψ coincides with the Newtonian potential Φ in the non-relativistic limit, S is a scalar function, Q_j is a transversal vector (i.e. $\partial_j Q_j = 0$), and Y_{ij} is a symmetric, transversal, and traceless tensor (i.e. $Y_{ij} = Y_{ji}$, $\partial_j Y_{ij} = 0$, and $Y_{ii} = 0$). The vector Q_j and the tensor Y_{ij} describe, respectively, vector and tensor perturbations and will be disregarded in what follows.

Using the freedom of the choice of coordinates, we can impose the conditions $f_j = 0$ and $S = 0$, see Sect. 12.2.3, so the perturbed metric due to scalar perturbations becomes

$$g_{tt} = 1 + 2\Phi, \quad g_{tj} = 0, \quad g_{ij} = -a^2\delta_{ij}(1-2\Psi). \tag{12.72}$$

This is the so-called Newtonian gauge. In cosmology, sometimes the so-called synchronous gauge is employed, where instead of (12.72) the conditions $g_{tt} = 1$, $g_{tj} = 0$, and $S \neq 0$ are imposed, or the gauge invariant approach (Mukhanov 2005; Mukhanov et al. 1992) is used.

As we can see from Eq. (12.42), the space-space components of the energy-momentum tensor are proportional to δ_{ij}, so the terms with the derivatives ∂_i and ∂_j in the equation for G_{ij} (12.35) must vanish, and we obtain

$$\Phi = \Psi. \tag{12.73}$$

From the equation for G_{tt} (12.33), using the expressions (12.44) or (12.72) for the metric coefficients and Eq. (12.41) for T_{tt}, we find

$$\frac{\Delta\Psi}{a^2} - 3H\dot{\Psi} - H^2\Psi = -\frac{1}{2}\delta\tilde{\rho}. \tag{12.74}$$

This equation is the cosmological counterpart of the Poisson equation (12.1). Of course, they coincide if $a \equiv 1$.

At this point, we have two equations and four unknowns: Ψ, Φ, $\delta\rho$ and $\mathbf{v}$. To close the system, we can use the Euler and the continuity equations, which can be derived from the covariant conservation of the energy-momentum tensor $\nabla_\mu T^\mu_\nu = 0$, respectively with $\nu = j$ and $\nu = t$. In the derivation of these equations we should keep in mind that, in contrast to the non-relativistic problem, the background pressure is not necessarily zero. Writing the the covariant conservation of $T_{\mu\nu}$ with the Christoffel symbols (12.28), we find the hydrodynamic continuity equation

$$\delta\dot{\rho} + (\rho_b + P_b)\frac{\partial_k v_k}{a} + 3H(\delta\rho + \delta P) - 3\dot{\psi}(\rho_b + P_b) = 0. \tag{12.75}$$

In the same way, we can obtain the Euler equation from $\nabla_\mu T^\mu_j = 0$ and we find

$$a\partial_t\left[v_j(\rho + P)\right] + 4Ha\,v_j(\rho_b + P_b) + \partial_j P + (\rho_b + P_b)\partial_j\Phi = 0. \tag{12.76}$$

Since here we are talking about scalar perturbations, the velocity vector should be a gradient of some scalar velocity potential, say $\mathbf{v}_j = \nabla\sigma$. This allows to eliminate the derivative ∂_j in Eq. (12.76), excluding modes with zero wave number. It is done below in conformal time in Eq. (12.92).

Even if we have already all the necessary equations, it is technically simpler if, instead of one of the above equations, we use the equations for the space-space components of the Einstein tensor, G_{ij}. More precisely, we consider the part proportional to δ_{ij} in (12.35)

$$\frac{\Delta A}{2A} + \frac{\Delta B}{2B} - \frac{\ddot{B}}{A} + \frac{\dot{B}^2}{4AB} + \frac{\dot{A}\dot{B}}{2A^2} = B\tilde{P}, \tag{12.77}$$

12.2 Density Perturbations in General Relativity

where we neglected the terms proportional to $\partial^k A \, \partial_k A$ and $\partial^k B \, \partial_k B$, as they are of the second order in the perturbations. From the expansions

$$\begin{aligned} A &= A_b(1+2\Phi) = 1+2\Phi, \\ B &= B_b(1-2\Psi) = a^2(t)(1-2\Psi), \\ P &= P_b + \delta P, \end{aligned} \quad (12.78)$$

we find, after some simple but rather tedious algebra, the expression for the background pressure

$$\tilde{P}_b = -3H^2 - 2\dot{H} \quad (12.79)$$

and the equation for the first order perturbations

$$\ddot{\Psi} + 4H\dot{\Psi} + \left(3H^2 + 2\dot{H}\right)\Psi = \frac{1}{2}\delta\tilde{P}, \quad (12.80)$$

where we used $\Phi = \Psi$.

For what follows, it is convenient to write all the equations in conformal time. The spatially flat FRW metric in conformal time takes the simple form

$$ds^2 = a^2(\eta)(d\eta^2 - d\mathbf{r}^2), \quad (12.81)$$

where the conformal time is defined by

$$d\eta = \frac{dt}{a(t)}. \quad (12.82)$$

In conformal time, the background metric satisfies the first Friedmann equation

$$H^2 = \frac{8\pi}{3M_{\text{Pl}}^2}\rho, \quad (12.83)$$

which is formally the same as Eq. (4.9), but the Hubble parameter is now

$$H = \frac{a'}{a^2}. \quad (12.84)$$

Here and in what follows, the prime means differentiation over η.

According to the discussion above, we also need the equation for the ij-components of the background Einstein tensor. In conformal time, this equation is

$$\frac{(a')^2}{a^2} - \frac{2a''}{a} = \frac{8\pi a^2 P_b}{M_{\text{Pl}}^2}. \quad (12.85)$$

Examples of expansion laws for typical cosmological regimes in conformal time have already been reported in Sect. 6.5, but here we present them again to make the chapter self-contained

$$a(\eta) \sim \eta \text{ in radiation dominated regime}, \tag{12.86}$$

$$a(\eta) \sim \eta^2 \text{ in matter dominated regime}, \tag{12.87}$$

$$a(\eta) \sim -1/\eta \text{ in de Sitter regime}. \tag{12.88}$$

Equations (12.73)–(12.76) govern the density perturbations and can be rewritten in conformal time as

$$\Phi - \Psi = 0, \text{ (the same as above)}, \tag{12.89}$$

$$\Delta \Psi - 3\frac{a'}{a}\Psi' - 3\left(\frac{a'}{a}\right)^2 \Psi = \frac{1}{2}a^2 \delta\tilde{\rho}, \tag{12.90}$$

$$\delta\rho' + (\rho_b + P_b)\Delta\sigma + 3\frac{a'}{a}(\delta\rho + \delta P) - 3(\rho_b + P_b)\Psi' = 0, \tag{12.91}$$

$$[\sigma(\rho_b + P_b)]' + 4\frac{a'}{a}\sigma(\rho_b + P_b) + \delta P + (\rho_b + P_b)\Phi = 0. \tag{12.92}$$

We have thus four equations for five unknowns (Ψ, Φ, $\delta\rho$, δP, and σ). The system can be solved if we add an equation of state, which is usually taken as $\delta P = c_s^2 \delta\rho$, where c_s is the speed of sound in the matter under scrutiny. In this way, we implicitly assumed that perturbations are adiabatic, namely entropy perturbations vanish.

As we have already mentioned above, it is more convenient to use the equation for G_{ij} (12.80). In conformal time, it becomes

$$\Phi'' + \frac{3a'}{a}\Phi' + \left[\frac{2a''}{a} - \left(\frac{a'}{a}\right)^2\right]\Phi = \frac{1}{2}a^2 \delta\tilde{P}. \tag{12.93}$$

Note that Eqs. (12.57), (12.90) and (12.93) contain the total energy-momentum tensor, namely the sum of all the matter components. In the case in which the cosmological plasma consists of several independent components, and every component is independently conserved, the continuity and Euler equations are fulfilled for every component. If different components are interacting with an exchange of energy and/or momentum, then one has to include their total energy-momentum tensor with the interaction term included.

Equations (12.93) and (12.90) lead to the following equation containing only one unknown function

$$\Phi'' + \frac{3a'}{a}\left(1 + c_s^2\right)\Phi' - c_s^2 \Delta\Phi + \left[\frac{2a''}{a} - \left(1 - 3c_s^2\right)\left(\frac{a'}{a}\right)^2\right]\Phi = 0. \tag{12.94}$$

Since the coefficients of the system of the linear differential Equations (12.89)–(12.93) and, in particular, of Eq. (12.94) do not depend on the space coordinates,

12.2 Density Perturbations in General Relativity

we can make the Fourier transformation without any problem and find a system of ordinary differential equations in which the Laplacian Δ is replaced by $-k^2$, where k is the comoving momentum. The Fourier amplitude of the metric perturbations in a cosmological background obeys an ordinary differential equation in time and can be easily analyzed.

Let us first consider the case of non-relativistic matter, which is characterized by negligible pressure, $P_b = 0$, and therefore $c_s = 0$. The expansion law for a spatially flat matter dominated universe is $a \sim \eta^2$, and Eq. (12.94) turns into the very simple form

$$\Phi'' + \frac{6\Phi'}{\eta} = 0. \tag{12.95}$$

The solution is

$$\Phi(\mathbf{x}, \eta) = \Phi_1(\mathbf{x}) + \frac{\Phi_2(\mathbf{x})}{\eta^5}. \tag{12.96}$$

Density perturbations can be found from Eq. (12.90)

$$\frac{\delta\rho}{\rho_b} = \frac{1}{6}\left(\eta^2 \Delta\Phi_1 + \frac{\Delta\Phi_2}{\eta^3}\right) - 2\Phi_2 + \frac{3\Phi_2}{\eta^5}. \tag{12.97}$$

Note that we have calculated the initial expression for $a^2\delta\rho$, so we have to normalize it to $a^2\rho_b$. In a spatially flat universe $a^2\tilde{\rho}_b = 3a^2H^2 = 3(a'/a)^2$. The evolution of perturbations is very much different for long and short wavelengths. If $k\eta \ll 1$, namely the physical wavelength $\lambda \sim a/k$ is much larger than the Hubble length $H^{-1} \sim a\eta$, we can neglect the first term proportional to the Laplacian in Eq. (12.97), and see that the density fluctuations stay essentially constant, $\delta\rho/\rho_b \approx -2\Phi_1$. In the opposite limit of waves shorter than the Hubble horizon, i.e. $k\eta \gg 1$, the density perturbations evolve as

$$\frac{\delta\rho}{\rho_b} \approx -k^2 \left(\Phi_1 \eta^2 + \frac{\Phi_2}{\eta^3}\right) \sim t^{2/3} + \frac{C}{t}. \tag{12.98}$$

The second term can be neglected and we conclude that, in a matter dominated regime, the short wave density perturbations rise as the cosmological scale factor.

In a radiation dominated universe, $P_b = \rho_b/3$ and the speed of sound is $c_s^2 = 1/3$. The cosmological expansion goes as $a(\eta) \sim \eta$. Now Eq. (12.94) for the Fourier modes Φ_k is

$$\Phi_k'' + \frac{4}{\eta}\Phi_k' + \frac{1}{3}k^2\Phi_k = 0. \tag{12.99}$$

This is a Bessel equation and it is solved as a linear superposition of $J_{\pm 3/2}\left(k\eta/\sqrt{3}\right)$. It is reduced to the elementary functions

$$\Phi_k = \frac{C_{k1}}{z}\left(\frac{\sin z}{z} - \cos z\right) + \frac{C_{k2}}{z}\left(\frac{\cos z}{z} + \sin z\right), \quad (12.100)$$

where $z = k\eta/\sqrt{3}$. In this case, the fractional density contrast is

$$\frac{\delta\rho_k}{\rho_b} = 2C_{k1}\left[\cos z\left(1 - \frac{2}{z^2}\right) - \frac{2\sin z}{z}\left(1 - \frac{1}{z^2}\right)\right]$$
$$+ 2C_{k2}\left[\sin z\left(\frac{2}{z^2} - 1\right) - \frac{2\cos z}{z}\left(1 - \frac{1}{z^2}\right)\right]. \quad (12.101)$$

Density perturbations in relativistic matter during a radiation dominated regime do not rise but form sound waves. This is intuitively evident since fast particles are reluctant to clump.

Let us consider now the evolution of density perturbations in subdominant non-relativistic matter during a radiation dominated regime with the expansion law $a \sim \eta$. The fluctuations of non-relativistic matter over the relativistic background are described by Eqs. (12.91) and (12.92), which now become

$$k^2\sigma\rho_{mb} = \delta\rho_m'' + \frac{3a'}{a}\delta\rho_m - 3\Phi'\rho_{mb}, \quad (12.102)$$

$$\left(k^2\rho_{mb}\sigma\right)' + \frac{4a'}{a}k^2\rho_{mb}\sigma + k^2\rho_{mb}\Phi = 0, \quad (12.103)$$

where ρ_{mb} is the energy density of the background non-relativistic matter which, by assumption, is much smaller than that of the relativistic one, $\rho_{mb} \ll \rho_{rb}$. In these equations, all the δ quantities above are the Fourier mode amplitudes. Excluding σ, we get

$$\delta\rho_m'' + \frac{7a'}{a}\delta\rho_m' + \left[\frac{3a''}{a} + \left(\frac{3a'}{a}\right)^2\right]\delta\rho_m$$
$$= 3\rho_{mb}\left(\Phi'' + \frac{4a'}{a}\Phi'\right) + 3\Phi'\rho_{mb}' - \rho_{mb}k^2\Phi. \quad (12.104)$$

Here the potential Φ is given by Eq. (12.100) and it is a decreasing function of time. Moreover, it always enters with a small factor ρ_{mb}. So this term is not essential for a possible rising mode of $\delta\rho_m$. The latter is determined by the free part of this equation, which has only a logarithmically rising solution

$$\frac{\delta\rho_m}{\rho_{mb}} \sim \ln\eta. \quad (12.105)$$

So matter fluctuations during a radiation dominated expansion are essentially constant.

12.2.6 Concluding Remarks

First, we would like to stress the role of a proper choice of the coordinate system. Even in the ideal homogeneous and isotropic FRW background, there is an infinite number of coordinate frames in which equal time surfaces have energy densities that depend on the space coordinates. In this case, the density perturbations are fictitiously created by a space dependent choice of the equal time surfaces. An opposite statement is also true, that real density perturbations can be formally removed by a coordinate freedom. To avoid this problem, we have to consider a full set of the relevant quantities, i.e. density, metric, and velocity perturbations, as it is done in the previous subsections. An unambiguous approach is based on the use of gauge (coordinate) independent (physical) quantities, see e.g. Mukhanov et al. (1992), Mukhanov (2005).

In this section, we have studied density perturbations of matter made of a single component or of matter consisting of some non-interacting components, as in the case of the subdominant non-relativistic matter in a radiation dominated universe discussed in the previous section. In a realistic cosmological scenario, this is not true. Non-relativistic electrons can strongly interact with the background photons. Protons are strongly coupled to electrons and so structures in a baryon dominated universe could not be developed due to large resistance from the photon pressure. The first structures could indeed develop in the dark matter sector, because dark matter particles do not interact with light. The onset of the structure formation in dark matter started at redshift $z \approx 10^4$, corresponding to the transition from the radiation dominated to the matter dominated stage of the Universe. Only later, after hydrogen recombination at redshift $z_{\text{rec}} \approx 1100$, the baryo-electron fluid essentially stopped interacting with the CMB and hydrogen and helium atoms were captured in the potential wells pre-created by dark matter. Since we know from the data on the angular fluctuations of the CMB temperature that $\delta\rho/\rho \sim 10^{-4}$ and that the density perturbations rose after that as the scale factor, see Eq. (12.98), we must conclude that the existence of dark matter is indeed obligatory for the structure formation in the Universe

Only adiabatic perturbations have been considered up to this point. There may exist entropy perturbations as well. Current CMB data provide strong constraints on entropy perturbations, though it is not excluded that they may be non-negligible at small scales.

It is also worth saying a few words about the spectrum of the density perturbations. The initial spectrum of perturbations is parametrized by a simple power law form. For the dimensionless perturbations of the FRW metric (i.e. for the gravitational potential of perturbations), it is taken in the form

$$\langle \phi(k)\phi(k') \rangle = \frac{P(k)}{(2\pi)^3} \delta^{(3)}(\mathbf{k} + \mathbf{k}') \sim k^{n-4} \delta^{(3)}(\mathbf{k} + \mathbf{k}'), \qquad (12.106)$$

where the left hand side is the statistical average of the Fourier transformed perturbations of the metric (the correlation function). Note that for $n = 1$ the power spectrum $P(k)$ is dimensionless, so it may not contain any dimensional parameter. Indeed we define the Fourier transformation as

$$\phi(\mathbf{x}) = \int d^3k \exp(i\mathbf{k}\mathbf{x}) \, \phi(\mathbf{k}), \qquad (12.107)$$

and hence the Fourier amplitude of the dimensionless potential has dimension k^{-3}. This spectrum is called flat or Harrison-Zeldovich spectrum (Harrison 1970; Zeldovich 1972), since Harrison and Zeldovich were the first to suggest that the primordial density fluctuations may have this form to avoid too much power at small and large scales.

The fluctuations of the energy density are expressed through the gravitational potential by the Poisson-like equation (12.90). Neglecting the cosmological expansion, we can conclude that $a^{-2}\Delta\delta\phi \sim 4\pi \delta\rho/M_{\text{Pl}}^2$ and obtain

$$\langle \delta\rho(x,t)\delta\rho(x',t) \rangle \sim \int d^3k k^n e^{i\mathbf{k}(\mathbf{x}-\mathbf{x}')} . \qquad (12.108)$$

It can be shown that, for the flat spectrum with $n = 1$, the perturbations with wavelength λ would have equal magnitude $\delta\rho/\rho = const$ at the horizon crossing $\lambda \sim t$.

12.3 Density Perturbations in Modified Gravity

12.3.1 General Equations

Let us consider the gravity theory described by the action (11.6). Since the Lagrangian is a non-linear function of R, the equation of motion is of higher (fourth) order and the evolution of perturbations may be different from that in General Relativity. In cosmology, this problem has been considered for different forms of $F(R)$, see e.g. Zhang (2006), Song et al. (2007), Tsujikawa (2007), Cruz-Dombriz et al. (2008), Ananda et al. (2011, 2009), Motohashi et al. (2009), Matsumoto (2015). An analysis of the Jeans instability for stellar-like objects in modified gravity was performed in Capozziello et al. (2011, 2012), Eingorn et al. (2014). In those papers, a perturbative expansion of $F(R)$ was performed around either $R = 0$ or $R = R_c$, where R_c is the present cosmological scalar curvature. Below we expand $F(R)$ around the curvature of the background metric R_b, which is typically much larger than R_c. It leads to some quantitative difference. Moreover, we study the associated instabilities not only in a quasi-stationary background, as it is usually done, but also in a quickly oscillating one. As it was mentioned in Sect. 11.1.3, such high frequency oscillations are induced in contracting systems with rising energy density.

12.3 Density Perturbations in Modified Gravity

We assume that the background spacetime weakly deviates from the Minkowski metric, while the corrections due to gravity modifications may be significantly different from those of General Relativity. In particular, R may be very different from $R_{GR} = -\tilde{T}$. We consider astronomical systems with $|R| \gg |R_c|$, but $R \ll m^2$. It is expected that in this limit $F(R) \ll R$ and $F'(R) \ll 1$. This is surely fulfilled for $F(R)$ given by Eq. (11.9), for which at $R \gg R_c$ we have

$$F(R) \approx -\lambda R_c \left[1 - \left(\frac{R_c}{R}\right)^{2n} \right] - \frac{R^2}{6m^2}. \tag{12.109}$$

In this modified gravity model, the new equations for the gravitational field have the form (11.7). For the particular choice of $F(R)$ (11.9), under the conditions specified above, the equations take the form

$$G_{\mu\nu} + \frac{1}{3\omega^2} \left(\nabla_\mu \nabla_\nu - g_{\mu\nu} \nabla^2 \right) R = \tilde{T}_{\mu\nu}, \tag{12.110}$$

where $G_{\mu\nu} = R_{\mu\nu} - g_{\mu\nu} R/2$ is still the usual Einstein tensor and $\omega^{-2} = -3F''_{RR}$.

As usually, the metric and the curvature tensor are expanded around their background values to the first order in infinitesimal perturbations

$$\begin{aligned} A &= A_b + \delta A, \\ B &= B_b + \delta B, \\ R &= R_b + \delta R. \end{aligned} \tag{12.111}$$

The background internal metric for a spherically symmetric distribution of matter, the analog of the Schwarzschild-type solution in modified gravity, has been studied by several authors. Here we use the form for the internal solution obtained in Arbuzova et al. (2014) (references to other papers can be found there)

$$B_b(r, t) = 1 + \frac{2M(r, t)}{M_{\text{Pl}}^2 r} \equiv 1 + B_1, \tag{12.112}$$

$$A_b(r, t) = 1 + \frac{R_b(t) r^2}{6} + A_1(r, t), \tag{12.113}$$

where

$$M(r, t) = \int_0^r d^3r \, T_{00}(r, t) = 4\pi \int_0^r dr \, r^2 \, T_{00}(r, t), \tag{12.114}$$

$$A_1(r, t) = \frac{r_g r^2}{2r_m^3} - \frac{3r_g}{2r_m} + \frac{\pi \ddot{\rho}_m}{3M_{\text{Pl}}^2} (r_m^2 - r^2)^2, \tag{12.115}$$

and $r_g = 2M/M_{Pl}^2$ with M being the total mass of the object under scrutiny. The functions A_1 and B_1 are the same as in standard General Relativity and the only deviation from General Relativity in this approximation comes from the second term in Eq. (12.113).

The evolution of the perturbations are studied in the approximation of constant ω. This is a good approximation if $R_b \approx R_{GR} = -\tilde{T}$, which is a slowly changing function of time. In the example (12.109)

$$\omega^2 = \left[\frac{1}{m^2} + \frac{6\lambda n(2n+1)}{|R_c|} \left(\frac{R_c}{R} \right)^{2n+2} \right]^{-1}. \tag{12.116}$$

Cosmological perturbations start rising at the onset of the matter dominated epoch, corresponding to the redshift $z_{eq} = 10^4$, when $R_c/R_{eq} \sim 10^{12}$. So for $m = 10^5$ GeV, which is the lower limit from the BBN (Arbuzova et al. 2012), ω may be treated as a constant if $n \geq 3$. If ω rises with time, the perturbations would rise even faster than it is obtained below. If Eq. (12.116) is dominated by the second term, then

$$\omega^2 = \frac{|R_c|}{6\lambda n(2n+1)} \left(\frac{R}{R_c} \right)^{2n+2}. \tag{12.117}$$

In Arbuzova et al. (2012, 2013), a high frequency oscillating solution with R strongly deviating from R_{GR} was found. In this case, the frequency (12.117) might crucially depend on time and the approximation of constant frequency would not be valid [though $R \gg R_{GR}$ diminishes the second term in Eq. (12.116). Nevertheless, in what follows we assume that $\omega = const$ and we study the development of instabilities described by the fourth order differential equation governing the evolution of perturbations in this model. In this case, the evolution of instabilities is quite different from the standard situation described by the second order equation of General Relativity. We will not dwell on a particular choice of the $F(R)$-function, but we will assume that the high frequency oscillations of the curvature are a generic phenomenon in such models. Indeed all known $F(R)$ scenarios would lead to a singularity with $R \to +\infty$, if the R^2/m^2-term is not purposely added to prevent from that. This term creates a repulsive effective potential for the evolution of R and so leads to an oscillatory behavior.

Equation (12.110) for the tt-component can be written as

$$-\frac{\Delta B}{B^2} + \frac{1}{3\omega^2} \left(\frac{\Delta R}{B} - \frac{3\dot{B}\dot{R}}{2AB} \right) = \tilde{\rho}, \tag{12.118}$$

because, according to (12.33) and (12.41), $G_{tt} = -A\Delta B/B^2$ and $\tilde{T}_{tt} = \tilde{\rho}A$.

Since the background quantities R_b and, according to Eq. (12.113), A_b are quickly oscillating functions of time with possibly large amplitude ("spikes"), which were found in Arbuzova et al. (2012, 2013), their time derivatives are large and we will keep the terms of the second order in ∂_t, such as $\partial_t^2 A$, $\partial_t A \, \partial_t R$, and so on. Correspondingly,

12.3 Density Perturbations in Modified Gravity

the equation for fluctuations takes the form

$$-\Delta\delta B - 2\tilde{\rho}_b\,\delta B + \frac{1}{3\omega^2}\left(\Delta\delta R - \frac{3}{2}\delta\dot{B}\dot{R}_b\right) = \delta\tilde{\rho}, \quad (12.119)$$

where the curvature fluctuation δR is [see Eq. (12.32)]

$$\delta R = \Delta\delta A - 3\delta\ddot{B} + 2\Delta\delta B + \frac{3}{2}\dot{A}_b\delta\dot{B}. \quad (12.120)$$

We assume that $\dot{B}_b$ is small in comparison with $\dot{A}_b$ and $\dot{R}_b$, see Eqs. (12.112) and (12.113), and put $A_b = B_b = 1$ in the denominators of Eq. (12.119).

Analogously, we find the $\partial_i\partial_j$-component of the equation for G_{ij}

$$\partial_i\partial_j(\delta A + \delta B - 2\omega^{-2}\delta R) = 0. \quad (12.121)$$

The continuity equation, derived from $\nabla_\mu T_t^\mu = 0$, takes the form

$$\delta\dot{\rho} + \rho_b\,\partial_j U^j + \frac{3}{2}\rho_b\,\delta\dot{B} = 0, \quad (12.122)$$

while the Euler equation, $\nabla_\mu T_j^\mu = 0$, becomes

$$\rho_b\,\delta\dot{U}_j + \partial_j P + \frac{1}{2}\rho_b\,\partial_j\delta A + \frac{1}{2}\dot{A}_b\,\rho_b\,U_j = 0. \quad (12.123)$$

Introducing $U^j = -U_j = -\partial_j\sigma$, $P = c_s^2\delta\rho$ and looking for a solution in the form $\sim \exp[-i\mathbf{k}\cdot\mathbf{x}]$, we obtain the following system of equations for the five unknown functions of time, δA, δB, δR, $\delta\tilde{\rho}$, and σ, which are the Fourier amplitudes of the original functions in the space coordinates

$$3\omega^2(k^2 - 2\tilde{\rho}_b)\delta B - \frac{3}{2}\delta\dot{B}\dot{R}_b - k^2\delta R - 3\omega^2\delta\tilde{\rho} = 0, \quad (12.124)$$

$$\delta R = \frac{3}{2}\omega^2(\delta A + \delta B), \quad (12.125)$$

$$\delta R = -k^2\delta A - 3\delta\ddot{B} - 2k^2\delta B + \frac{3}{2}\dot{A}_b\delta\dot{B}, \quad (12.126)$$

$$\delta\dot{\tilde{\rho}} + \tilde{\rho}_b k^2\sigma + \frac{3}{2}\tilde{\rho}_b\,\delta\dot{B} = 0, \quad (12.127)$$

$$\tilde{\rho}_b\dot{\sigma} - c_s^2\delta\tilde{\rho} + \frac{1}{2}\tilde{\rho}_b(\dot{A}_b\sigma - \delta A) = 0. \quad (12.128)$$

At this point, it is proper to comment on the application of the Fourier transformation in the derivation of Eqs. (12.124)–(12.128). The use of Fourier transformations is typically applied if the coefficients in the equations to solve are space independent.

In our case, the metric function A_b explicitly depends upon the space coordinates, $A = 1 + r^2 R(t)/6$, and the approach of the Fourier amplitudes may thus look valid only in the limit in which $(kr)^2$ is small, see the discussion in Sect. 12.1, after Eq. (12.21). This is the truth, but not the whole truth. The description of perturbations in terms of Fourier transformed quantities is applicable also in the limit of large kr. We make Fourier transformation of the equations multiplying them by $\exp(ikr)$ and integrating over d^3k. If the coefficients of the linear equations for infinitesimal δA and δB do not depend upon the coordinates, we come to an algebraic system of linear equations for the Fourier modes of the fluctuations. Let us assume now that some coefficients in the original differential equations depend upon r. Still in this case we can transform our equations taking the integral $d^3 r \exp(ikr)$ not in the infinite limit but in some finite limit around a fixed value $r = r_0$ with the linear size of the volume Δr. If $\Delta r k \gg 1$, then such integral is close to the real Fourier transformation with the infinite integration limit. If the cloud under scrutiny permits separation into some pieces with $k \Delta r \gg 1$ and such that $\Delta r/r_0 \ll 1$, we can thus make the approximate Fourier transformation taking r_0 out of the integral. Such adiabatic limit can be of practical interest. Moreover, there may be examples of time-oscillating but spatially homogeneous backgrounds.

Keeping this in mind, we can derive from the system of five low order equations (12.124)–(12.128) the fourth order equation for the function δB

$$\overset{....}{\delta B} - \overset{...}{\delta B}\left(1 + \frac{2k^2}{3\omega^2}\right)\frac{\dot{R}_b}{2k^2}$$

$$+ \delta\ddot{B}\left[\omega^2 - \frac{\tilde{\rho}_b \omega^2}{2k^2}\left(1 + \frac{8k^2}{3\omega^2}\right) + k^2(1 + c_s^2) - \ddot{A}_b \right.$$

$$\left. - \frac{1}{k^2}\left(1 + \frac{2k^2}{3\omega^2}\right)\left(\ddot{R}_b + \frac{\dot{A}_b \dot{R}_b}{4}\right) - \frac{\dot{A}_b^2}{4}\right]$$

$$+ \delta\dot{B}\left[-\frac{\ddot{A}_b}{2} - \frac{1}{4k^2}\left(1 + \frac{2k^2}{3\omega^2}\right)\left(2\overset{...}{R}_b + \dot{A}_b \ddot{R}_b + 2\dot{R}_b c_s^2 k^2\right) - \frac{\ddot{A}_b \dot{A}_b}{4}\right.$$

$$\left. + \frac{\dot{A}_b}{2}\left(\omega^2 + k^2(1 - c_s^2) + \frac{2\tilde{\rho}_b}{3} - \frac{\tilde{\rho}_b \omega^2}{2k^2}\right)\right]$$

$$+ \delta B\left[c_s^2 k^2(k^2 + \omega^2) - 2c_s^2 \tilde{\rho}_b \omega^2\left(1 + \frac{2k^2}{3\omega^2}\right) - \frac{\tilde{\rho}_b \omega^2}{2}\left(1 + \frac{4k^2}{3\omega^2}\right)\right] = 0. \quad (12.129)$$

One can see that in this example $\dot{A}_b^2 \ll \ddot{A}_b$, $\dot{A}_b \dot{R}_b \ll \ddot{R}_b$, $\dot{A}_b \ddot{A}_b \ll \overset{...}{A}_b$, and $\dot{A}_b \ddot{R}_b \ll \overset{...}{R}_b$, so the corresponding terms in Eq. (12.129) can be neglected. In accordance with Eq. (12.113) we take $A_b = 1 + R_b r^2/6$.

Let us now estimate the factor $(kr)^2$ near its Jeans value $k = k_J = \sqrt{\tilde{\rho}/(2c_s^2)}$. The mass of the object under scrutiny is $M_{tot} = 4\pi \rho r_m^3/3$, where r_m is the maximum radius. So

12.3 Density Perturbations in Modified Gravity

$$(rk_J)^2 = \frac{3}{2c_s^2}\frac{r_g r^2}{r_m^3} \ll 1, \qquad (12.130)$$

where $r_g = 2M/M_{\rm Pl}^2 = \tilde\rho r_m^2/3$.

Now we introduce the dimensionless time $\tau = \omega t$, the dimensionless parameters

$$a \equiv \frac{\tilde\rho_b}{k^2}, \quad b \equiv \frac{k^2}{\omega^2}, \quad c \equiv c_s^2, \qquad (12.131)$$

and

$$\alpha = \frac{a}{2}\left(1 + \frac{2b}{3}\right), \qquad (12.132)$$

$$\Omega^2 = 1 - \frac{a}{2}\left(1 + \frac{8b}{3}\right) + b(1+c), \qquad (12.133)$$

$$\mu = b\left[c(1+b) - \frac{a}{2}\left(1 + \frac{4b}{3}\right) - 2ac\left(1 + \frac{2b}{3}\right)\right]. \qquad (12.134)$$

Denoting $\delta B \equiv z$, $R_b = -\tilde\rho_b y$, and taking the limit $(kr)^2 \ll 1$, we rewrite Eq. (12.129) in the following very simple form

$$z'''' + \alpha y' z''' + (\Omega^2 + 2\alpha y'')z'' + \alpha(y''' + bcy')z' + \mu z = 0. \qquad (12.135)$$

Since the physically interesting quantity is the magnitude of the density perturbations, we present $\delta\rho/\rho_b$ expressed through $z \equiv \delta B$

$$\frac{\delta\rho}{\rho_b} = z\left[\frac{1+b}{a(1+2b/3)} - 2\right] + \frac{1}{2}z'y' + \frac{z''}{a(1+2b/3)}. \qquad (12.136)$$

According to Eqs. (12.12) and (12.131), when k is close to its Jeans value $a \sim \tilde\rho_b/k_J^2 \sim c_s^2$ and, correspondingly, the first term in the square brackets dominates if $c_s^2 < 1/2$, which is true in practically all the physically realizable cases.

12.3.2 Modified Jeans Instability

For very small amplitudes of curvature oscillations, we can neglect $y(\tau)$ in Eq. (12.135), which is thus reduced to a simple equation with constant coefficients. It can be solved by employing the substitution $z = \exp(i\lambda\tau)$. The eigenvalue λ is determined by the algebraic equation

$$\lambda^4 - \Omega^2\lambda^2 + \mu = 0, \qquad (12.137)$$

which gives

$$\lambda_\pm^2 = \frac{\Omega^2}{2} \pm \sqrt{\frac{\Omega^4}{4} - \mu}. \tag{12.138}$$

If $\mu < 0$, $\lambda_+^2 > 0$ and therefore one of the eigenvalues is negative imaginary. It corresponds to the usual exponential Jeans instability, though the values of the Jeans wave vector in modified gravity and in General Relativity are different. The magnitude of the Jeans wave number is found from the equation $\mu = 0$, which, in the case of small speed of sound, gives

$$a = \frac{2c(1+b)}{1+4b/3}, \tag{12.139}$$

where a and b depend upon k according to the definitions (12.131). So it is a quadratic equation with respect to the square of the Jeans wave number in modified gravity, k_J^{MG}. We present an explicit solution for large ω

$$(k_J^{MG})^2 = (k_J^{GR})^2 \left[1 + \frac{(k_J^{GR})^2}{3\omega^2} \right], \tag{12.140}$$

which turns into the result of General Relativity in the limit $\omega \to +\infty$. Equation (12.140) shows that in modified gravity the Jeans wave number is larger than in General Relativity. This corresponds to a reduced minimum length scale associated to structure formation. The correction is typically quite small, but, for models in which k_J^{GR}/ω is non-negligible, it could lead to a significant deviations from General Relativity.

If μ is positive, but $\mu < \Omega^4/4$, the eigenvalues λ^2 in (12.138) are real and positive, so all λs are real, which corresponds to acoustic oscillations with constant amplitude. So these two cases of negative and positive μ are in one-to-one correspondence to the usual Jeans analysis. For a very large μ, namely for $\mu > \Omega^4/4$, there exists a new type of unstable oscillating solution with exponentially rising amplitude. Indeed, in this case the solutions for λ^2 become complex conjugate numbers and two out of the four eigenvalues λ would have negative imaginary part, so the factor $\exp(i\lambda t)$ would rise with time. This is a new phenomenon, present only in modified gravity. However, in the model based on $F(R)$, given by Eq. (11.9), the parameter μ cannot exceed $\Omega^4/4$. It is unclear if this is a general property of all reasonable models of modified gravity and hence this new type of gravitational instability is always absent or models possessing such exciting property can be found. There is another unusual possibility for instability in modified gravity, which would appear if $\Omega^2 < 0$. In the frameworks of the chosen model (11.9), the acceptable values of the parameters do not allow to realize such an unusual possibility, but the above question remains, if this is possible with further gravity modifications.

12.3.3 Effects of Time Dependent Background

If $y(\tau)$ is non-negligible, quite new interesting effects can show up. The function $y(\tau)$ entering Eq. (12.135) is an oscillating function of the "time" τ. It can induce an analogue of the parametric resonance instability resulting in a very fast rise of perturbations at a certain set of frequencies. Another new effect can be called "anti-friction". It appears at sufficiently large amplitudes of oscillations of y such that the coefficients in front of the odd derivative terms in Eq. (12.135) become periodically negative. This phenomenon leads to an explosive rise of z in a wide range of frequencies. Both effects do not exist in standard General Relativity and, if discovered, would be a proof of modified gravity. On the contrary, the non-observation of these effects would allow to put stringent restrictions on the parameters of $F(R)$ theories.

The anti-friction behavior can be demonstrated in the limit of large derivatives of $y(\tau)$, when Eq. (12.135) can be solved analytically. In this case, this equation turns into

$$z'''' + \alpha y' z''' + 2\alpha z'' y'' + \alpha z' y''' = 0. \tag{12.141}$$

The last three terms in this equation can be written as $\alpha(z'y')''$, so the equation is easily integrated

$$z'' + \alpha z' y' = C_1 + C_2 \tau, \tag{12.142}$$

leading, in turn, to the solution

$$z' = C_0 e^{-\alpha y(\tau)} + C_1 e^{-\alpha y(\tau)} \int_0^\tau d\tau' e^{\alpha y(\tau')} + C_2 e^{-\alpha y(\tau)} \int_0^\tau d\tau' \tau' e^{\alpha y(\tau')}. \tag{12.143}$$

We take for illustration $y(\tau) = y_0 \cos(\Omega_1 \tau)$. From Eq. (12.143) it is clear that the derivative z' is small when $\alpha y < 0$, while z' is positive and large for $\alpha y > 0$. So the function z remains constant during the first period and rises during the second one.

The effect of parametric resonance can be observed for smaller $y(\tau)$ and only for frequencies close to an integer fraction Ω/n. The theoretical description of this effect is quite similar to that of the usual parametric resonance exhibited by the Matheau equation, see Sect. 6.4.3.

Problems

12.1 Evaluate numerically the evolution of perturbations in the time dependent background (12.17) and (12.18).

12.2 Check that for $P_b = w\rho_b$, with $w = 0, 1/3$, and -1, ρ_b indeed satisfies the first Friedmann equation (4.9).

12.3 Check that the covariant conservation of $T_{\mu\nu}$ (4.12) in conformal time becomes

$$\rho' = -3\frac{a'}{a}(\rho + P). \tag{12.144}$$

12.4 Show that during a quasi-de Sitter regime, namely in a period dominated by a vacuum-like energy, matter fluctuations decrease as the cube of the scale factor.

12.5 Solve Eq. (12.135) numerically for $y = y_0 \cos(\Omega_1 \tau)$ for different y_0 and Ω_1 to observe the effects of parametric resonance and anti-friction (mentioned in Sect. 12.3.3).

References

K.N. Ananda, S. Carloni, P.K.S. Dunsby, Class. Quant. Grav. **26**, 235018 (2009). arXiv:0809.3673 [astro-ph]
K.N. Ananda, S. Carloni, P.K.S. Dunsby, Springer Proc. Phys. **137**, 165 (2011). arXiv:0812.2028 [astro-ph]
E.V. Arbuzova, A.D. Dolgov, L. Reverberi, JCAP **1202**, 049 (2012). arXiv:1112.4995 [gr-qc]
E.V. Arbuzova, A.D. Dolgov, L. Reverberi, Eur. Phys. J. C **72**, 2247 (2012). arXiv:1211.5011 [gr-qc]
E.V. Arbuzova, A.D. Dolgov, L. Reverberi, Phys. Rev. D **88**(2), 024035 (2013). arXiv:1305.5668 [gr-qc]
E.V. Arbuzova, A.D. Dolgov, L. Reverberi, Astropart. Phys. **54**, 44 (2014). arXiv:1306.5694 [gr-qc]
E.V. Arbuzova, A.D. Dolgov, L. Reverberi, Phys. Lett. B **739**, 279 (2014). arXiv:1406.7104 [gr-qc]
S. Capozziello, M. De Laurentis, I. De Martino, M. Formisano, S.D. Odintsov, Phys. Rev. D **85**, 044022 (2012). arXiv:1112.0761 [gr-qc]
S. Capozziello, M. De Laurentis, S.D. Odintsov, A. Stabile, Phys. Rev. D **83**, 064004 (2011). arXiv:1101.0219 [gr-qc]
A. de la Cruz-Dombriz, A. Dobado, A.L. Maroto, Phys. Rev. D **77**, 123515 (2008). arXiv:0802.2999 [astro-ph]
M. Eingorn, J. Novk, A. Zhuk, Eur. Phys. J. C **74**, 3005 (2014). arXiv:1401.5410 [astro-ph.CO]
D.S. Gorbunov, V.A. Rubakov, *Introduction to the Theory of the Early Universe: Cosmological Perturbations and Inflationary Theory* (World Scientific, Hackensack, 2011)
E.R. Harrison, Phys. Rev. D **1**, 2726 (1970)
J.H. Jeans, Phil. Trans. Roy. Soc. A **199**, 1 (1902)
E.M. Lifshitz, Zh Eksp. Teor. Fiz. **16**, 587 (1946)
A.P. Lightman, W.H. Press, R.H. Price, S.A. Teukolsky, *Problem Book in Relativity and Gravitation* (Princeton University Press, Princeton, 1975)
J. Matsumoto. arXiv:1401.3077 [astro-ph.CO]
H. Motohashi, A.A. Starobinsky, J. Yokoyama, Int. J. Mod. Phys. D **18**, 1731 (2009). arXiv:0905.0730 [astro-ph.CO]
V. Mukhanov, *Physical Foundations of Cosmology* (Cambridge University Press, Cambridge, 2005)
V.F. Mukhanov, H.A. Feldman, R.H. Brandenberger, Phys. Rept. **215**, 203 (1992)
Y.S. Song, W. Hu, I. Sawicki, Phys. Rev. D **75**, 044004 (2007) [astro-ph/0610532]
S. Tsujikawa, Phys. Rev. D **76**, 023514 (2007). arXiv:0705.1032 [astro-ph]
S. Weinberg, *Cosmology*, 1st edn. (Oxford University Press, Oxford, 2008)
Y.B. Zeldovich, Mon. Not. Roy. Astron. Soc. **160**, 1P (1972)
Y.B. Zeldovich, I.D. Novikov, *Relativistic Astrophysics: The Structure And Evolution Of The Universe* (University of Chicago Press, Chicago, 1983)
P. Zhang, Phys. Rev. D **73**, 123504 (2006) [astro-ph/0511218]

Correction to: Introduction to Particle Cosmology

Correction to:
C. Bambi and A. D. Dolgov, *Introduction to Particle Cosmology*, **UNITEXT for Physics,**
https://doi.org/10.1007/978-3-662-48078-6

In the original version of the book, typographical errors and equations in the following sections have been corrected: Sections 4.5, 5.2.1, 6.4.1, 6.4.2, 12.1 and 12.2.6.

The book and the chapters have been updated with the changes.

The updated version of these chapters can be
https://doi.org/10.1007/978-3-662-48078-6_4
https://doi.org/10.1007/978-3-662-48078-6_5
https://doi.org/10.1007/978-3-662-48078-6_6
https://doi.org/10.1007/978-3-662-48078-6_12

Appendix A
Natural Units

In particle physics, it is common—and very convenient—to use the so-called *natural units*. The three fundamental dimensional quantities are energy, action, and velocity. Energy is measured in eV or keV (10^3 eV), MeV (10^6 eV), GeV (10^9 eV), TeV (10^{12} eV), etc. Action is measured in units of $\hbar$ and velocity is measured in units in of c, where $\hbar = h/(2\pi)$ is the reduced Planck constant and c is the speed of light. Setting $\hbar = 1$ and $c = 1$, we can simplify many equations.

In fact, the system of units with $c = 1$ is used in astronomy for a long time. Distances can be measured in units of time needed for the light propagation along the distance. It is a matter of convention and convenience to measure distances in centimeters or in seconds.

The relation between the particle energy and the wave frequency of a quantum mechanical, or even classical, electromagnetic wave is $E = \hbar\omega$. It is again a matter of convention: one can set $\hbar = 1$ and measure energy in units of inverse time or, vice versa, measure time in units of inverse energy. In this units, an electromagnetic plane wave can be written as $\exp(-iEt + i\mathbf{k}\mathbf{x})$.

The relation between the average particle energy and the temperature in a thermal bath is $E \sim kT$, where T is the temperature and k is the Boltzmann constant. The numerical value of k is determined by our definition of the unit of temperature. Taking $k = 1$, we measure temperature in units of energy, and so the unit of temperature would be eV or GeV or any other energy unit.

Eventually, all dimensional quantities have dimensions of a power of energy. For instance, masses have dimensions of energy, lengths and times have dimensions of $1/E$:

$$M = \frac{E}{c^2}, \quad L = \frac{\hbar c}{E}, \quad T = \frac{\hbar}{E}. \tag{A.1}$$

The numerical values of $\hbar$ and $\hbar c$ in one of the usual systems of units are

$$\hbar = 6.582 \cdot 10^{-25} \text{ GeV} \cdot \text{s},$$
$$\hbar c = 1.973 \cdot 10^{-14} \text{ GeV} \cdot \text{cm}, \tag{A.2}$$

© Springer-Verlag Berlin Heidelberg 2016
C. Bambi and A.D. Dolgov, *Introduction to Particle Cosmology*,
UNITEXT for Physics, DOI 10.1007/978-3-662-48078-6

and therefore the conversion factors GeV ↔ cm and GeV ↔ s are

$$\frac{1}{\text{GeV}} = 6.582 \cdot 10^{-25} \text{ s},$$
$$\frac{1}{\text{GeV}} = 1.973 \cdot 10^{-14} \text{ cm}. \quad (A.3)$$

From the Newton gravitational constant G_N, we can obtain the Planck mass M_{Pl}, the Planck length L_{Pl}, and the Planck time T_{Pl}

$$M_{Pl} = \sqrt{\frac{\hbar c}{G_N}} = 1.222 \cdot 10^{19} \text{ GeV},$$

$$L_{Pl} = \sqrt{\frac{\hbar G_N}{c^3}} = 1.615 \cdot 10^{-33} \text{ cm},$$

$$T_{Pl} = \sqrt{\frac{\hbar G_N}{c^5}} = 0.539 \cdot 10^{-45} \text{ s}. \quad (A.4)$$

In cosmology, it is common to write $1/M_{Pl}^2$ instead of G_N. For instance, the Einstein equations becomes

$$G^{\mu\nu} = \frac{8\pi}{M_{Pl}^2} T^{\mu\nu} \quad \text{instead of} \quad G^{\mu\nu} = \frac{8\pi G_N}{c^4} T^{\mu\nu}. \quad (A.5)$$

Appendix B
Gauge Theories

A *group* is a set G with an operation $m : G \times G \to G$ such that:

1. $m(x, m(y, z)) = m(m(x, y), z) \;\forall\; x, y, z \in G$ (associative property).
2. $\exists\; u \in G$ such that $m(u, x) = m(x, u) = x \;\forall\; x \in G$ (existence of the identity element).
3. $\forall\; x \in G, \exists\; x^{-1}$ such that $m(x, x^{-1}) = m(x^{-1}, x) = u$ (existence of the inverse element).

Moreover, G is called *abelian group* if $m(x, y) = m(y, x) \;\forall\; x, y \in G$ (commutative property). The unitary group of degree n is usually indicated by $U(n)$ and it is the group of $n \times n$ unitary matrices with the group operation of matrix multiplication. In the case $n = 1$, the group is $U(1)$ and its elements are all complex numbers with norm 1, so they have the form $e^{i\alpha}$ with α real.

Let us now consider a complex scalar field ϕ. Its Lagrangian is

$$\mathscr{L} = \frac{1}{2}\eta^{\mu\nu}\partial_\mu\phi^*\partial_\nu\phi + m^2\phi^*\phi. \tag{B.1}$$

Such a Lagrangian is clearly invariant under the "global" $U(1)$ transformation

$$\phi \to \phi' = e^{i\alpha}\phi, \tag{B.2}$$

where α is a real constant. The name global is because α does not depend on the spacetime coordinates. However, the Lagrangian in (B.1) is not invariant under a "local" $U(1)$ transformation, namely a transformation in which $\alpha = \alpha(x)$. We can anyway "promote" the global $U(1)$ symmetry in the Lagrangian (B.1) to a local symmetry by introducing an auxiliary field A_μ

$$\mathscr{L} = \frac{1}{2}\eta^{\mu\nu}D_\mu\phi^*D_\nu\phi + m^2\phi^*\phi, \tag{B.3}$$

where

$$D_\mu = \partial_\mu + ig A_\mu. \tag{B.4}$$

The auxiliary field A_μ must transform as

$$A_\mu \to A'_\mu = A_\mu - \frac{1}{g}\partial_\mu \alpha. \tag{B.5}$$

In this way, the Lagrangian (B.1) is invariant under a local $U(1)$ transformation and there is a conserved current

$$J_\mu = i\left(\partial_\mu \phi^*\right)\phi, \quad \partial_\mu J^\mu = 0. \tag{B.6}$$

In order to preserve the symmetry, the kinetic term of the field A_μ must have the form $F_{\mu\nu}F^{\mu\nu}$, where $F_{\mu\nu} = \partial_\mu A_\nu - \partial_\nu A_\mu$ and no mass-term $m^2 A^2$ is allowed (the auxiliary field must be associated to massless particles).

The procedure to introduce the new field A_μ is called *gauge principle* and A_μ is a gauge field. In the case of the $U(1)$ symmetry, we can introduce the electromagnetic force in the Standard Model of particle physics, $U_{em}(1)$. The approach can be extended to more complicated groups and the Standard Model of particle physics is described by $U_Y(1) \times SU_L(2) \times SU(3)$. We note that not all the global symmetries can be promoted to local symmetries and only experiments can tell us if a symmetry is global or local. The Standard Model of particle physics has two global symmetries associated to the baryon and lepton numbers. There is thus a conserved current and (classically) the baryon and lepton numbers cannot be violated. However, there is no force associated with these symmetries and therefore no gauge theory.

Appendix C
Field Quantization

The transition from classical field theory to quantum field theory is similar to the transition from classical to quantum mechanics. In the case of mechanics, we assume that the classical coordinates and momenta are no more just numbers (the so-called C-numbers), but operators satisfying certain commutation relations, such as $[x, p_x] = i\hbar$. Instead of the 3-dimensional coordinate $\mathbf{x}$, in field theory a multi-dimensional quantity $\chi(\mathbf{x}, t)$ plays the role of coordinate, where $\mathbf{x}$ labels continuous number of χ. In other words, $\chi(\mathbf{x}, t)$ for fixed $\mathbf{x}$ plays the role of the coordinate $\mathbf{x}(t)$ in classical or quantum mechanics, while $\mathbf{x}$ in field theory is just a label, like $j = 1, 2, 3$ for the 3-dimensional coordinate x_j in mechanics.

To quantize a field, one needs to introduce an analogue of the particle momenta and postulate the corresponding commutation relations. In the case of scalar field theory, the analogue of the momentum is $\dot\chi$ and the canonical commutator takes the form

$$[\dot\chi(\mathbf{x}, t), \chi(\mathbf{x}', t)] = (2\pi)^3 \delta(\mathbf{x} - \mathbf{x}'). \quad (C.1)$$

Usually the proper field operators satisfying the commutation relation (C.1) are introduced through an expansion in terms of creation-annihilation operators:

$$\chi(\mathbf{x}, t) = \int \widetilde{dk} \left[a_\mathbf{k} f_k(t) \exp(i\mathbf{k} \cdot \mathbf{x}) + b_\mathbf{k}^\dagger f_k^*(t) \exp(-i\mathbf{k} \cdot \mathbf{x}) \right], \quad (C.2)$$

where

$$\widetilde{dk} = \frac{d^3 k}{2\omega_k (2\pi)^3}. \quad (C.3)$$

Here $\omega_k = \sqrt{\mathbf{k}^2 + m^2}$, which reduces to $\omega_k = |\mathbf{k}|$ for massless particles. The creation-annihilation operators commutes as

$$[a_{\mathbf{k}_1}, a_{\mathbf{k}_2}^\dagger] = 2\omega_{k_1} (2\pi)^3 \delta^{(3)}(\mathbf{k}_1 - \mathbf{k}_2), \quad (C.4)$$

and a similar relation holds for the antiparticle creation and annihilation operators $b_\mathbf{k}$. All the other pairs of the operators have zero commutators.

Note that, by construction, the creation operator $a_k^\dagger$ acting on the vacuum state $|0\rangle$ creates the one-particle state with momentum k, $|\mathbf{k}\rangle$. The annihilation operator a_k acting on the one-particle state with momentum k creates the vacuum state $|0\rangle$. If the annihilation operator acts on the vacuum state, it kills it.

This quantization procedure is usually done for a free, non-interacting field, while interactions are taken into account perturbatively. Correspondingly, in flat spacetime the operator χ satisfies the free Klein-Gordon equation

$$\ddot{\chi} - \Delta\chi + m_\chi^2 \chi = 0. \tag{C.5}$$

Its Fourier amplitudes, which are C-numbers, obey the simple harmonic oscillator equation

$$\ddot{f}_k + \left(\mathbf{k}^2 + m_\chi^2\right) f_k = 0. \tag{C.6}$$

Equation (C.6) has the following two linear independent solutions

$$f_k = \exp(-i\omega_k t), \quad f_k^* = \exp(i\omega_k t). \tag{C.7}$$

Note that in the expansion (C.2) the coefficients in front of a_k and $b_k^\dagger$ are, respectively, f_k and f_k^*. This ensures positive definiteness of energy.

The vacuum state is defined as a no-particle state, so, as we have already mentioned, the annihilation operator acting on this state "kills" it

$$a_\mathbf{k}|0\rangle = b_\mathbf{k}|0\rangle = 0. \tag{C.8}$$

The vacuum state is normalized as $\langle 0|0\rangle = 1$. The one-particle (antiparticle) state with momentum $\mathbf{k}$ is defined as

$$|\mathbf{k}\rangle = a_\mathbf{k}^\dagger|0\rangle \quad \left(|\bar{\mathbf{k}}\rangle = b_\mathbf{k}^\dagger|0\rangle\right). \tag{C.9}$$

In the more general case, when the field χ evolves in a time dependent background, e.g. in curved spacetime, the measure of integration instead of (C.3) is taken as

$$\widetilde{dk} = \frac{d^3k}{|W_k|(2\pi)^2}, \tag{C.10}$$

where W_k is the Wronskian of the equation of motion obeyed by $f_k(t)$, in particular of Eq. (6.48)

$$W_k = \dot{f}_k^* f_k - f_k^* \dot{f}_k. \tag{C.11}$$

Appendix C: Field Quantization

Using the equation of motion, we find that $W_k = const$. The choice of the Wronskian as a normalization factor ensures that the equal time commutator of χ and $\dot\chi$ has the canonical form

$$[\dot\chi(t, \mathbf{x}'), \chi(t, \mathbf{x})] = (2\pi)^3 \delta(\mathbf{x} - \mathbf{x}'). \tag{C.12}$$

Sometimes a different form of the measure is taken, with $(2\pi)^{3/2}\sqrt{W_k}$ instead of $(2\pi)^3 W_k$. In this case, the creation-annihilation operators are renormalized in such a way that the factor in front of the δ-function becomes also $(2\pi)^{3/2}\sqrt{W_k}$. The first choice is probably preferable, because the factor d^3k/ω_k is Lorenz invariant. Anyhow, both definitions lead to the same canonical commutator (C.12).

MIX
Papier aus verantwortungsvollen Quellen
Paper from responsible sources
FSC® C105338

If you have any concerns about our products,
you can contact us on
ProductSafety@springernature.com

In case Publisher is established outside the EU,
the EU authorized representative is:
**Springer Nature Customer Service Center GmbH
Europaplatz 3, 69115 Heidelberg, Germany**

Printed by Libri Plureos GmbH
in Hamburg, Germany